Modern Birkhäuser Classics

Many of the original research and survey monographs in pure and applied mathematics published by Birkhäuser in recent decades have been groundbreaking and have come to be regarded as foundational to the subject. Through the MBC Series, a select number of these modern classics, entirely uncorrected, are being re-released in paperback (and as eBooks) to ensure that these treasures remain accessible to new generations of students, scholars, and researchers.

Loop Spaces, Characteristic Classes and Geometric Quantization

Jean-Luc Brylinski

Reprint of the 1993 Edition

Birkhäuser
Boston • Basel • Berlin

Jean-Luc Brylinski
Brylinski Research
PO Box 1329
Hyannis, MA 02601-1329
U.S.A.

Originally published as Volume 107 in the series *Progress in Mathematics*

Cover design by Alex Gerasev.

Mathematics Subject Classification: 14-xx, 14F05, 14D21, 18F20, 20L05, 46M20, 55-xx, 55-02, 53Cxx, 55N30

Library of Congress Control Number: 2007936854

ISBN-13: 978-0-8176-4730-8   ISBN 978-0-8176-4731-5 (eBook)

Printed on acid-free paper.

9 8 7 6 5 4 3 2 1

www.birkhauser.com   (IBT)

Jean-Luc Brylinski

Loop Spaces,
Characteristic Classes
and Geometric Quantization

Birkhäuser

Boston · Basel · Berlin

Jean-Luc Brylinski
Department of Mathematics
The Penn State University
University Park, PA 16802

Library of Congress Cataloging-in-Publication Data

Brylinski, J.-L. (Jean-Luc)
　　Loop spaces, characteristic classes, and geometric quantization /
Jean-Luc Brylinski, 1951-
　　　　p.　　cm. -- (Progress in mathematics : 107)
　　Includes bibliographical references and index.
　　ISBN 0-8176-3644-7　(acid free)
　　1. Loop spaces.　2. Characteristic classes.　3. Homology theory.
I. Title.　II. Series : Progress in mathematics　(Boston, Mass.) ;
vol. 107.
QA612.76.B79　　1992　　　　　　　　　　　　　92-34421
514' .24--dc20　　　　　　　　　　　　　　　　CIP

Printed on acid-free paper

ISBN 0-8176-3644-7
ISBN 3-7643-3644-7

Camera-ready copy prepared by the Author in TeX.

9 8 7 6 5 4 3 2

To Ranee
with love

Table of Contents

Introduction

Characteristic classes are certain cohomology classes associated either to a vector bundle or a principal bundle over a manifold. For instance, there are the Chern classes $c_i(E) \in H^{2i}(M, \mathbb{Z})$ of a complex vector bundle E over a manifold M. These Chern classes can be described concretely in two ways. In the de Rham theory, a connection on the bundle is used to obtain an explicit differential form representing the cohomology class. In singular cohomology, the Chern class is the obstruction to finding a certain number of linearly independent sections of the vector bundle; this is Pontryagin's original method. The relation between the two approaches is rather indirect and in some ways still mysterious. A better understanding requires a geometric theory of cohomology groups $H^p(M, \mathbb{Z})$ for all p.

In the past 25 years, characteristic classes have been used to construct some of the basic objects in infinite-dimensional geometry and loop groups, geometric quantization, knot invariants and gauge theory. In fact, a number of recent developments, especially Witten's treatment of the Jones polynomial for knots [**Wi2**], has focused on so-called secondary characteristic classes, for instance, the Chern-Simons classes of a vector bundle with connection. These secondary classes are more refined invariants than the usual classes, and live in more sophisticated cohomology groups, like the Cheeger-Simons group of differential characters, or Deligne cohomology groups. It is a crucial task to elucidate the geometric significance of these cohomology theories and of the characteristic classes which take values in them. The approach in this book is based on generalizations of the notion of bundles over a manifold. Such generalizations are not pure abstractions. In fact, experience shows that they are just as natural as ordinary bundles and occur quite frequently in geometric problems.

For degree 2 cohomology, there is a very satisfactory theory of A. Weil [**Weil1**] and B. Kostant [**Ko1**], which identifies $H^2(M, \mathbb{Z})$ with the group of isomorphism classes of line bundles over M. Furthermore, the isomorphism admits a completely explicit description in terms of Čech cocycles associated to some open covering of the manifold. By picking an open covering (U_i) and a trivialization of the line bundle over each open set U_i, one obtains the transition functions g_{ij} which take values in $\mathbb{C}^*$. Then, choosing logarithms $Log(g_{ij})$ of the g_{ij}, one can obtain a Čech 2-cocycle with values in $\mathbb{Z}$, which is an alternating sum of the logarithms of the g_{ij}, divided by $2\pi\sqrt{-1}$. Furthermore, given a connection on a line bundle L, its curvature K is a closed 2-form. One can see directly that the closed 2-form $\dfrac{K}{2\pi\sqrt{-1}}$ has integral cohomology class equal to that of the Čech 2-cocycle described above. This is discussed in Chapter 2.

One way of looking at the theory of Weil and Kostant is to see that it gives a *quantization condition* for a closed 2-form K on a manifold: K is the curvature of some line bundle if and only if its cohomology class, divided by $2\pi\sqrt{-1}$, is integral. This quantization condition is very important in symplectic geometry, since in the geometric quantization program of Kostant and Souriau, a line bundle on a manifold is used to construct representations of the group of symplectomorphisms. It also is central in algebraic geometry, where ample line bundles are used to embed a complex manifold into projective space.

It would of course be highly desirable to have a theory for integer cohomology of any degree, which is as clear, geometric and as explicit as the Weil-Kostant theory for line bundles. This book intends to be a first step in that direction.

We develop here a Chern-Weil theory of characteristic classes of gerbes. Gerbes (invented by J. Giraud) are fiber bundles over a manifold, whose fibers are groupoids (that is to say, categories in which every morphism is invertible). Gerbes occur very frequently in geometry, for instance, if there exists a family of symplectic manifolds or a bundle of projective Hilbert spaces. We think of gerbes as a receptacle for the obstruction to constructing some sort of global bundle with some properties. In the case of a family of symplectic manifolds, the gerbe is the obstruction to finding a line bundle on the total space with given restriction to the fibers. Such a global line bundle may not exist, but the categories of line bundles over all possible open sets organize into a gerbe. We claim this is a very fruitful way of looking at this obstruction problem.

The gerbes that we need are connected with the group of smooth functions with values in $\mathbb{C}^*$. We will call them Dixmier-Douady sheaves of groupoids, since they are a generalization of some infinite-dimensional algebra bundles considered by the authors [D-D]. This book lays the foundation of the theory of Dixmier-Douady sheaves of groupoids and of their differential geometry, leading to the notion of curvature of a gerbe, which turns out to be a closed degree 3 differential form (called the "3-curvature"). This differential form is "quantized," that is, all of its periods are integers. One of our main results is the converse theorem, stating that any closed 3-form with integer periods appears as the 3-curvature of some gerbe. Thus the differential geometry of gerbes allows us to construct all the degree 3 cohomology with integer coefficients. We then prove that the group $H^3(M,\mathbb{Z})$ identifies with the group of equivalence classes of Dixmier-Douady sheaves of groupoids.

One important application of our theory has to do with the degree 3 cohomology of a compact Lie group. For a simple simply-connected compact Lie group G, it is clear from [P-S] that the canonical generator of $H^3(G, \mathbf{Z}) \simeq \mathbf{Z}$ is what gives rise to the whole theory of loop groups, their central extensions, and their representations. Furthermore, work by Gawedzki [Ga2] suggests that the generator of $H^3(G, \mathbf{Z})$ is in some sense accountable for the existence of quantum groups. Our geometric interpretation of $H^3(G, \mathbf{Z})$ is based on a canonical gerbe over any compact simply-connected Lie group G, which is associated to the path fibration of Serre, viewed as a family of symplectic manifolds over the group itself. The 2-form on a fiber is therefore a natural 2-form on the pointed loop group, although not invariant. Then, by transgression from this gerbe on G, we derive the central extension of the loop group LG. The gerbe on G is a central feature of the geometric theory of degree four characteristic classes, developed by D. McLaughlin and the author [Br-ML1] [Br-ML2]. In the case of $G = SU(2) = S^3$, the gerbe on G has another description in terms of the magnetic monopole of Dirac. The resulting quantization condition for the volume form on S^3 gives the Dirac quantization of the electric charge, which thus appears as a truly 3-dimensional phenomenon.

Another application of our description of $H^3(M, \mathbf{Z})$ is the construction of line bundles on the free loop space LM of a manifold M. From a topological point of view, line bundles on LM are classified by the group $H^2(LM, \mathbf{Z})$, and there is a natural transgression map $H^3(M, \mathbf{Z}) \to H^2(LM, \mathbf{Z})$. Therefore any class in $H^3(M, \mathbf{Z})$ gives rise to an isomorphism class of line bundles. It is, however, better to actually construct the line bundle itself. Corresponding to any given Dixmier-Douady sheaf of groupoids $\mathcal{C}$ over M, we give a direct construction of a line bundle over LM, using a geometric analog of the topological transgression. The line bundle is described symbolically as a sort of tautological bundle; the fiber at a loop γ is described in terms of the holonomy of objects of the sheaf of the groupoid around γ. The curvature of the line bundle is the 2-form on LM obtained from the 3-curvature of the sheaf of groupoids by transgression. The line bundle is used to construct so-called "action functionals" for mappings of a surface with boundary to the ambient manifold. In fact, the action functional is not a number in $\mathbf{C}^*$, but is an element of the fiber of the line bundle at the boundary of the surface. Thus, following physicists, the line bundle may be called the "anomaly line bundle."

The same method gives the line bundle on the space of knots in a 3-manifold. This space of knots, which was studied in [Bry3], has several remarkable structures. It has a riemannian structure, a complex structure, and a symplectic structure, all of which are related in the same way as the corresponding structures on a Kähler manifold. Our methods enable us to construct a holomorphic line bundle over the space of knots, whose curvature is $2\pi\sqrt{-1}$ times the symplectic form.

We will present two levels of the theory of sheaves of groupoids and its relation with degree 3 cohomology. The first level is an extension of the *Dixmier-Douady theory*, in [D-D], these authors described $H^3(M, \mathbb{Z})$ in terms of *projective Hilbert space bundles* over M, that is, fiber bundles whose fiber is the projective space associated to a separable Hilbert space. After introducing suitable analogs of the notions of connection and curvature, it is found that the curvature is a degree 3 differential form. At this point, we obtain a theory of degree 3 cohomology which is as natural and geometric as the Kostant-Weil theory for degree 2 cohomology.

This viewpoint, however, has two limitations. First, it is very difficult in general to construct explicitly a projective Hilbert space bundle associated to a given degree 3 cohomology class, and, second, it is necessary to consider infinite-dimensional bundles for a problem which is intrinsically finite-dimensional in nature. Therefore we develop a second level, in which the infinite-dimensional projective Hilbert space is replaced by the more abstract notion of a *groupoid*. Given a projective Hilbert space P, there is indeed a natural groupoid C, such that an object of C is a Hilbert space E whose associated projective space identifies with P. We then prove that the group $H^3(M, \mathbb{Z})$ is equal to the group of equivalence classes of Dixmier-Douady sheaves of groupoids. Analogous notions of connection and curvature are then very natural in this context. The first level theory is a special case of the more abstract second level, but the second level is more flexible, and more natural in many cases.

Very interesting phenomena occur when we try to make a sheaf of groupoids equivariant under some group of transformations. To explain this, we recall what happens in the case of a line bundle over a symplectic manifold (M, ω). Assume that a Lie group G acts on M, preserving the 2-form ω. As Kostant showed, when one tries to make the line bundle L equivariant under G, one finds an obstruction. What happens is that L is equivariant under a new Lie group $\tilde{G}$, which is a central extension of G by $\mathbb{C}^*$. The central extension $\tilde{G}$ is crucial in the geometric quantization program of Kostant [Ko1] and Souriau [So1].

Just as there is the notion of equivariant line bundles over a space, one can speak of an equivariant sheaf of groupoids. Let a Lie group G act on a manifold M, preserving the 3-curvature of a sheaf of groupoids C. Then the obstruction to lifting the G-action to the sheaf of groupoids is a group cohomology class in $H^3(G, \mathbb{C}^*)$. One of our main theorems states that in the case where $G = SU(2)$ acting on itself by translation and C is the sheaf of groupoids corresponding to the Dirac monopole, the class in $H^3(SU(2), \mathbb{T})$ is the Chern-Simons class $\hat{c}_2$ for flat $SU(2)$-bundles. This is close in spirit to the author's work on the first Pontryagin class in collaboration with D. McLaughlin [Br-ML1] [Br-ML2]. We have no doubt that all secondary characteristic classes for flat bundles can be constructed as the obstruction to making some natural geometric object over the group equivariant. The object will become more intricate and hard to pin down as the degree of the characteristic class increases.

In developing the theory of geometric quantization, Kostant was led to study the group of isomorphism classes of pairs (L, ∇) of a line bundle with a connection, and to describe it via a 2-term complex of sheaves. Deligne later did a similar computation in the holomorphic case, and the upshot is that the group in question is isomorphic to the Deligne cohomology group $H^2(M, \mathbb{Z}(1)_D)$. Such a cohomological description is also well-suited to studying the central extension of a group of symplectomorphisms. It is crucial in some computations of the regulator map of Beilinson, from algebraic K-theory to Deligne cohomology (more precisely, Beilinson's modified version of Deligne cohomology) [Be1]. We find a similar geometric interpretation of the Deligne cohomology group $H^3(M, \mathbb{Z}(2)_D)$ as a group of equivalence classes of sheaves of groupoids over M, equipped with some extra differential structures. Such a description may be of use in other computations involving Beilinson's regulator. We feel that the method employed here will lead to geometric interpretation of all relevant Deligne cohomology groups. This may help to relate the point of view of this book to several works in algebraic geometry and number theory [Be1] [Den] [Ku], which address relations between L-functions, characteristic classes and groupoids (and their generalizations).

The book is meant to be accessible to topologists, geometers, Lie theorists and mathematical physicists, as well as to operator-algebraists. Basic knowledge of point-set topology, manifolds, differential geometry and graduate algebra is necessary, as is some familiarity with some basic facts regarding Lie groups, Hilbert spaces and categories. There are many discussions of open problems and possible avenues for research to be found in the later sections of the book, which we will not list here. Suffice it to say that the full theory of geometric quantization is not yet available for sheaves of groupoids. This is not surprising, since such a theory should include quantization of loop spaces viewed as symplectic manifolds. The case of gerbes with a non-commutative band has barely been studied in a geometric context. The full applicability of this circle of ideas to topological quantum field theory remains largely open; we refer the reader to [Br-ML2] for a conjecture in this direction.

We now give a more detailed description of the seven chapters. Chapter 1 is an introduction to sheaves, complexes of sheaves, and their cohomology, covering standard aspects, and also a few less standard ones which are needed in the rest of the book. We start from scratch, developing both sheaf cohomology from injective resolutions and Čech cohomology using open coverings. We discuss de Rham cohomology, Deligne cohomology and the Cheeger-Simons differential characters, as well as the Leray spectral sequence for a mapping and a complex of sheaves.

Chapter 2 introduces line bundles, their relation with H^2, connections and curvature. The theory of Kostant and Deligne relating line bundles with connection to Deligne cohomology is detailed, both in the smooth and in the holomorphic case. We explain the Kostant-Souriau central extension of the Lie algebra of hamiltonian vector fields and the Kostant central extension of the group of symplectomorphisms. We give the action of this central extension on sections of the line bundle; we also prove that the central extension splits if the group has a fixed point, and give numerical consequences. Our coverage includes the case of a symplectic form with non-integral cohomology class, following Weinstein [**Wein**].

Chapter 3 presents a number of geometric results on the space $\hat{Y}$ of singular knots in a smooth 3-manifold M. The striking fact is that $\hat{Y}$ looks exactly like a Kähler manifold. It has a complex structure, which, by a theorem of Lempert, is integrable in a weak sense. It has a symplectic structure that is familiar from fluid mechanics [**M-W**]. The symplectic form has integral cohomology class. Finally, there is a riemannian structure, which is compatible with both the symplectic and complex structures. Of great interest is the action on $\hat{Y}$ of the group G of volume-preserving diffeomorphisms of M. As it turns out, the space of knots of given isotopy type is a coadjoint orbit of the connected component of G.

Chapter 4 exposes the first level of degree 3 cohomology theory. We start with the theory of Dixmier and Douady [**D-D**], based on infinite-dimensional algebra bundles, and formulate the general obstruction of lifting the structure group of a principal bundle to a central extension of the group. We then proceed to develop the relevant geometric concepts leading to the notion of 3-curvature. Examples include families of symplectic manifolds (in the spirit of [**G-L-S-W**]), and a class in $H^3(M, \mathbb{Z})$, obtained by cup-product. In the latter case, we give a natural construction of a bundle of projective Hilbert spaces, using the generalized Heisenberg group associated to the circle group.

Chapter 5, really the heart of the book, exposes the second level of this theory, using sheaves of groupoids. We begin with a discussion of the glueing properties of sheaves with respect to surjective local homeomorphisms, and introduce the notion of torsors, which is a generalization of principal bundles. Then we define sheaves of groupoids and gerbes using glueing axioms that mimic the glueing properties of sheaves. The gerbe associated to a G-bundle over M and a group extension $\tilde{G}$ of G is discussed. Given a gerbe $\mathcal{C}$, there is an associated a sheaf of groups, called the band of $\mathcal{C}$. Gerbes with band equal to the sheaf of smooth $\mathbb{C}^*$-valued functions are called Dixmier-Douady sheaves of groupoids. The differential-geometry of such sheaves of groupoids is developed, for which the exotic names "connective structure" and "curving" were invented. A differential form of degree 3 is defined, and called the 3-curvature of the sheaf of groupoids. The group of equivalence classes of sheaves of groupoids of Dixmier-Douady type, with connective structure and curving, is identified with the Deligne cohomology group $H^3(M, \mathbb{Z}(2)_D)$. Many examples and applications are given. The case of a family of symplectic manifolds is explained, and the important special case of the path-fibration $\Omega M \to P_1 M \to M$ is treated in some detail. The sheaf of groupoids arising from a central extension of the structure group of a principal bundle is studied; this allows the first level of the theory to be recovered inside the second level. Sheaves of groupoids occur, at least implicitly, in the study of the group of isomorphism classes of equivariant line bundles for a compact Lie group acting on a smooth manifold. This leads us to an accurate description of this important group, which did not seem to be known.

Chapter 6 gives the construction of a line bundle on LM associated to a sheaf of groupoids on M with connective structure. The construction is in some sense very explicit, in a spirit similar to the construction of a tautological line bundle. The curvature of the line bundle on LM is the 2-form obtained by transgressing the 3-curvature of the sheaf of groupoids over M. This construction corresponds to a transgression map $H^3(M, \mathbb{Z}(2)_D) \to H^2(LM, \mathbb{Z}(1)_D)$ in Deligne cohomology. We show how the central extension of a loop group can be understood in this way (following [Br-ML 2]). We then construct a holomorphic line bundle over the space of knots in a smooth 3-manifold. Finally, we discuss the notion of parallel transport in the context of sheaves of groupoids.

Chapter 7 treats the Dirac magnetic monopole, and the resulting quantization condition for a closed 3-form on the 3-sphere. We first explain Dirac's construction and its classical mathematical interpretation using a 2-sphere centered at the monopole. Then we give our purely 3-dimensional approach, in which the monopole is viewed as a sheaf of groupoids on the 3-sphere. The quantization condition follows from the theory of Chapter 5. We then address the problem of making the sheaf of groupoids equivariant under $SU(2)$ and find that the obstruction to so doing is the Chern-Simons class $\hat{c}_2$ for flat bundles, which is a degree 3 group cohomology class with coefficients in the circle group. In fact, this obstruction class extends to the whole group of volume-preserving diffeomorphisms of S^3.

A number of the methods and results of this book were first exposed in a preprint circulated in 1990 [**Bry 3**]. The idea of using general concepts from [**Gi**] rather than infinite-dimensional algebra bundles arose from discussions with Pierre Deligne, and I am very grateful to him. I have benefited a great deal from correspondence and discussions with Alan Weinstein, especially concerning the relation of sheaves of groupoids with families of symplectic manifolds, and the line bundle over the space of knots. I am also grateful to him for his constant interest in this book, for his encouragement as editor, and for his many useful comments. I would like to thank Ann Kostant for her support throughout the whole writing and publishing process, and for her invaluable help and expert advice.

I would like to express my gratitude to Sean Bates, who proofread two versions of the book and made a number of very useful comments and suggestions, some of which resulted in improvements of the exposition, and saved me from many mistakes. I would also like to thank Ethel Wheland for her aid in the "final" version of the book.

Finally, I thank my wife Ranee for her support and encouragement. The author was supported in part by grants from the National Science Foundation.

Chapter 1

Complexes of Sheaves and their Hypercohomology

1.1. Injective resolutions and sheaf cohomology

The word *sheaf* is a translation of the French word *faisceau*. Sheaves were first introduced in the 1940s by H. Cartan [**Ca**] in complex analysis (following works by K. Oka [**Ok**] and H. Cartan himself), and by J. Leray in topology [**Le**]. Since then sheaves have become important in many other branches of mathematics. Whenever one wants to investigate global properties of geometric objects like functions, bundles, and so on, sheaves come into play. There are many excellent introductions to the theory of sheaves [**Go**] [**Iv**] [**B-T**]; we will develop here only those aspects which will be useful throughout the book. We will begin with the notion of presheaf.

1.1.1. Definition. *Let X be a topological space. A <u>presheaf</u> $\mathcal{F}$ of sets on X (or over X) consists of the following data:*

(1) For every open set U of X, a set $\mathcal{F}(U)$. One assumes $\mathcal{F}(\emptyset)$ is a set with one element.

(2) For every inclusion $V \subseteq U$ between open sets, a mapping of sets $\rho_{V,U} : \mathcal{F}(U) \to \mathcal{F}(V)$. $\rho_{V,U}$ is called the <u>restriction map</u> (from U to V).

One requires the following transitivity condition. If $W \subseteq V \subseteq U$ are open sets, then

$$\rho_{W,U} = \rho_{W,V}\,\rho_{V,U}. \qquad (1-1)$$

One also requires that $\rho_{U,U}$ be the identity. An element of $\mathcal{F}(U)$ is called a <u>section</u> of $\mathcal{F}$ over U.

Here are some examples. If Y is some space, let $\mathcal{F}(U)$ be the set of continuous mappings from U to Y. For $V \subseteq U$, let $\rho_{V,U} : \mathcal{F}(U) \to \mathcal{F}(V)$ be the restriction of mappings to an open set, that is, $\rho_{V,U}(f) = f_{/V}$ for $f : U \to Y$. Another example is the constant presheaf obtained from a set S; for this presheaf $\mathcal{G}$, we put $\mathcal{G}(\mathcal{U}) = S$ for any open set U, and $\rho_{V,U} = Id_S$, where Id_S is the identity map.

When all $\mathcal{F}(U)$ are groups and the $\rho_{U,V}$ are group homomorphisms, we say that $\mathcal{F}$ is a presheaf of groups. Similarly, when all $\mathcal{F}(U)$ are rings ($\mathbb{C}$-algebras, and so on) and the $\rho_{V,U}$ are ring (resp. $\mathbb{C}$-algebra, and so on) homomorphisms, we say that $\mathcal{F}$ is a presheaf of rings (resp. $\mathbb{C}$-algebras).

If X is a set of cardinality one, a presheaf $\mathcal{F}$ of sets on X simply amounts to the set $\mathcal{F}(X)$; similarly, a presheaf of groups on X is just a group.

In this book, we will often use the word "space" to mean "topological space." We wish to formalize the nice properties of the presheaf of continuous mappings with values in a space Y. The essential property is that which allows us to go from sections of $\mathcal{F}$ over open sets in a covering of X to a global section.

1.1.2. Definition. *Let $\mathcal{F}$ be a presheaf over a space X. One says that $\mathcal{F}$ is a <u>sheaf</u> if for every open set V of X, for every open covering $\mathcal{U} = (U_i)_{i \in I}$ of V and for every family $(s_i)_{i \in I}$ where $s_i \in \mathcal{F}(U_i)$, such that*

$$\rho_{U_i \cap U_j, U_i}(s_i) = \rho_{U_i \cap U_j, U_j}(s_j),$$

there exists a unique $s \in \mathcal{F}(V)$ such that $\rho_{U_i, V}(s) = s_i$.

For a sheaf $\mathcal{F}$, we will use the notation $\Gamma(U, \mathcal{F}) = \mathcal{F}(U)$. With the notations of 1.1.2, we say that the section s of $\mathcal{F}$ over X is obtained by glueing the sections s_i of $\mathcal{F}$ over U_i. The condition $\rho_{U_i \cap U_j, U_i}(s_i) = \rho_{U_i \cap U_j, U_j}(s_j)$ is called the glueing condition.

Given a presheaf $\mathcal{F}$ on X and an open subset U of X, one gets a presheaf $\mathcal{F}_{/U}$ on U, where $\mathcal{F}_{/U}(V) = \mathcal{F}(V)$ for V an open subset of U (which is also of course open in X). We call this the *restriction* of $\mathcal{F}$ to the open set U. If $\mathcal{F}$ is a sheaf, then $\mathcal{F}_{/U}$ is a sheaf. For convenience we use the notation $U_{ij} = U_i \cap U_j$, $U_{ijk} = U_i \cap U_j \cap U_k$, and so on.

1.1.3. Lemma. *Let Y be another space, and let $\mathcal{F}$ be the presheaf of continuous maps from open sets of X to Y. Then $\mathcal{F}$ is a sheaf.*

Proof. Let V be an open subset of X, and let (U_i) be an open covering of V. Any $f_i \in \mathcal{F}(U_i)$ is a continuous mapping from U_i to Y. The condition $\rho_{U_{ij}, U_i}(f_i) = \rho_{U_{ij}, U_j}(f_j)$ means that f_i and f_j have the same restriction to U_{ij}. Then there exists a unique function $f : V \to Y$ such that $f_{/U_i} = f_i$ for any $i \in I$. Since each f_i is continuous and $V = \cup_{i \in I} U_i$, it is clear that f is continuous. ∎

If Y is a group, then the sheaf $\mathcal{F}$ of Lemma 1.1.3 is a sheaf of groups. Some of the most important sheaves are of this type; the cases $Y = \mathbb{C}, \mathbb{Z}, \mathbb{C}^*$

are especially significant. If X is a smooth manifold, one has the sheaf $\underline{\mathbb{C}}_X$ of smooth complex-valued functions (on open sets of X). If X is a complex manifold, one has the sheaf $\mathcal{O}_X$ of holomorphic functions.

Given two presheaves of sets $\mathcal{F}$ and $\mathcal{G}$ on the same space X, we have a natural notion of morphism of presheaves $\phi : \mathcal{F} \to \mathcal{G}$. Such a morphism consists of a mapping of sets $\phi_U : \mathcal{F}(U) \to \mathcal{G}(U)$ for each open set U, such that whenever $V \subseteq U$ the following diagram commutes

$$
\begin{array}{ccc}
\mathcal{F}(U) & \xrightarrow{\phi_U} & \mathcal{G}(U) \\
\downarrow{\scriptstyle \rho_{V,U}} & & \downarrow{\scriptstyle \rho_{V,U}} \\
\mathcal{F}(V) & \xrightarrow{\phi_V} & \mathcal{G}(V)
\end{array}
$$

Similarly, one defines the notion of morphisms of sheaves of groups over X. If $\mathcal{F}$ and $\mathcal{G}$ are sheaves of groups over X, we will denote simply by $Hom_X(\mathcal{F}, \mathcal{G})$ the set of morphisms of presheaves of groups from $\mathcal{F}$ to $\mathcal{G}$. If $\mathcal{G}$ is a sheaf of abelian groups, $Hom_X(\mathcal{F}, \mathcal{G})$ is an abelian group.

One can compose morphisms of presheaves (of sets, or groups) in an obvious fashion. Furthermore, there is some associativity built into this composition of morphisms. The upshot is that we have defined the category of presheaves of sets; similarly one defines the category of presheaves of groups, rings, and so on. One defines a morphism of sheaves exactly in the same way as a morphism of presheaves. The resulting category of sheaves is a full subcategory of the category of presheaves. For the basic notions about categories, [McL1] is an excellent reference.

We have the notion of *subsheaf* $\mathcal{F}$ of a sheaf $\mathcal{G}$; each $\mathcal{F}(U)$ is a subset of $\mathcal{G}(U)$, and the restriction maps for $\mathcal{F}$ are induced by those of $\mathcal{G}$.

As sheaves are so much better behaved than presheaves, it is important to have a procedure which straightens a presheaf into a sheaf. Before going to this, we introduce the notion of *stalk* $\mathcal{F}_x$ of a presheaf $\mathcal{F}$ on X at a point x. Intuitively, an element of $\mathcal{F}_x$ is a section of $\mathcal{F}$ defined over a neighborhood of x. Two such sections are identified if they agree on a (possibly smaller) neighborhood.

1.1.4. Definition. *Let $\mathcal{F}$ be a presheaf of sets on a space X, and let $x \in X$. The __stalk__ $\mathcal{F}_x$ of $\mathcal{F}$ at x is the quotient of the set* $\coprod\limits_{U \text{ open}, U \ni x} \mathcal{F}(U)$
by the following equivalence relation: $s \in \mathcal{F}(U) \simeq s' \in \mathcal{F}(V)$ if and only if there exists an open set $W \subseteq U \cap V$ containing x such that s and s' have the same restriction to W.

The elements of $\mathcal{F}_x$ are called *germs* at x of sections of $\mathcal{F}$. This is familiar in complex analysis: for a complex manifold X, the $\mathbb{C}$-algebra $\mathcal{O}_{X,x}$

of germs at x of holomorphic functions is a local ring. For $x \in U$ and $s \in \mathcal{F}(U)$, we will denote by $s_x \in \mathcal{F}_x$ the corresponding element of the stalk at x. It is called the germ of f at x.

One can define $\mathcal{F}_x$ more compactly in terms of direct limits. The set of open neighborhoods of x is an ordered set, where $U < V$ means $V \subset U$. This ordered set is *directed*, i.e., for U_1, U_2 in the set, there exists an element V such that $U_1 < V$ and $U_2 < V$. Given $U < V$, we have the restriction map $\rho_{V,U} : \mathcal{F}(U) \to \mathcal{F}(V)$, with the transitivity property (1-1). Then the stalk $\mathcal{F}_x$ is the *direct limit*

$$\mathcal{F}_x = \varinjlim_{U \ni x} \mathcal{F}(U). \qquad (1-2)$$

In fact the construction in Definition 1.1.4 amounts to the standard description of this direct limit. The stalk of a presheaf of groups (resp. abelian groups) is itself a group (resp. an abelian group).

A morphism $\phi : \mathcal{F} \to \mathcal{G}$ of presheaves of sets (resp. groups) over X induces for each $x \in X$ a mapping of sets (resp. a group homomorphism) $\phi_x : \mathcal{F}_x \to \mathcal{G}_x$ on the stalks at x.

By construction, if $x \in U$, there is a restriction map $\rho_{x,U} : \mathcal{F}(U) \to \mathcal{F}_x$. Given a presheaf $\mathcal{F}$, we want to introduce the notion of a continuous section $x \mapsto s_x \in \mathcal{F}_x$ of $\mathcal{F}$ over some open set U. We say that $x \mapsto s_x$ is *continuous* if for every $x \in U$ there exists an open neighborhood V of x in U, and an element s of $\mathcal{F}(V)$ such that $\rho_{y,V}(s) = s_y$ for any $y \in V$.

1.1.5.   Proposition and Definition.   *Let $\mathcal{F}$ be a presheaf on a space X. For any open set U of X, let $\tilde{\mathcal{F}}(U)$ denote the set of continuous sections $(s_x)_{x \in U}$ of F over U. For $V \subseteq U$, let $\rho_{V,U} : \tilde{\mathcal{F}}(U) \to \tilde{\mathcal{F}}(V)$ be the restriction map $\rho_{V,U}(s_x)_{x \in U} = (s_x)_{x \in V}$. Then the presheaf $\tilde{\mathcal{F}}$ is a sheaf, called the <u>associated sheaf</u> to $\mathcal{F}$. If $\mathcal{F}$ is a presheaf of groups, $\tilde{\mathcal{F}}$ is a sheaf of groups.*

Proof. The proof is very similar to that of Lemma 1.1.3, so we will not give the details. ∎

Note the following properties of the construction of the associated sheaf. For any presheaf $\mathcal{F}$, there is a canonical morphism of presheaves $\phi : \mathcal{F} \to \tilde{\mathcal{F}}$ from $\mathcal{F}$ to the associated sheaf; for U an open set and $s \in \mathcal{F}(U)$, $\phi(s)$ is the continuous section $s_x = \rho_{x,U}(s)$.

1.1.6. Proposition. (1) *For any presheaf $\mathcal{F}$ on X, the morphism $\phi : \mathcal{F} \to \tilde{\mathcal{F}}$ induces an isomorphism of stalks $\phi_x : \mathcal{F}_x \xrightarrow{\sim} \tilde{\mathcal{F}}_x$.*

(2) *If $\mathcal{F}$ is a sheaf on X, the morphism of presheaves $\phi : \mathcal{F} \to \tilde{\mathcal{F}}$ is an isomorphism.*

(3) *For any presheaf $\mathcal{F}$ of groups and any sheaf $\mathcal{G}$ of groups on X, there is a canonical bijection $Hom_X(\tilde{\mathcal{F}}, \mathcal{G}) \xrightarrow{\sim} Hom_X(\mathcal{F}, \mathcal{G})$. A similar result holds for presheaves and sheaves of sets.*

Proof. Let $\mathcal{F}$ be a presheaf on X. The map $\phi_x : \mathcal{F}_x \to \tilde{\mathcal{F}}_x$ is surjective because an element σ of $\tilde{\mathcal{F}}_x$ is of the form $\sigma = s_x$, for some $s \in \mathcal{F}(U)$, where U is a neighborhood of x. The map ϕ_x is injective because, by construction of $\tilde{\mathcal{F}}$, there is a map $\psi_x : \tilde{\mathcal{F}}_x \to \mathcal{F}_x$ such that $\psi_x \phi_x = Id$. This proves (1).

Let $\mathcal{F}$ be a sheaf on X, U an open set in X, and let $\phi_U : \mathcal{F}(U) \to \tilde{\mathcal{F}}(U)$ be the canonical map. We have to show ϕ_U is a bijection. First consider s, s' in $\mathcal{F}(U)$ which have the same image in $\tilde{\mathcal{F}}(U)$. This means that $s_x = s'_x$ (as elements of the stalk $\mathcal{F}_x$) for any $x \in U$; since $\mathcal{F}_x$ is the direct limit of the $\mathcal{F}(V)$, taken over open neighborhoods V of x in U, there exists a neighborhood V_x such that $s_{/V_x} := \rho_{V_x,U}(s)$ is equal to $s'_{/V_x} := \rho_{V_x,U}(s')$. Then the $(V_x)_{x \in U}$ form an open covering of U, such that s and s' have the same restriction to each open set of the covering. Since $\mathcal{F}$ is a sheaf, this implies $s = s'$. Now let $(s_x)_{x \in U}$ be a continuous section of $\mathcal{F}$ over U. For every $x \in U$, there exists an open neighborhood V_x of x in U and an element $s(V_x)$ of $\mathcal{F}(V_x)$ such that $\rho_{x,V_x}(s(V_x)) = s_x$. For $x, y \in U$, the restrictions of $s(V_x)$ and of $s(V_y)$ to $V_x \cap V_y$ are equal, by the first part of the proof of (2), since they have the same image in $\mathcal{F}_z$ for any $z \in V_x \cap V_y$. As $\mathcal{F}$ is a sheaf, there exists $s \in \mathcal{F}(U)$ such that $\rho_{V_x,U}(s) = s(V_x)$ for any $x \in U$. This in particular implies that $\rho_{x,U}(s) = \rho_{x,V_x}(s(V_x)) = s_x$ for any x. This proves (2).

Let $\mathcal{F}$ be a presheaf of groups, and let $\phi_{\mathcal{F}} : \mathcal{F} \to \tilde{\mathcal{F}}$ be the canonical morphism of presheaves of groups. Then, for a sheaf $\mathcal{G}$ of groups, there is a map $\alpha : Hom_X(\tilde{\mathcal{F}}, \mathcal{G}) \to Hom_X(\mathcal{F}, \mathcal{G})$ defined by $\alpha(g) = g \circ \phi$, where $\phi = \phi_{\mathcal{F}}$. Let us show that α is injective. Let $f, f' : \tilde{\mathcal{F}} \to \mathcal{G}$ be two morphisms of sheaves such that $f \circ \phi = f' \circ \phi$. For U an open set and $s \in \tilde{\mathcal{F}}(U)$, we have to show that $f(s) = f'(s)$. As $\mathcal{G}$ is a sheaf, it is enough to show that $f(s)$ and $f'(s)$ have the same image in any stalk $\mathcal{G}_x$ (this will suffice because of (2)). But there exists some open neighborhood V of x, and a section s' of $\mathcal{F}$ over V such that the section s of $\tilde{\mathcal{F}}$ and the section s' of $\mathcal{F}$ have the same image in $\mathcal{F}_x$. Therefore $f(s_x) = f \circ \phi(s'_x) = f' \circ \phi(s'_x) = f'(s_x)$ in the stalk $\mathcal{G}_x$. Hence $f(s)$ and $f'(s)$ have the same image in every stalk; since $\mathcal{G}$ is a sheaf, they must coincide.

It is easy to show that each morphism of presheaves of groups $f : \mathcal{F} \to \mathcal{G}$

factors through a morphism $\tilde{\mathcal{F}} \to \mathcal{G}$, since f induces a morphism $\tilde{\mathcal{F}} \to \tilde{\mathcal{G}}$ of the associated sheaves, which one can compose with the inverse of the isomorphism $\mathcal{G} \xrightarrow{\sim} \tilde{\mathcal{G}}$. This gives the required morphism of sheaves of groups. ∎

For a set S and a space X, the associated sheaf to the constant presheaf S on X is denoted S_X (instead of $\tilde{S}$). Such a sheaf S_X is called a *constant sheaf*. Each stalk $S_{X,x}$ of S_X identifies with S. For U an open set of X, a section $x \in U \mapsto s_x \in S$ is continuous if and only if it is *locally constant*. Hence $S_X(U)$ is the set of locally constant functions $U \to S$.

The notion of associated sheaf is used in defining the *image* of a morphism of sheaves $\phi : \mathcal{F} \to \mathcal{G}$. First we have the presheaf $\mathcal{H}(U) = Im(\phi_U : \mathcal{F}(U) \to \mathcal{G}(U))$. Then we take the associated sheaf, which we denote by $Im(\phi)$, and call the image of ϕ. This is a subsheaf of $\mathcal{G}$. A section s of $\mathcal{G}$ over U belongs to $Im(\phi)$ if and only if, for every $x \in U$, the element s_x of $\mathcal{G}_x$ belongs to the image of ϕ_x.

We are mostly interested in sheaves of groups. For $\phi : \mathcal{F} \to \mathcal{G}$ a morphism of sheaves of groups, its *kernel* $Ker(\phi)$ is the subsheaf of $\mathcal{F}$ such that $(Ker(\phi))(U) = Ker(\phi_U)$. A sequence $A \xrightarrow{f} B \xrightarrow{g} C$ of sheaves of groups is *exact* if and only if the two subsheaves $Im(f)$ and $Ker(g)$ of B coincide. The following result gives a purely local characterization of exact sequences of sheaves.

1.1.7. Proposition. *A sequence $A \xrightarrow{f} B \xrightarrow{g} C$ of morphisms of sheaves of groups is exact if and only if, for every $x \in X$, the sequence of groups $A_x \xrightarrow{f_x} B_x \xrightarrow{g_x} C_x$ is exact.*

Proof. This follows immediately from

1.1.8. Lemma. *Two subsheaves $\mathcal{F}$ and $\mathcal{F}'$ of a sheaf $\mathcal{G}$ coincide if and only if, for every $x \in X$, the stalks $\mathcal{F}_x$ and $\mathcal{F}'_x$ coincide, as subsets of $\mathcal{G}_x$.* ∎

Let us give one of the most important examples of an exact sequence of sheaves—the exponential exact sequence. For X a smooth manifold (finite or infinite-dimensional) and for a Lie group G, denote by $\underline{G}_X$ the sheaf such that $\underline{G}_X(U)$ is the group of smooth functions $U \to G$. The kernel of the exponential mapping $exp : \mathbb{C} \to \mathbb{C}^*$ is the subgroup $2\pi\sqrt{-1} \cdot \mathbb{Z}$ of $\mathbb{C}$. Following Deligne [De1], we denote this group by $\mathbb{Z}(1)$. This is a cyclic group, which has a preferred generator only after one chooses a square root of -1.

1.1.9. Corollary. *For a smooth manifold X, we have an exact sequence of sheaves*

$$0 \to \mathbb{Z}(1) \to \underline{\mathbb{C}}_X \xrightarrow{exp} \underline{\mathbb{C}}_X^* \to 0 \qquad\qquad (1-3)$$

Proof. The only non-trivial point is that the morphism of sheaves $exp :$ $\underline{\mathbb{C}}_X \to \underline{\mathbb{C}}_X^*$ is onto. For every $x \in X$ and every $f_x \in \underline{\mathbb{C}}_{X,x}^*$, there exists a contractible open neighborhood U of x such that f_x is the germ at x of a smooth function $f : U \to \mathbb{C}^*$. There exists a smooth function $g : U \to \mathbb{C}$ such that $exp(g) = f$. Then for $g_x \in \underline{\mathbb{C}}_{X,x}$, the germ at x of g, we have $exp(g_x) =$ f_x. So exp gives a surjective map on the stalks at x. By Proposition 1.1.7, exp is onto. ∎

We now focus on the category of sheaves $\mathcal{AB}(X)$ of abelian groups over X. Recall that we denote by $Hom_X(A, B)$ the abelian group of morphisms in this category. There is a natural notion of direct sum $\mathcal{F} \oplus \mathcal{G}$ of two such sheaves. One may define direct limits $\varinjlim_{i \in I} \mathcal{F}_i$ of sheaves of abelian groups, where I is a directed ordered set. This is the sheaf associated to the presheaf $U \mapsto \varinjlim_{i \in I} \mathcal{F}_i(U)$. Products and inverse limits are easier to define, as one already obtains directly a sheaf by taking the product or inverse limit for each open set.

We have the notion of the kernel and of the image of a morphism $\phi : \mathcal{F} \to \mathcal{G}$ in $\mathcal{AB}(X)$. We also have the notion of cokernel $Coker(\phi)$, which is the sheaf associated to the presheaf $U \mapsto Coker(\phi_U)$. Given $\phi : \mathcal{F} \to \mathcal{G}$, the *coimage* $Coim(\phi)$ is the quotient of $\mathcal{F}$ by $Ker(\phi)$. The natural morphism $Coim(\phi) \to Im(\phi)$ is an isomorphism, and thus $\mathcal{AB}(X)$ is an abelian category in which one can do the usual sort of homological algebra. Recall that a morphism $\phi : \mathcal{F} \to \mathcal{G}$ is called a *monomorphism* if $Ker(\phi) = 0$, an *epimorphism* if $Coker(\phi) = 0$.

We wish to study the functor $\Gamma(X, -) : \mathcal{AB}(X) \to \mathcal{AB}$, such that $\Gamma(X, \mathcal{F}) = \mathcal{F}(X)$. $\Gamma(X, -)$ is called the functor of *global sections*.

1.1.10. Lemma. *The functor $\Gamma(X, -)$ is* left-exact, *i.e., for every exact sequence*

$$0 \to A \to B \to C \to 0$$

in $\mathcal{AB}(X)$, the corresponding sequence

$$0 \to \Gamma(X, A) \to \Gamma(X, B) \to \Gamma(X, C)$$

is exact.

The functor $\Gamma(X, -)$ is not exact in general. For instance, the exponential map $exp : \underline{\mathbb{C}}(X) \to \underline{\mathbb{C}}^*(X)$ is not surjective if $H^1(X, \mathbb{Z}) \neq 0$. In this situation, one introduces the *right derived functors* of $\Gamma(X, -)$ [C-E] [Gr2] [McL2].

1.1.11. Definition. *A sheaf $\mathcal{F}$ of abelian groups over X is called <u>injective</u> if for any diagram in $\mathcal{AB}(X)$*

$$A \xrightarrow{\ i\ } B$$
$$\downarrow f$$
$$\mathcal{F}$$

with $Ker(i) = 0$, there exists a morphism $g : B \to \mathcal{F}$ in $\mathcal{AB}(X)$ such that $g \circ i = f$.

If X is a point, we recover the notion of injective abelian group. An abelian group A is injective if and only it is *divisible*, i.e., for every $n \in \mathbb{N}$ the homomorphism $x \mapsto n \cdot x$ from A to itself, is surjective.

1.1.12. Lemma. *Let $0 \to A \xrightarrow{\ i\ } B \to C \to 0$ be an exact sequence of sheaves of abelian groups over a space X, in which A is an injective sheaf. Then the sequence splits, i.e., there exists a morphism $p : B \to A$ in $\mathcal{AB}(X)$ such that $p \circ i = Id_A$.*

Proof. Since A is injective, the identity morphism Id_A can be extended to a morphism $p : B \to A$. ∎

1.1.13. Lemma. *For any sheaf $\mathcal{F}$ of abelian groups on a space X, there exists an injective sheaf I of abelian groups and a monomorphism $f : \mathcal{F} \to I$.*

Proof. Recall that for any abelian group B, there exists an injection $B \hookrightarrow J$, where J is an injective abelian group [C-E]. For every $x \in X$, take an injective resolution $\mathcal{F}_x \hookrightarrow J_x$ of the abelian group $\mathcal{F}_x$. Let $\tilde{J}_x$ be the so-called "skyscraper sheaf" which has all stalks equal to 0, except the stalk at x, which is equal to J_x. In other words

$$\tilde{J}_x(U) = \left\{ \begin{array}{l} J_x \text{ if } x \in U \\ 0 \text{ if } x \notin U \end{array} \right\}.$$

For any sheaf A of abelian groups, we have $Hom_X(A, \tilde{J}_x) \xrightarrow{\sim} Hom(A_x, J_x)$.

It follows that $\tilde{J}_x$ is an injective sheaf, and that we have a natural morphism $\mathcal{F} \to \tilde{J}_x$. Therefore we have a morphism $\mathcal{F} \to \prod_{x \in X} \tilde{J}_x$. The product of any family of injective sheaves is clearly injective. The morphism $\mathcal{F} \to \prod_{x \in X} \tilde{J}_x$ is then a monomorphism of $\mathcal{F}$ into an injective sheaf. ∎

A *complex of sheaves*

$$\cdots \xrightarrow{d^{n-1}} K^n \xrightarrow{d^n} K^{n+1} \xrightarrow{d^{n+1}}$$

is a sequence (indexed by $n \in \mathbb{Z}$ or by $n \in \mathbb{Z}_+$, according to circumstances) of sheaves K^n of abelian groups and of morphisms $d^n : K^n \to K^{n+1}$ such that $d^n \circ d^{n-1} = 0$. The complex (K^n, d^n) is often denoted $K^\bullet$ to simplify notation. The *cohomology sheaf* $\underline{H}^j(K^\bullet)$ is the sheaf associated to the presheaf $Ker(d^j)/Im(d^{j-1})$. A morphism $\psi : K^\bullet \to L^\bullet$ of complexes of sheaves consists of morphisms $\psi^n : K^n \to L^n$ of sheaves of groups such that the following diagram commutes

$$
\begin{array}{ccc}
K^n & \xrightarrow{d^n_K} & K^{n+1} \\
\downarrow{\psi^n} & & \downarrow{\psi^{n+1}} \\
L^n & \xrightarrow{d^n_L} & L^{n+1}
\end{array}
$$

Any such morphism ϕ induces morphisms of cohomology sheaves $\underline{H}^j(K^\bullet) \to \underline{H}^j(L^\bullet)$. Given two such morphisms of complexes of sheaves ψ and ρ from $K^\bullet$ to $L^\bullet$, a *homotopy* H from ψ to ρ is a sequence of morphisms $H^n : K^n \to L^{n-1}$ of sheaves of abelian groups such that $d^{n-1}_L H^n + H^{n+1} d^n_K = \psi^n - \rho^n$ for all n. Two morphisms of complexes are called homotopic if there exists a homotopy between them, in which case the induced maps on cohomology sheaves coincide. A morphism of complexes of sheaves $f : K^\bullet \to L^\bullet$ is called a *homotopy equivalence* if there exists a morphism $g : L^\bullet \to K^\bullet$ such that fg and gf are homotopic to the identity maps.

An exact sequence $A^\bullet \xrightarrow{f} B^\bullet \xrightarrow{g} C^\bullet$ of complexes of sheaves consists of morphisms of complexes $f : A^\bullet \to B^\bullet$ and $g : B^\bullet \to C^\bullet$ such that for all n the sequence $A^n \xrightarrow{f^n} B^n \xrightarrow{g^n} C^n$ is exact.

1.1.14. Proposition. *Let $0 \to A^\bullet \xrightarrow{f} B^\bullet \xrightarrow{g} C^\bullet \to 0$ be an exact sequence of complexes of sheaves. Then we have a long exact sequence of sheaves*

$$\cdots \to \underline{H}^n(A^\bullet) \xrightarrow{H^n(f)} \underline{H}^n(B^\bullet) \xrightarrow{H^n(g)} \underline{H}^n(C^\bullet) \xrightarrow{\delta} \underline{H}^{n+1}(A^\bullet) \to \cdots, \quad (1-4)$$

where δ is the morphism of sheaves defined on stalks as follows. For $s \in \underline{H}^n(C^\bullet)$ and $x \in X$, let σ be an element of B^n_x such that $g_x(\sigma) = s$. Then

$d\sigma$ has zero image in C_x^{n+1}, and so belongs to A_x^{n+1}; $\delta(s)_x$ is then the class of $d\sigma$ in $\underline{H}^{n+1}(A)_x = Ker(d_x^{n+1})/Im(d_x^n)$.

A *resolution* of a sheaf $A \in \mathcal{AB}(X)$ is a complex of sheaves $K^\bullet$ together with a morphism $i : A \hookrightarrow K^0$ in $\mathcal{AB}(X)$ such that

(1) i is a monomorphism with image equal to $Ker(d^0)$.

(2) For $n \geq 1$, the kernel of d^n coincides with the image of d^{n-1}.

Then $\underline{H}^0(K^\bullet) = A$ and $\underline{H}^j(K^\bullet) = 0$ for $j > 0$. The resolution will be denoted by $A \xrightarrow{i} K^\bullet$. Given two resolutions $A \xrightarrow{i_1} K_1^\bullet$ and $A \xrightarrow{i_2} K_2^\bullet$ of A, a morphism of resolutions is a morphism of complexes $\phi : K_1^\bullet \to K_2^\bullet$ such that $\phi^0 i_1 = i_2$ in $Hom_X(A, K_2^0)$.

A complex $K^\bullet$ of sheaves is called *acyclic* if all cohomology sheaves $\underline{H}^j(K^\bullet)$ are zero. Note that given a complex of sheaves $K^\bullet$ and $i : A \to K^0$, $K^\bullet$ is a resolution of A if and only if the complex $A \xrightarrow{i} K^0 \to K^1 \cdots \to$ is acyclic.

1.1.15. Proposition. (1) *For any sheaf A of abelian groups, there exists a resolution $A \xrightarrow{i} I^\bullet$ of A in which each sheaf I^n is injective. This is called an* injective resolution *of A.*

(2) *Let $\phi : A \to B$ be a morphism in $\mathcal{AB}(X)$. Given a resolution $A \xrightarrow{f} I^\bullet$ of A and an injective resolution $B \xrightarrow{g} J^\bullet$ of B, there exists a morphism of complexes $\psi : I^\bullet \to J^\bullet$ such that $\psi^0 \circ f = g \circ \phi$. This morphism of complexes is unique up to homotopy.*

Proof. To prove (1), we construct by induction on $n \in \mathbb{N}$ an exact sequence $A \xrightarrow{i} I^0 \xrightarrow{d^0} I^1 \xrightarrow{d^1} \cdots \xrightarrow{d^{n-1}} I^n$ with $I^0, \cdots, I^n$ injective sheaves. The case $n = 0$ follows easily from Lemma 1.1.13. Having chosen I^n, we then choose an injective sheaf I^{n+1} and a monomorphism $I^n/Im(d^{n-1}) \xrightarrow{d^n} I^{n+1}$ with I^{n+1} an injective sheaf.

To prove (2), we construct $\psi^n : I^n \to J^n$ by induction on n. The existence of ψ^0 follows from the fact that J^0 is injective. If $\psi^0, \psi^1, \ldots, \psi^n$ have been constructed (for some $n \geq 1$), the map $d^n \circ \psi^n : I^n \to J^{n+1}$ has zero composition with d^{n-1}, since $d^n \psi^n d^{n-1} = d^n d^{n-1} \psi^{n-1} = 0$. Hence it factors through a morphism $f : I^n/Im(d^{n-1}) \to J^{n+1}$. We have $I^n/Im(d^{n-1}) = I^n/Ker(d^n)$ since $I^\bullet$ is a resolution of A. Note that d^{n+1} induces a monomorphism $I^n/Im(d^{n-1}) \hookrightarrow I^{n+1}$. Using the fact that J^{n+1} is an injective sheaf to get a morphism $\psi^{n+1} : I^{n+1} \to J^{n+1}$ extending f, we then have $\psi^{n+1} d^n = d^n \psi^n$.

Now let $\rho = (\rho^n)$ be another morphism of complexes of sheaves with the same properties. We construct by induction on $n \geq 0$ a morphism

$H^n : I^n \to J^{n-1}$ such that

$$d^{n-1}H^n + H^{n+1}d^n = \psi^n - \rho^n \qquad (E_n)$$

(by convention, we have $d^{-1} = 0$). Assume that $H^0, H^1, \cdots, H^n$ have been constructed, so that equation E_i is satisfied for $i \leq n-1$. Then $\psi^n - \rho^n - d^{n-1}H^n$ vanishes on $Ker(d^n)$, according to equation E_{n-1}. Therefore there exists a morphism of sheaves $h : Im(d^n) \to J^{n+1}$ such that $hd^n = \psi^n - \rho^n - d^{n-1}H^n$. Then if we take $H^{n+1} : I^{n+1} \to J^{n+1}$ to be any extension of h (which exists since J^{n+1} is injective), E_n will be satisfied. ∎

1.1.16. Corollary. *Given two injective resolutions $I_1^\bullet$ and $I_2^\bullet$ of a sheaf A of abelian groups, there exists a morphism ϕ of resolutions from $I_1^\bullet$ to $I_2^\bullet$, which is a homotopy equivalence. This morphism is unique up to homotopy.*

We are now ready to define sheaf cohomology.

1.1.17. Proposition and Definition. *For a sheaf A of abelian groups in a space X, the __sheaf cohomology groups__ $H^j(X, A)$ are defined as follows. Let $I^\bullet$ be an injective resolution of A. We define $H^j(X, A)$ to be the j-th cohomology group of the complex*

$$\cdots \to \Gamma(X, I^j) \to \Gamma(X, I^{j+1}) \to \cdots$$

Up to canonical isomorphism, these groups are independent of the injective resolution. If A is an injective sheaf, $H^j(X, A) = 0$ for $j > 0$. Any morphism $\phi : A \to B$ induces well-defined group homomorphisms $H^j(X, A) \to H^j(X, B)$. The cohomology group $H^0(X, A)$ is canonically isomorphic to $\Gamma(X, A)$.

Proof. Given two injective resolutions $I_1^\bullet$ and $I_2^\bullet$ of A, according to Proposition 1.1.15, there exists a morphism of resolutions, which gives a morphism of the complex $\Gamma(X, I_1^\bullet)$ to $\Gamma(X, I_2^\bullet)$. This morphism is unique up to homotopy, hence the induced map on cohomology groups is uniquely defined and is an isomorphism. This shows that $H^j(X, A)$ is independent of the resolution. If A is injective, we can choose the injective resolution $I \xrightarrow{Id} I \to 0 \to 0 \cdots$, hence $H^p(X, A) = 0$ for $p > 0$. Given a morphism $\phi : A \to B$ in $\mathcal{AB}(X)$, an injective resolution $I^\bullet$ of A and an injective resolution $J^\bullet$ of B, one can find a morphism $\psi : I^\bullet \to J^\bullet$ of complexes of sheaves as in Proposition 1.1.15. This induces a morphism of the complexes of global sections, hence homomorphisms $H^j(X, A) \to H^j(X, B)$. As the morphism ψ is unique up to homotopy, the induced homomorphisms on cohomology are well-defined. Finally, one has the exact sequence

$$0 \to \Gamma(X, A) \to \Gamma(X, I^0) \to \Gamma(X, I^1)$$

by Lemma 1.1.10. Hence $H^0(X, I^\bullet) = \Gamma(X, A)$. ∎

This definition of sheaf cohomology using injective resolutions is of almost no use to actually compute these groups. It does however permit one to prove a basic result, the long exact sequence for sheaf cohomology.

1.1.18. Proposition. *Let* $0 \to A \xrightarrow{f} B \xrightarrow{g} C \to 0$ *be an exact sequence of sheaves of abelian groups. Let* $A \xrightarrow{u} I^\bullet$ *and* $C \xrightarrow{v} J^\bullet$ *be injective resolutions. Then there exists an injective resolution* $B \xrightarrow{w} K^\bullet$ *and a commutative diagram*

$$
\begin{array}{ccccccccc}
& & 0 & & 0 & & 0 & & \\
& & \downarrow & & \downarrow & & \downarrow & & \\
0 & \to & A & \xrightarrow{u} & I^0 & \to & I^1 & \to & \cdots \\
& & \downarrow f & & \downarrow a^0 & & \downarrow a^1 & & \downarrow \\
0 & \to & B & \xrightarrow{w} & K^0 & \to & K^1 & \to & \cdots \\
& & \downarrow g & & \downarrow b^0 & & \downarrow b^1 & & \downarrow \\
0 & \to & C & \xrightarrow{v} & J^0 & \to & J^1 & \to & \cdots \\
& & \downarrow & & \downarrow & & \downarrow & & \\
& & 0 & & 0 & & 0 & &
\end{array}
$$

with each sequence $0 \to I^n \to K^n \to J^n \to 0$ *exact.*

Proof. We put $K^n = I^n \oplus J^n$. We will define $d_K^{-1} = w : B \to I^0 \oplus J^0$ as $w = \alpha \oplus vg$, where $\alpha : B \to I^0$ satisfies $\alpha f = u$. Then we construct $d_K^n : I^n \oplus J^n \to I^{n+1} \oplus J^{n+1}$ by induction on n (for $n \geq 0$) so that the following conditions are verified for all $n \geq -1$:

(a) d_K^n is represented by a "block matrix" $\begin{pmatrix} d_I^n & h^n \\ 0 & d_J^n \end{pmatrix}$ (we set $d_I^{-1} = u$ and $d_J^{-1} = v$).

(b) $d_K^{n+1} d_K^n = 0$.

What we need to define by induction is $h^n : J^n \to I^{n+1}$. We explain the construction of h^n, assuming that h^{n-1} (hence d_K^{n-1}) has been constructed. The only condition on h^n to be verified is that $h^n d_J^{n-1} = -d_I^n h^{n-1}$. The existence of h^n with this restriction follows from the hypothesis that I^{n+1} is injective and the fact that $(-d_I^n h^{n-1})d_J^{n-2} = d_I^n d_I^{n-1} h^{n-2} = 0$.

Note that conditions (a) and (b) imply that the complex $(I^n \oplus J^n, d_K^n)$ is a resolution of B, using Proposition 1.1.14. Hence we have fulfilled all the required conditions. ■

1.1.19. Corollary. *If $0 \to A \to B \to C \to 0$ is an exact sequence of sheaves of abelian groups, we have a long exact sequence of cohomology groups*

$$\cdots \to H^n(X, A) \to H^n(X, B) \to H^n(X, C) \xrightarrow{\delta} H^{n+1}(X, A) \to \cdots \quad (1-5)$$

Proof. Choose injective resolutions $A \to I^\bullet$ and $C \to J^\bullet$, and construct an injective resolution $B \to K^\bullet$, as in Proposition 1.1.18. Then for each n we have an exact sequence $0 \to I^n \to K^n \to J^n \to 0$. Since I^n is injective, this gives an exact sequence $0 \to \Gamma(X, I^n) \to \Gamma(X, K^n) \to \Gamma(X, J^n) \to 0$. We now have an exact sequence of complexes of abelian groups $0 \to \Gamma(X, I^\bullet) \to \Gamma(X, K^\bullet) \to \Gamma(X, J^\bullet) \to 0$. By Proposition 1.1.14, we get a long exact sequence involving the cohomology groups. ■

1.2. Spectral sequences and complexes of sheaves

In this section we will develop some tools for computing sheaf cohomology, and we will introduce sheaf hypercohomology for a complex of sheaves. We will also introduce spectral sequences and apply them to sheaf hypercohomology.

First we will develop the spectral sequence of a filtered complex. References on this topic are [Go] [McCl] [Spa]. We will have the occasion to consider not only complexes of abelian groups or of modules over some ring, but also complexes of objects in an abelian category (like the category $\mathcal{AB}(X)$ of sheaves of abelian groups on X). If $K^\bullet$ is a complex, a *subcomplex* $L^\bullet$ consists of a family $L^p \subseteq K^p$ of subobjects such that $d(L^p) \subseteq L^{p+1}$. Then $L^\bullet$ itself is a complex, and we can form the quotient complex $K^\bullet/L^\bullet$. A *filtration* of a complex $K^\bullet$ consists of a nested family $\cdots \subseteq F^{p+1} \subseteq F^p \subseteq \cdots$ of subcomplexes F^p of $K^\bullet$. The filtration is called *regular* if for every i there exists n_i such that $F^p \cap K^i = 0$ for $p > n_i$. It is called *exhaustive* if for every i there exists an integer m_i such that $K^i \subseteq F^{m_i}$.

One of the most frequent examples of a filtered complex comes from double complexes. A *double complex* $K^{\bullet\bullet}$ consists of objects $K^{p,q}$ for $(p, q) \in \mathbb{Z}^2$, together with two sorts of differentials:

   (1) The *horizontal differential* δ or δ_K, which maps $K^{p,q}$ to $K^{p+1,q}$.
   (2) The *vertical differential* d or d_K, which maps $K^{p,q}$ to $K^{p,q+1}$.

One requires the following properties:

$$\delta\delta = dd = 0 \quad \text{and} \quad \delta d = d\delta. \qquad (1-6)$$

A double complex $K^{\bullet\bullet}$

Each row $(K^{\bullet,q}, \delta)$ is then a complex, as is each column $(K^{p,\bullet}, d)$, and the differentials δ and d commute. We say that p is the *first degree* of an element of $K^{p,q}$, q is its *second degree*, and $p+q$ is the *total degree*.

Later we will see many examples of double complexes. One of the most well-known comes from complex analysis. If X is a complex manifold, and $\underline{A}^n_{X,\mathbb{C}} = \underline{A}^n_X \otimes \mathbb{C}$ denotes the sheaf of n-forms with complex coefficients, then we have a decomposition $\underline{A}^n_{X,\mathbb{C}} = \oplus_{0\leq p\leq n}\underline{A}^{p,n-p}_X$, where $\underline{A}^{p,n-p}_X$ denotes those n-forms which have local expressions

$$\omega = \sum_{i_1,\ldots,i_p,j_1,\ldots,j_{n-p}} f_{i_1,\ldots,j_{n-p}} \cdot dz_{i_1} \wedge \cdots \wedge dz_{i_p} \wedge \overline{dz}_{j_1} \wedge \cdots \wedge \overline{dz}_{j_{n-p}}.$$

For $\alpha \in \underline{A}^{p,n-p}_X$, the exterior derivative can be written $d\alpha = d'\alpha + d''(\alpha)$, for $d'(\alpha) \in \underline{A}^{p+1,n-p}_X$ and $d''(\alpha) \in \underline{A}^{p,n-p+1}_X$. Then the $\underline{A}^{p,q}_X$ form a double complex of sheaves with horizontal differential d' and vertical differential $(-1)^p d''$ in degree (p,q). The global sections $\Gamma(X, \underline{A}^{p,q}_X) = A^{p,q}(X)$ form a double complex of vector spaces.

This example suggests how to get an ordinary complex $K^{\bullet}$ from a double complex $K^{\bullet\bullet}$. One puts $K^n = \oplus_{p+q=n}K^{p,q}$, with the differential $D = \delta + (-1)^p d$ in bidegree (p,q). This sign is necessary to have a complex. The complex $K^{\bullet}$ is called the *total complex* of $K^{\bullet\bullet}$ and is often denoted by $Tot(K^{\bullet\bullet})$. The cohomology $H^n(K^{\bullet\bullet})$ is simply the cohomology of the total complex. So a degree n cycle c consists of a finite family $c_p \in K^{p,n-p}$ such that

$$\delta(c_p) = (-1)^p d(c_{p+1}) \in K^{p+1,n-p}.$$

Clearly cycles may be hard to construct because of the presence of the two differentials d and δ. One would like in first approximation to neglect one of the differentials. This is what the theory of spectral sequences accomplishes in this case.

There are two natural filtrations on the total complex $K^\bullet$. The *first filtration* $F_H^\bullet$ is given by the subcomplexes $F_H^p = \oplus_{i \geq p} K^{i,j}$, with degree n component $\oplus_{i \geq p} K^{i,n-i}$. The *second filtration* F_V is given by the subcomplexes $F_V^p = \oplus_{j \geq p} F^{i,j}$, with degree n component $\oplus_{j \geq p} F^{n-j,j}$.

A morphism $\phi : K^{\bullet\bullet} \to L^{\bullet\bullet}$ of double complexes consists of homomorphisms $\phi^{p,q} : K^{p,q} \to L^{p,q}$ such that $\delta_L \phi = \phi \delta_K$ and $d_L \phi = \phi d_K$. Then ϕ induces morphisms of the corresponding total complexes, as well as homomorphisms $H^n(K^{\bullet\bullet}) \to H^n(L^{\bullet\bullet})$.

The spectral sequence for a filtered complex is an algebraic tool used to extract information about the cohomology of the complex $K^\bullet$ from knowledge of the cohomology of the *graded quotients* $Gr_F^p(K^\bullet) = F^p/F^{p+1}$ of the filtration, which is often much easier to compute. As a matter of notation, we will denote by D the differential of $K^\bullet$, as this is our notation for the total differential of a double complex. For any integer $r \geq 1$, we define a new filtration $\cdots \subseteq Z_r^{p+1} \subseteq Z_r^p \subseteq \cdots$ by $Z_r^p = \{x \in F^p : Dx \in F^{p+r}\}$. This is a graded group. We put $Z_r^{p,q}$ to be the degree $p + q$-component of Z_r^p. We define the E_r-term of the spectral sequence to be $E_r^{p,q} = Z_r^{p,q}/(DZ_{r-1}^{p-r+1,q+r-1} + Z_{r-1}^{p+1,q-1})$. This definition may seem rather arbitrary at first. The subscript r tells one how far one is in the spectral sequence. As we will see, one starts by computing the E_1-term, then the E_2-term, and so on. The first degree (the *filtering degree*) p tells one in which step F^p of the filtration one is located, and the *complementary degree* q is such that $p + q$ is the *total degree* in the complex $K^\bullet$. We next construct the differential

$$d_r : E_r^{p,q} \to E_r^{p+r,q-r+1}. \qquad (1-7)$$

Let $a \in E_r^{p,q}$ be represented by some $x \in F^p$ such that $Dx \in F^{p+r}$. Then $Dx \in F^{p+r}$ will be a representative of $d_r(x)$. To see that this is well-defined independently of the choice of x, one needs to verify that for $x \in Z_{r-1}^{p+1}$, the class of Dx in $E_r^{p+r,q-r+1}$ is trivial. This is clear, since $Dx \in dZ_{r-1}^{p+1}$ has zero class in $E_r^{p+r,q-r+1}$.

1.2.1. Proposition. *We have $d_r d_r = 0$, and the cohomology*

$$Ker(d_r : E_r^{p,q} \to E_r^{p+r,q-r+1})/Im(d_r : E_r^{p-r,q+r-1} \to E_r^{p,q})$$

identifies canonically with $E_{r+1}^{p,q}$.

Proof. Let us describe the group $A = Ker(d_r : E_r^{p,q} \to E_r^{p+r,q-r+1})$. It consists of the classes in $E_r^{p,q}$ represented by some $x \in F^p(K^{p+q})$ such that Dx is of the form $Dz + y$, where $y \in F^{p+r+1}$ and $z \in F^{p+1}$ with $Dz \in F^{p+r}$. This class is represented also by $x - z$, which satisfies $D(x-z) = y \in F^{p+r+1}$. Thus one has a well-defined map $f : A \to E_{r+1}^{p,q}$, which maps the class of x to the class of $x - z$. The kernel of this map consists of the classes of $x \in F^p$ such that $x - z$ is of the form $x - z = u + Dv$, for $u \in Z_r^{p+1}$ and $v \in Z_r^{p-r}$. So $u \in F^{p+1}$, $Du \in F^{p+r+1}$, and $v \in F^{p-r}$, $Dv \in F^p$. Therefore $[x] \in E_r^{p,q}$ is equal to $d_r([v])$, where $[v] \in E_r^{p-r,q+r-1}$. So the kernel of f is contained in the image of d_r. The same line of reasoning proves the reverse inclusion. ∎

We have the terms $E_1^p = \oplus_q E_1^{p,q}$, $E_2^p = \oplus_q E_2^{p,q}$, and so on of the spectral sequence; each of them gives a complex, with a differential d_r of degree r. The cohomology of the complex $E_r^{\bullet}$ is equal to E_{r+1}. What we are really interested in is the cohomology $H^n(K^{\bullet})$ of the complex $K^{\bullet}$ itself. This cohomology acquires a filtration $F^p(H^n(K^{\bullet}))$ from the filtration F^p of $K^{\bullet}$. The term E_{∞}^p of the spectral sequence is the quotient $F^p(H^n(K^{\bullet}))/F^{p+1}(H^n(K^{\bullet}))$. In analogy with the definition of E_r, one may also define it as follows. Let $Z_{\infty}^p = \{x \in F^p : Dx = 0\}$ be the group of cycles in F^p, and let $B_{\infty}^p = F^p \cap Im(D)$ be the group of boundaries in F^p. We now have $E_{\infty}^p = Z_{\infty}^p / B_{\infty}^p + Z_{\infty}^{p+1}$.

The sense in which the E_r approximate E_{∞} is made precise in the next proposition.

1.2.2. Proposition. *Let F^p be a filtration of the complex $K^{\bullet}$. Assume that for every n we have integers $m_n \le q_n$ such that $K^n \cap F^{q_n} = 0$ and $K^n \subseteq F^{m_n}$. Let $r = r(n) = \max(q_{n+1} - m_n, q_n - m_{n-1})$. Then for any p we have*

$$E_r^{p,n-p} = E_{r+1}^{p,n-p} = \cdots = E_{\infty}^{p,n-p}.$$

Proof. The only values of p for which $E_s^{p,n-p}$ may be non-zero are in the interval $\{m_n, \cdots, q_n - 1\}$. The differential $d_s : E_s^{p,n-p} \to E_s^{p+s,n-s+1}$ is therefore zero if $s + m_n \ge q_{n+1}$. This shows that for r as in the statement this differential, as well as the differential $d_r : E_s^{p-r,n-p+r-1} \to E_s^{p,n-p}$, are zero. We have $E_{s+1}^{p,n-p} = E_s^{p,n-p}$ for $s \ge r$. The same argument shows that $E_s^{p,n-p} = E_{\infty}^{p,n-p}$ for $s >> 0$. ∎

When the conclusion of Proposition 1.2.2 holds for all p and n (for suitable r, which may depend on (p,n)), we say that the spectral sequence *converges* to E_∞^p. One calls E_∞ the *abutment* of the spectral sequence. Proposition 1.2.2 says that if the filtration F^p is regular and exhaustive, the spectral sequence is convergent. It is commonly said that the spectral sequence "converges to $H^n(K^\bullet)$." We will avoid this terminology in order to emphasize that the spectral sequence does not allow one to compute the cohomology groups $H^n(K^\bullet)$ themselves, but only the graded quotients $E^{p,n-p}(K^\bullet)$ of the filtration $F^p(H^n(K^\bullet))$. It may be a non-trivial matter to compute $H^n(K^\bullet)$ from these graded quotients.

We say that a spectral sequence *degenerates at* E_r if we have $E_r = E_{r+1} = \cdots = E_\infty$. Here is a frequently used case of degeneracy.

1.2.3. Proposition. *Consider a complex with a regular and exhaustive filtration. Assume that for some integers p and $r \geq 2$, the following holds. For every n we have $E_r^{n-l,l} = 0$ for $l \neq p$. Then the spectral sequence degenerates at E_r, and for every n we have a canonical isomorphism $H^n(K^\bullet) \xrightarrow{\sim} E_r^{n-l,l}$.*

Proof. We know that the spectral sequence converges. It therefore suffices to show that $E_s = E_{s+1}$ for any $s \geq r$. Assume it is true for $r \leq s' < s$. Then we have $E_s^{n-l,l} = E_r^{n-l,l}$ by the inductive assumption, and this is zero unless $l = p$. The differential $d_s : E_s^{n-l,l} \rightarrow E_s^{n-l+s,l-s+1}$ could only be non-zero if $l = p$ and $l - s + 1 = p$, which is impossible as $s \geq 2$. Hence $E_s = E_{s+1}$. ■

The beginning terms of a spectral sequence are easy to understand in principle. We have $E_1^{p,q} = H^{p+q}(F^p/F^{p+1})$. The differential $d_1 : H^{p+q}(F^p/F^{p+1}) \rightarrow H^{p+q+1}(F^{p+1}/F^{p+2})$ is the connecting homomorphism in the exact sequence of complexes

$$0 \rightarrow F^{p+1}/F^{p+2} \rightarrow F^p/F^{p+2} \rightarrow F^p/F^{p+1} \rightarrow 0.$$

We have the notion of morphism $\phi : E_r^{p,q} \rightarrow (E')_r^{p,q}$ of spectral sequences. This means that for each r we have a homomorphism $\phi_r : E_r^{p,q} \rightarrow (E')_r^{p,q}$, such that $d_r\phi_r = \phi_r d_r$. Given filtered complexes $(K^\bullet, F^p)$ and $(L^\bullet, G^p)$, a morphism $\phi : K^\bullet \rightarrow L^\bullet$ is said to be compatible with the filtrations if $\phi(F^p) \subseteq G^p$ for all p. Such ϕ induces a morphism of spectral sequences.

1.2.4. Proposition. *Let $E_r^{\bullet,\bullet}$ and $(E')_r^{\bullet,\bullet}$ be convergent spectral sequences, and let $\phi_r : E_r^{p,q} \rightarrow (E')_r^{p,q}$ be a morphism of spectral sequences. If for some s, ϕ_s is an isomorphism, ϕ_r is an isomorphism for all $r \geq s$, including $r = \infty$.*

This is very useful in comparing the cohomology of various complexes.

We now consider the important case of the first filtration of a double complex $K^{\bullet\bullet}$. The corresponding spectral sequence is called the first spectral sequence. Recall that $F^p = \oplus_{i \geq p} K^{i,j}$. The E_1-term is the cohomology of the graded quotients of the filtration. The complex F^p/F^{p+1} is the "column complex"

$$
\begin{array}{c}
\vdots \\
\uparrow {\scriptstyle (-1)^{j+1} \cdot d} \\
K^{p,j+1} \\
\uparrow {\scriptstyle (-1)^{j} \cdot d} \\
K^{p,j} \\
\uparrow {\scriptstyle (-1)^{j-1} \cdot d} \\
\vdots
\end{array}
$$

with $K^{p,j}$ in degree $p+j$. So $E_1^{p,q}$ is the bidegree (p,q)-vertical cohomology (i.e., the cohomology of $K^{\bullet\bullet}$ with respect to the vertical differential alone). Then the differential $d_1 : E_1^{p,q} \to E_1^{p+1,q}$ is the map on vertical cohomology induced by the horizontal differential δ. As for the E_2-term, we have

$$E_2 = H_H(H_V(K^{\bullet\bullet})) \qquad (1-8)$$

(horizontal cohomology of the vertical cohomology).

Convergence of the first spectral sequence poses no problem for a double complex $K^{\bullet\bullet}$ such that $K^{p,q} \neq 0$ implies $p \geq 0$ and $q \geq 0$. Such a double complex is often said to be concentrated in the first quadrant. The corresponding spectral sequence is said to be a "first quadrant spectral sequence." For such a spectral sequence, there is a homomorphism $H^n(K^{\bullet}) \to E_2^{n,0}$, with image equal to the subgroup $E_\infty^{n,0}$ of $E_2^{n,0}$. This is called an *edge homomorphism*. Using this edge homomorphism, one obtains the exact sequence for low degree cohomology

$$0 \to E_2^{1,0} \to H^1(K^{\bullet}) \to E_2^{0,1} \to E_2^{2,0} \to H^2(K^{\bullet})$$

The second spectral sequence of a double complex, corresponding to the second filtration, is analyzed similarly. The $E_1^{p,q}$-term is the horizontal cohomology in bidegree (p,q). The E_2-term is

$$E_2 = H_V(H_H(K^{\bullet\bullet})). \qquad (1-9)$$

The following result is very useful.

1.2.5. Lemma. *Let $K^{\bullet\bullet}$ and $L^{\bullet\bullet}$ be double complexes such that there exists k for which $K^{p,q} = L^{p,q} = 0$ unless $p \geq k$ and $q \geq k$. Let $\phi : K^{\bullet\bullet} \to L^{\bullet\bullet}$ be a morphism of double complexes. If ϕ induces an isomorphism of horizontal cohomology groups $H_H^{p,q}(K^{\bullet\bullet}) \xrightarrow{\sim} H_H^{p,q}(L^{\bullet\bullet})$, then ϕ induces an isomorphism $H^j(K^{\bullet\bullet}) \xrightarrow{\sim} H^j(L^{\bullet\bullet})$. The same conclusion holds if ϕ induces an isomorphism of the vertical cohomology groups.*

Proof. This is an easy consequence of Proposition 1.2.4. The spectral sequences for the second filtrations of $K^{\bullet\bullet}$ and $L^{\bullet\bullet}$ are convergent. ϕ induces a morphism between these spectral sequences. It is an isomorphism on the E_1-terms. Hence we also have an isomorphism of the E_∞-terms. Now, for a given n, the $E_\infty^{p,n-p}(K^{\bullet\bullet})$-terms are the graded quotients for a finite (i.e., regular and exhaustive) filtration F^p of $H^n(K^{\bullet\bullet})$. A similar statement holds true for $E_\infty^{p,n-p}(L^{\bullet\bullet})$. Since we get an isomorphism of the graded quotients, the statement follows. ∎

An important case of a spectral sequence is that of a *spherical* spectral sequence. This means that for some $k \geq 1$, we have $E_2^{p,q} = 0$ unless $q = 0$ or $q = k$ (we will see examples of this in §6). Then we have $E_r^{p,q} = E_2^{p,q}$ for $r \leq k + 1$, followed by

$$E_{k+2}^{p,0} = E_2^{p,0}/d_{k+1}\,(E_2^{p-k-1,k})$$
$$E_{k+2}^{p,k} = Ker(d_{k+1} : E_2^{p,k} \to E_2^{p+k+1,0})$$

The spectral sequence degenerates at E_{k+2}. We can then put together the short exact sequences

$$0 \to E_\infty^{n,0} \to H^n(K^\bullet) \to E_\infty^{n-k,k} \to 0$$

into a long exact sequence

$$\cdots \to E_2^{n-k-1,k} \xrightarrow{d_{k+1}} E_2^{n,0} \to H^n(K^\bullet)$$
$$\to E_2^{n-k,k} \xrightarrow{d_{k+1}} E_2^{n+1,0} \cdots \to \qquad (1-10)$$

We will need double complexes in order to define and compute hypercohomology of a complex of sheaves. Many natural complexes of sheaves occur in geometry, for instance the *de Rham complex* $\underline{A}_M^\bullet$ on a smooth manifold:

$$\underline{A}_M^0 \xrightarrow{d} \underline{A}_M^1 \xrightarrow{d} \cdots,$$

where the sheaf $\underline{A}^p_M$ of p-forms is put in degree p, and d denotes exterior differentiation. A complex of sheaves $K^\bullet$ is said to be *bounded below* if there exists some integer $k \in \mathbb{Z}$ such that $K^p = 0$ for $p < k$.

1.2.6. Proposition. *Let $(K^\bullet, d_K)$ be a complex of sheaves on a space X, which is bounded below. There exists a double complex $(I^{\bullet\bullet}, \delta, d)$, with $I^{p,q} = 0$ for $p < 0$, and a morphism of complexes $u : K^\bullet \to (I^{0,\bullet}, d)$ such that*

(1) *For each $q \in \mathbb{Z}$, the complex of sheaves $I^{\bullet,q}$ (with differential δ) is an injective resolution of K^q.*

(2) *For each $q \in \mathbb{Z}$, the complex of sheaves $d(I^{\bullet,q-1}) \subseteq I^{\bullet,q}$ is an injective resolution of $d_K(K^{q-1})$.*

(3) *For each $q \in \mathbb{Z}$, the complex of sheaves $Ker(d) \subseteq I^{\bullet,q}$ is an injective resolution of $Ker(d_K : K^{q-1} \to K^q)$.*

(4) *For each $q \in \mathbb{Z}$, the complex of sheaves $\underline{H}^{\bullet,q}_H(I^{\bullet\bullet})$ (horizontal cohomology) is an injective resolution of $\underline{H}^q(K^\bullet)$.*

Furthermore, if $f : K^\bullet \to L^\bullet$ is a morphism of bounded below complexes of sheaves, and if $I^{\bullet\bullet}$ and $J^{\bullet\bullet}$ are double complexes as described above, f can be extended to a morphism of double complexes, unique up to homotopy.

Proof.   We will use Proposition 1.1.15 repeatedly. Let k be such that $K^q = 0$ for $q < k$. First we construct an injective resolution $H^{\bullet,q}$ of $\underline{H}^q(K^\bullet)$ for all q, and an injective resolution $R^{\bullet,q}$ of $\underline{d}_K(K^{q-1})$. Then we use Proposition 1.1.15 to find an injective resolution $S^{\bullet,q}$ of $Ker(d : K^q \to K^{q+1})$, by induction on $q \geq k$, such that we have a commutative diagram

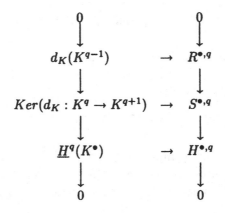

We find, by induction on q, an injective resolution $(I^{\bullet,q}, \delta)$ of K^q and a

commutative diagram

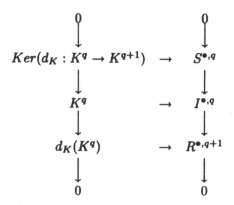

We define the vertical differential $d : I^{\bullet,q} \to I^{\bullet,q+1}$ to be the composition of the maps $I^{\bullet,q} \to R^{\bullet,q+1}$, $R^{\bullet,q+1} \to S^{\bullet,q+1}$ and $S^{\bullet,q+1} \to I^{\bullet,q+1}$. The composition dd is 0, because it factors through a composition $S \to I \to R$, which is 0 by construction. Because d is a morphism of complexes, we have a double complex. All other properties of $I^{\bullet\bullet}$ are then easily verified. The last part of the statement is proved in the same way as Proposition 1.1.15. ∎

The double complex $I^{\bullet\bullet}$ is called an injective resolution of $K^{\bullet}$. We will use injective resolutions to define the hypercohomology groups of a complex of sheaves.

1.2.7. Definition and Proposition. *Let $K^{\bullet}$ be a bounded below complex of sheaves on a space X. The <u>hypercohomology group</u> $H^p(X, K^{\bullet})$ is the p-th cohomology group of the double complex $\Gamma(X, I^{p,q})$, where $I^{\bullet\bullet}$ is an injective resolution of $K^{\bullet}$. These cohomology groups are defined independently of the injective resolution. Any morphism $\phi : K^{\bullet} \to L^{\bullet}$ of complexes of sheaves induces group homomorphisms $H^p(X, K^{\bullet}) \to H^p(X, L^{\bullet})$.*

Proof. This follows easily from Proposition 1.2.6. ∎

Note that a sheaf A may be viewed as a complex of sheaves, which is A in degree 0 and 0 in all other degrees. Then for such a complex we recover sheaf cohomology as defined in §1. The following is a generalization of Proposition 1.1.18 and Corollary 1.1.19 to complexes of sheaves.

1.2.8. Proposition. *Let $0 \to A^{\bullet} \to B^{\bullet} \to C^{\bullet} \to 0$ be an exact sequence of bounded below complexes of sheaves. Let $I^{\bullet\bullet}$ be an injective resolution of $A^{\bullet}$, and $J^{\bullet\bullet}$ an injective resolution of $C^{\bullet}$. Then there exists an injective* *resolution $K^{\bullet\bullet}$ of $B^{\bullet}$ and a commutative diagram*

$$
\begin{array}{ccccccccc}
0 & \to & I^{\bullet\bullet} & \to & K^{\bullet\bullet} & \to & J^{\bullet\bullet} & \to & 0 \\
& & \uparrow & & \uparrow & & \uparrow & & \\
0 & \to & A^{\bullet} & \to & B^{\bullet} & \to & C^{\bullet} & \to & 0
\end{array}.
$$

1.2.9. Corollary. *Under the assumptions of Proposition 1.2.8, there exists a long exact sequence*

$$
\cdots \to H^p(X, A^{\bullet}) \to H^p(X, B^{\bullet}) \to H^p(X, C^{\bullet}) \xrightarrow{\delta} H^{p+1}(X, A^{\bullet}) \to \cdots
$$

At first sight, hypercohomology looks like a formidable thing. But one can often get a handle on it using spectral sequences.

1.2.10. Proposition. *Let $K^{\bullet}$ be a bounded below complex of sheaves on a space X. Then there is a convergent spectral sequence with $E_1^{p,q} = H^q(X, K^p)$ and $E_{\infty}^{p,q} = Gr_F^p(H^{p+q}(X, K^{\bullet}))$ for some filtration F. The differential $d_1 : H^q(X, K^p) \to H^q(X, K^{p+1})$ is induced by the morphism of sheaves $d_K^p : K^p \to K^{p+1}$.*

Proof. Take an injective resolution $I^{\bullet\bullet}$ of $K^{\bullet}$ as in Proposition 1.2.6 (it must only satisfy property (1) in 1.2.6). Then $H^{\bullet}(K^{\bullet})$ is the cohomology of the double complex $\Gamma(X, I^{\bullet\bullet})$. Consider the second filtration of this double complex. The E_1-term is the vertical cohomology, i.e., $E_1^{p,q}$ is the degree q-cohomology of the complex $\Gamma(X, I^{\bullet,p})$. Since $I^{\bullet,p}$ is an injective resolution of the sheaf K^p, we have $E_1^{p,q} = H^q(X, K^p)$. The d_1-differential is induced by the horizontal differential, which is a morphism of complexes $\Gamma(X, I^{\bullet,p}) \to \Gamma(X, I^{\bullet,p+1})$ induced by the morphism of complexes $I^{\bullet,p} \xrightarrow{d} I^{\bullet,p+1}$ from an injective resolution of K^p to an injective resolution of K^{p+1}. Hence $d_1 : H^q(X, K^p) \to H^q(X, K^{p+1})$ is induced by $d_K : K^p \to K^{p+1}$. As $K^{\bullet\bullet}$ is bounded below, there exists some k such that $I^{p,q} = 0$ unless $p \geq 0$ and $q \geq k$. Therefore the spectral sequence converges. The E_{∞}-terms are then the graded quotients for the filtration of $H^{\bullet}(\Gamma(X, I^{\bullet\bullet})) = H^{\bullet}(X, K^{\bullet})$ induced by the filtration of $K^{\bullet}$ by the subcomplexes

$$
F^p(K^{\bullet}) = 0 \to \cdots \to 0 \to K^p \to K^{p+1} \to \cdots. \qquad \blacksquare
$$

1.2.11. Corollary *Let $K^{\bullet}$ be a bounded below complex of sheaves on a space X. Assume that each sheaf $K^{\bullet}$ is <u>acyclic</u>, i.e., satisfies $H^q(X, K^p) = 0$ for*

$q > 0$. *Then we have a canonical isomorphism between the hypercohomology groups* $H^n(X, K^\bullet)$ *and the cohomology groups of the complex*

$$\cdots \to \Gamma(X, K^p) \to \Gamma(X, K^{p+1}) \to \cdots.$$

Proof. The assumption means that $E_1^{p,q} = 0$ for $q \neq 0$. The same holds for the E_2-term, and by Proposition 1.2.3 we have $H^n(K^\bullet) = E_2^{n,0}$. Now the E_2-terms $E_2^{n,0}$ are the cohomology groups of the complex of groups with $E_1^{n,0} = \Gamma(X, K^n)$ in degree n. ∎

In effect this means that acyclic sheaves are as good as injective sheaves for sheaf cohomology. To justify this fully, we need to introduce the so-called hypercohomology spectral sequence. Let $K^\bullet$ be a bounded below complex of sheaves on X, such that each sheaf K^p is acyclic. Let $I^{\bullet\bullet}$ be an injective resolution of $K^\bullet$ which satisfies the assumptions of Proposition 1.2.6. We consider the double complex of abelian groups $\Gamma(X, I^{\bullet\bullet})$. The horizontal cohomology is the cohomology of the complex $\Gamma(X, I^{\bullet,p})$, i.e., the sheaf cohomology $H^q(X, K^p)$, which is zero for $q > 0$. Hence the second spectral sequence degenerates at E_2, and the total cohomology of the double complex $\Gamma(X, I^{\bullet\bullet})$ identifies with the cohomology of the complex $\Gamma(X, K^\bullet)$.

On the other hand, the first spectral sequence has as E_1-term the vertical cohomology

$$E_1^{p,q} = Ker\ (d : \Gamma(X, I^{p,q}) \to \Gamma(X, I^{p,q+1}))/d\ \Gamma(X, I^{p,q-1}).$$

Recall, with the notations of the proof of Proposition 1.2.6, that we have
(1) $S^{\bullet,q} = Ker\ (d : I^{\bullet,q} \to I^{\bullet,q+1})$ is an injective resolution of $Ker(d_K^q)$.
(2) $R^{\bullet,q} = I^{\bullet,q}/S^{\bullet,q}$ is an injective resolution of $Im(d_K^{q-1})$.
(3) $H^{\bullet,q} = S^{\bullet,q}/R^{\bullet,q}$ is an injective resolution of $\underline{H}^q(K^\bullet)$.
We therefore get

$$E_1^{p,q} = \Gamma(X, S^{p,q})/\Gamma(X, R^{p,q}) = \Gamma(X, H^{p,q}).$$

The differential $d_1 : \Gamma(X, H^{p,q}) \to \Gamma(X, H^{p+1,q})$ is induced by the differential $H^{p,q} \to H^{p+1,q}$ in the injective resolution $H^{\bullet,q}$ of $\underline{H}^q(K^\bullet)$. Hence we have

$$E_2^{p,q} = H^p(X, \underline{H}^q(K^\bullet)). \qquad (1-11)$$

1.2.12. Proposition. *For any bounded below complex of sheaves* $K^\bullet$ *on a space* X, *there is a convergent spectral sequence* $E_2^{p,q} = H^p(X, \underline{H}^q(K^\bullet))$, *with E_∞-term the graded quotients of some filtration of $H^\bullet(X, K^\bullet)$. Any morphism of complexes of sheaves induces a morphism of the corresponding spectral sequences.*

Proof. We have proved this above in case $K^\bullet$ is a complex of acyclic sheaves. Given any bounded below complex of sheaves $K^\bullet$, there exists an injective resolution $I^{\bullet\bullet}$ of $K^\bullet$. Let $I^\bullet$ be the total complex of $I^{\bullet\bullet}$. Then the hypercohomology of $K^\bullet$ is by definition that of $I^\bullet$, and we consider the spectral sequence that we obtained for $I^\bullet$, which gives what we want, since $\underline{H}^p(K^\bullet) \xrightarrow{\sim} \underline{H}^p(I^\bullet)$. ∎

This is the so-called *hypercohomology spectral sequence*, which has very nice consequences. We say that a morphism $\phi : K^\bullet \to L^\bullet$ of complexes of sheaves is a *quasi-isomorphism* if for every n, ϕ induces an isomorphism $\underline{H}^n(K^\bullet) \xrightarrow{\sim} \underline{H}^n(L^\bullet)$ of cohomology sheaves.

1.2.13. Proposition. *Let $K^\bullet$ and $L^\bullet$ be bounded below complexes of sheaves, and let $\phi : K^\bullet \to L^\bullet$ be a quasi-isomorphism. For each n the induced homomorphism $H^n(X, K^\bullet) \to H^n(X, L^\bullet)$ is an isomorphism. In particular, the cohomology $H^n(X, A)$ of a sheaf A identifies with the hypercohomology $H^n(X, K^\bullet)$ of any resolution $K^\bullet$ of A. In case $K^\bullet$ is an acyclic resolution of A, then $H^j(X, A)$ is the j-th cohomology group of the complex*

$$\cdots \to \Gamma(X, K^p) \to \Gamma(X, K^{p+1}) \to \cdots$$

Proof. Let $E_r^{\bullet,\bullet}(K^\bullet)$ and $E_r^{\bullet,\bullet}(L^\bullet)$ be the hypercohomology spectral sequences for $K^\bullet$ and $L^\bullet$. According to Proposition 1.2.12, we have a natural morphism of filtered complexes for the injective resolutions; hence we have a morphism of spectral sequences $\phi_r : E_r^{p,q}(K^\bullet) \to E_r^{p,q}(L^\bullet)$. For $r = 1$, ϕ_1 is an isomorphism by the assumption. Since both spectral sequences are convergent, we see from Proposition 1.2.5 that ϕ_∞ is an isomorphism. Now we have a homomorphism $\phi : H^n(X, K^\bullet) \to H^n(X, L^\bullet)$, which is compatible with the finite filtrations of both groups, and induces isomorphisms on the quotient groups. It is then easy to see that it is an isomorphism. ∎

1.3. Čech cohomology and hypercohomology

This section is very much inspired by [**Go, Chapter 5**]. We have also incorporated some ideas from [**Mi**].

We will define the Čech cohomology of a presheaf A of abelian groups on a space X with respect to an open covering $\mathcal{U} = (U_i)_{i \in I}$. Recall that

$A(\emptyset) = \{0\}$. As opposed to sheaf cohomology, it will be defined explicitly in terms of the sections of A over finite intersections of open sets U_i. For $(i_0, \ldots, i_p) \in I$, $U_{i_0,\ldots,i_p}$ denotes the intersection $U_{i_0} \cap \cdots \cap U_{i_p}$. For $p \geq 0$, let $C^p(\mathcal{U}, A) = \prod_{i_0,\ldots,i_p} A(U_{i_0,\ldots,i_p})$, the product ranging over $(p+1)$-tuples of elements of I. An element $\underline{\alpha}$ of $C^p(\mathcal{U}, A)$ is a family $\alpha_{i_0,\ldots,i_p} \in A(U_{i_0,\ldots,i_p})$. Define a homomorphism $\delta : C^p(\mathcal{U}, A) \to C^{p+1}(\mathcal{U}, A)$ as follows:

$$\delta(\underline{\alpha})_{i_0,\ldots,i_{p+1}} = \sum_{j=0}^{p+1} (-1)^j \, (\alpha_{i_0,\ldots,i_{j-1},i_{j+1},\ldots,i_{p+1}})/U_{i_0,\ldots,i_{p+1}}. \tag{1-12}$$

Note that since $A(\emptyset) = \{0\}$, we need consider only those p-tuples $(i_0, \ldots, i_p)$ for which $U_{i_0,\ldots,i_p} \neq \emptyset$ in the definition of $C^p(\mathcal{U}, A)$.

1.3.1. Proposition and Definition. *We have $\delta\delta = 0$. The degree p cohomology of the complex of groups*

$$\cdots \xrightarrow{\delta} C^p(\mathcal{U}, A) \xrightarrow{\delta} C^{p+1}(\mathcal{U}, A) \xrightarrow{\delta} \cdots \tag{1-13}$$

is denoted by $\check{H}^p(\mathcal{U}, A)$ and called the degree p Čech cohomology of the covering $\mathcal{U}$ with coefficients in the presheaf A. If A is a sheaf, we have $\check{H}^0(\mathcal{U}, A) = \Gamma(X, A)$. An element $\underline{\alpha} = (\alpha_{i_0,\ldots,i_p})$ of $C^p(\mathcal{U}, A)$ is called a Čech p-cochain of the covering $\mathcal{U}$ with coefficients in A. If $\delta(\underline{\alpha}) = 0$, $\underline{\alpha}$ is called a Čech p-cocycle with coefficients in A. $\delta(\underline{\alpha})$ is called the coboundary of $\underline{\alpha}$. The image of the p-cocycle $\underline{\alpha}$ in $\check{H}^p(\mathcal{U}, A)$ is called the cohomology class of $\underline{\alpha}$. The complex (1-13) is called the Čech complex of $\mathcal{U}$ with coefficients in A and is denoted $C^\bullet(\mathcal{U}, A)$.

Proof. We have

$$\delta\delta(\underline{\alpha})_{i_0,\ldots,i_{p+2}} = \sum_{1 \leq j,k \leq p+2, j \neq k} (-1)^{j+k+1+H(k-j)} \, \alpha_{i_0,\ldots,\hat{i}_j,\ldots,\hat{i}_k,\ldots,i_{p+2}}$$

with the convention that

$$H(k-j) = \left\{ \begin{array}{l} 1 \text{ if } k-j > 0 \\ 0 \text{ if } k-j < 0 \end{array} \right\}.$$

It follows that each term appears twice, with signs canceling. The group $\check{H}^0(\mathcal{U}, A)$ is the kernel of the map $\delta : \prod_i A(U_i) \to A(U_{ij})$. A family $(s_i \in A(U_i))$ is in the kernel of δ if and only if for all $i, j \in I$, we have $s_i = s_j$ on U_{ij}. Since A is a sheaf, this means that the s_i glue together to form a (unique) element of $\Gamma(X, A)$. ∎

Degree 1 Čech cohomology appears already implicitly in the nineteenth century work by Cousin on finding a meromorphic function f on a Riemann surface S with poles at most at a finite set of points $x_1, \ldots, x_n$, with given polar part g_k at each x_k. By polar part at x_k, we mean the "negative part" of the Laurent series of f at x_k. There is locally no obstruction; hence one may find an open covering U_i of S, and a meromorphic function f_i over U_i, with the required polar parts. Then $\alpha_{ij} = f_j - f_i$ is holomorphic on U_{ij}, because the polar parts at each x_k cancel out. We have

$$\delta(\alpha)_{ijk} = \alpha_{jk} - \alpha_{ik} + \alpha_{ij} = f_k - f_j + f_i - f_k + f_j - f_i = 0$$

over U_{ijk}. So α_{ij} is a Čech 1-cocycle with values in the sheaf $\mathcal{O}_S$ of holomorphic functions. If the cohomology class of α is zero, there exist $h_i \in \Gamma(U_i, \mathcal{O}_S)$ such that $h_j - h_i = \alpha_{ij}$ over U_{ij}. Then $f_i - h_i = f_j - h_j$ over U_{ij}, so we obtain a global meromorphic function over S with the given polar parts at $x_1, \ldots, x_n$. Conversely, if such a meromorphic function f exists, then $h_i = f_i - f$ is a Čech 0-cochain of (U_i) with coefficients in $\mathcal{O}_S$, with coboundary equal to α. In conclusion, the problem of Cousin has a positive solution if and only if the cohomology class of α_{ij} in $\check{H}^1(\mathcal{U}, \mathcal{O}_S)$ is zero. It should be pointed out that it is largely the Cousin problem (in higher dimensions) which led H. Cartan and Oka to the concept of sheaf.

We will also see in Chapter 2 that the transition functions for a line bundle lead to a Čech 1-cocycle with coefficients in the sheaf $\mathbb{C}^*$ of smooth $\mathbb{C}^*$-valued functions. We want to view Čech cohomology as a very hands-on and geometrically significant way to construct classes in sheaf cohomology, which itself is rather abstract and nebulous so far. We start with a simple result.

1.3.2. Lemma. *Let* $\mathcal{U} = (U_i)_{i \in I}$ *be an open covering of the space* X *such that* $X = U_i$ *for some* $i \in I$. *Then for any presheaf* A *of abelian groups on* X, *we have* $\check{H}^p(\mathcal{U}, A) = 0$ *for* $p > 0$.

Proof. We construct a homotopy H, i.e., a homomorphism $H : C^p(\mathcal{U}, A) \to C^{p-1}(\mathcal{U}, A)$ such that $\delta H + H\delta = Id$ on $C^p(\mathcal{U}, A)$ for $p > 0$. This will imply the statement. Indeed, let $\underline{\alpha}$ be a degree p Čech cocycle. Then $\underline{\alpha} = \delta(H\underline{\alpha})$. We put $H(\underline{\alpha})_{i_0, \ldots, i_{p-1}} = \alpha_{i, i_0, \ldots, i_{p-1}}$. It is easy to verify that H is a homotopy. ∎

Now we consider the Čech complex $C^\bullet(\mathcal{U}, A)$. For any open set V of X, we have an induced open covering $\mathcal{U}_{/V} = (U_i \cap V)_{i \in I}$ of V. We can form the Čech complex $C^\bullet(\mathcal{U}_{/V}, A)$ for this covering and for $A_{/V}$. There are natural restriction maps to smaller open sets, so we get a complex $C^\bullet(\mathcal{U}, A)$

of presheaves on X. We have a natural morphism $j : A \to C^0(\mathcal{U}, A)$. If A is a sheaf, we obtain a complex of sheaves on X.

1.3.3. Proposition. *Let $\mathcal{U}$ be an open covering of X. Then for any sheaf A of abelian groups on X, $C^\bullet(\mathcal{U}, A)$ is a resolution of A.*

Proof. It is enough to show that each $x \in X$ has an open neighborhood V such that the sequence

$$0 \to A(W) \xrightarrow{j} C^0(\mathcal{U}_{/W}, A) \xrightarrow{\delta} C^1(\mathcal{U}_{/W}, A) \xrightarrow{\delta} \cdots$$

is exact, for every open set W in V. Indeed, since a direct limit of exact sequences is still exact, we will then get an exact sequence for the stalks at x. By Proposition 1.1.7, we will then have an exact sequence of sheaves. Take $V = U_{i_0}$, for some U_{i_0} which contains x. Then for $W \subseteq V$, the open covering $(U_i \cap W)$ of W has $W = W \cap U_{i_0}$ among its ranks, hence the above sequence is exact by Lemma 1.3.2. ∎

The consequence is as follows:

1.3.4. Proposition. *For any sheaf A of abelian groups on a space X and any open covering $\mathcal{U}$ of X, there is a canonical group homomorphism $\check{H}^j(\mathcal{U}, A) \to H^j(X, A)$ from Čech cohomology to sheaf cohomology, induced by a morphism of resolutions from $C^\bullet(\mathcal{U}, A)$ to some injective resolution of A.*

Proof. Let $I^\bullet$ be an injective resolution of A. By Proposition 1.3.3, $C^\bullet(\mathcal{U}, A)$ is a resolution of A. By Proposition 1.1.15 (2), there exists a morphism of resolutions $C^\bullet(\mathcal{U}, A) \to I^\bullet$, which is unique up to homotopy. Then the induced map on the complexes of global sections gives group homomorphisms $\check{H}^p(\mathcal{U}, A) \to H^p(X, A)$. ∎

Therefore one can use Čech cohomology to construct classes in sheaf cohomology. We are interested in the question of whether all sheaf cohomology classes can be captured in this fashion. First of all, we consider the Čech cohomology of injective sheaves.

1.3.5. Lemma. *Let I be a sheaf on X which has the property that for every open set V, the sheaf $U \mapsto I(V \cap U)$ on X is injective. Then for any open covering $\mathcal{U}$ of X, one has $\check{H}^p(\mathcal{U}, I) = 0$ for $p > 0$.*

Proof. The sheaf $C^p(\mathcal{U}, I)$ is a product of sheaves of the type $U \mapsto I(V \cap U)$, hence is injective. So $C^\bullet(\mathcal{U}, I)$ is an injective resolution of I. The complex of global sections is $C^\bullet(\mathcal{U}, I)$, and its cohomology is $H^p(X, I)$. So we have

$\check{H}^p(\mathcal{U}, I) = H^p(X, I)$, which is 0 for $p > 0$. ∎

One can define the *Čech hypercohomology* $\check{H}^p(\mathcal{U}, K^\bullet)$ of a bounded below complex of sheaves $K^\bullet$ on X. For this, one forms the double complex

$$
\begin{array}{ccccccc}
& & \uparrow d_K & & \uparrow d_K & & \\
\cdots \xrightarrow{\delta} & C^p(\mathcal{U}, K^{q+1}) & \xrightarrow{\delta} & C^{p+1}(\mathcal{U}, K^{q+1}) & \xrightarrow{\delta} & \cdots & \\
& & \uparrow d_K & & \uparrow d_K & & (1-14) \\
\cdots \xrightarrow{\delta} & C^p(\mathcal{U}, K^q) & \xrightarrow{\delta} & C^{p+1}(\mathcal{U}, K^q) & \xrightarrow{\delta} & \cdots & \\
& & \uparrow d_K & & \uparrow d_K & &
\end{array}
$$

and one takes the total cohomology of this double complex. The double complex is denoted $C^\bullet(\mathcal{U}, K^\bullet)$.

We will study this Čech hypercohomology for a resolution $I^\bullet$ of a sheaf A which has the property of Lemma 1.3.5. The injective resolution constructed in the proof of Proposition 1.1.15 is of this type. It also has the property that for U an open set of X, $I^\bullet_{/U}$ is an injective resolution of A_U.

The cohomology of the double complex (1-14) may be approached from the convergent spectral sequence for the second filtration. The E_1-term $E_1^{p,q}$ is the horizontal cohomology of (1-14), i.e., the Čech cohomology group $\check{H}^q(\mathcal{U}, I^p)$. This is 0 for $q > 0$. Hence by Proposition 1.2.4, the spectral sequence degenerates at E_2, and we have $\check{H}^n(\mathcal{U}, I^\bullet) = E_2^{n,0}$. This E_2-term is the cohomology of the complex $\Gamma(X, I^n)$, i.e., the sheaf cohomology $H^n(X, A)$. So we have $\check{H}^n(\mathcal{U}, I^\bullet) = H^n(X, A)$.

We will now consider the spectral sequence for the first filtration. The term E_1 is the column cohomology, i.e., $E_1^{p,q} = \prod_{i_0,\ldots,i_p} H^q(U_{i_0,\ldots,i_p}, A)$ (since the restriction of $I^\bullet$ to $U_{i_0,\ldots,i_p}$ is an injective resolution of $A_{/U_{i_0,\ldots,i_p}}$). We will not study this spectral sequence in general as in [Go]. Rather, we will assume that the restriction of A to an open set $U_{i_0,\ldots,i_p}$ is acyclic. Then we have $E_1^{p,q} = 0$ for $q > 0$, and this spectral sequence also degenerates. Thus we obtain the important

1.3.6. Theorem. *Let A be a sheaf on X, and let $\mathcal{U} = (U_i)_{i \in I}$ be an open covering of X such that $H^q(U_{i_0,\ldots,i_p}, A) = 0$ for all $q > 0$ and $i_0,\ldots,i_p \in I$. Then the canonical homomorphism $\check{H}^p(\mathcal{U}, A) \to H^p(X, A)$ of Proposition 1.3.4 is an isomorphism for all $p \geq 0$.*

Consider the case of the constant sheaf B_X over X, for X a finite-dimensional manifold and B some abelian group. Following A. Weil [We1], an open covering $\mathcal{U} = (U_i)_{i \in I}$ is called a *good covering* if all non-empty intersections $U_{i_0,\ldots,i_p}$ are contractible. One can show that the hypothesis of Theorem 1.3.6 is satisfied (for $B = \mathbb{R}$, this follows from Theorem 1.4.15, which does not use Theorem 1.3.6). Then Theorem 1.3.6 means that the sheaf cohomology $H^p(X, B_X)$ is the cohomology of a simplicial set, the *nerve* of the covering, with coefficients in B. This nerve has a vertex for each element of I, and a p-simplex with vertices $i_0, \ldots, i_p$ is filled whenever the intersection $U_{i_0, \ldots, i_p}$ is non-empty. If X is a riemannian manifold, a good covering can be constructed as follows. A subset S of X is called geodesically convex if given two points x and y of S, there is a unique geodesic from x to y, and this geodesic is contained in S. A geodesic subset is contractible. One can cover X by geodesically convex subsets; such a covering is good.

Čech cohomology has the advantage of allowing an easy and explicit construction of a cup-product

$$\check{H}^p(\mathcal{U}, F) \otimes \check{H}^q(\mathcal{U}, G) \rightarrow \check{H}^{p+q}(\mathcal{U}, F \otimes G) \qquad (1-15)$$

Here F and G are sheaves of abelian groups on X, and the *tensor product-sheaf* $F \otimes G$ is the sheaf associated to the presheaf $U \mapsto F(U) \otimes_{\mathbf{Z}} G(U)$. The stalk at x of $F \otimes G$ is $F_x \otimes G_x$. The cup-product will be defined from a morphism of complexes. We first need the notion of tensor product $A^\bullet \otimes B^\bullet$ of two complexes; this is the total complex of the double complex $A^p \otimes B^q$. So the degree n-term of $A^\bullet \otimes B^\bullet$ is $\oplus_p A^p \otimes B^{n-p}$. The differential in $A^\bullet \otimes B^\bullet$ is $d(a \otimes b) = (da) \otimes b + (-1)^p a \otimes db$ for $a \in A^p$, $b \in B^q$. We have the obvious tensor product $\otimes : H^p(A^\bullet) \otimes H^q(B^\bullet) \rightarrow H^{p+q}(A^\bullet \otimes B^\bullet)$.

We now return to the sheaves F and G of abelian groups on X. We have the complexes $C^\bullet(\mathcal{U}, F)$ and $C^\bullet(\mathcal{U}, G)$. The interesting part is the construction of a morphism of complexes

$$\phi : C^\bullet(\mathcal{U}, F) \otimes C^\bullet(\mathcal{U}, G) \rightarrow C^\bullet(\mathcal{U}, F \otimes G). \qquad (1-16)$$

For $\underline{\alpha} \in C^p(\mathcal{U}, F)$ and $\underline{\beta} \in C^q(\mathcal{U}, G)$, we put

$$\phi(\underline{\alpha} \otimes \underline{\beta})_{i_0, \ldots, i_{p+q}} = \alpha_{i_0, \ldots, i_p} \otimes \beta_{i_p, \ldots, i_{p+q}}. \qquad (1-17)$$

One checks easily that ϕ is indeed a morphism of complexes. The proof is the same as for the cup-product on the complex of singular cochains on a space. The induced map on cohomology gives the cup-product on Čech cohomology. For $\underline{\alpha}$ a degree p Čech cocycle with coefficients in F and $\underline{\beta}$ a degree q Čech cocycle with coefficients in G, we have

$$(\underline{\alpha} \cup \underline{\beta})_{i_0, \ldots, i_{p+q}} = \alpha_{i_0, \ldots, i_p} \otimes \beta_{i_p, \ldots, i_{p+q}}. \qquad (1-18)$$

The cup-product has the following properties.

1.3.7. Proposition. (1) *The cup-product is associative, i.e., for* $\alpha \in$

$\check{H}^p(\mathcal{U}, F)$, $\beta \in \check{H}^q(\mathcal{U}, G)$, $\gamma \in \check{H}^r(\mathcal{U}, H)$, we have

$$\alpha \cup (\beta \cup \gamma) = (\alpha \cup \beta) \cup \gamma \in \check{H}^{p+q+r}(\mathcal{U}, F \otimes G \otimes H).$$

(2) *The cup-product is graded-commutative. If* $\alpha \in \check{H}^p(\mathcal{U}, F)$, $\beta \in \check{H}^q(\mathcal{U}, G)$, *we have*

$$\alpha \cup \beta = (-1)^{pq} \beta \cup \alpha \in \check{H}^{p+q}(\mathcal{U}, F \otimes G).$$

Proof. (1) is immediate. The proof of (2), although classical, is rather complicated, as it amounts to proving the graded commutativity of the cup-product in the cohomology of a simplicial set with values in a coefficient system. We refer to [G-H] for details of the proof, which uses the technique of acyclic carriers. ∎

We will also have use for the notion of the tensor product $K^\bullet \otimes L^\bullet$ of two complexes of sheaves of abelian groups. The degree n component of $K^\bullet \otimes L^\bullet$ is $\oplus_p K^p \otimes L^{n-p}$. There is a morphism of complexes

$$\phi : \mathrm{Tot}\ (C^\bullet(\mathcal{U}, K^\bullet)) \otimes \mathrm{Tot}\ (C^\bullet(\mathcal{U}, L^\bullet)) \to \mathrm{Tot}\ (C^\bullet(\mathcal{U}, K^\bullet \otimes L^\bullet)).$$

For $\underline{\alpha}$ a Čech p-cochain with values in K^l, and $\underline{\beta}$ a Čech q-cochain with values in L^m, $\phi(\underline{\alpha} \otimes \underline{\beta})$ is the $p + q$-cochain with values in $K^l \otimes L^m$ given by formula (1-17).

Now there are two drawbacks of Čech cohomology. The first one is that there is no reason that an exact sequence of sheaves should lead to a long exact sequence of Čech cohomology groups. The second is that it depends on the choice of a covering. There is a remedy to both problems, at least for X paracompact. We will consider all open coverings of X. Recall that there is an order relation on the set of open coverings of X. An open covering $\mathcal{V} = (V_j)_{j \in J}$ is *finer* than an open covering $\mathcal{U} = (U_i)_{i \in I}$ if there exists a map $f : J \to I$ such that $V_j \subseteq U_{f(j)}$ for all $j \in J$. We then write $\mathcal{U} \prec \mathcal{V}$. Given such a map f, there exists a morphism of complexes $f_* : C^\bullet(\mathcal{U}, A) \to C^\bullet(\mathcal{V}, A)$ such that for α a Čech p-cochain for the covering $\mathcal{U}$, $f_*(\underline{\alpha})$ is the p-cochain of the covering $\mathcal{V}$ defined by

$$f_*(\underline{\alpha})_{j_0, \ldots, j_p} = (\alpha_{f(j_0), \ldots, f(j_p)}) / V_{j_0, \ldots, j_p} \qquad (1-19)$$

Note that $\alpha_{f(j_0), \ldots, f(j_p)}$ is a section of A over $U_{f(j_0), \ldots, f(j_p)}$, so we can restrict it to the smaller open set $V_{j_0, \ldots, j_p}$.

1.3.8. Lemma. [Go] *Given two maps* $f, f' : J \to I$ *such that* $V_j \subseteq U_{f(j)}$, $V_j \subseteq U_{f'(j)}$, *there is a homotopy between* f_* *and* f'_*.

Proof. We will only write down the formula for the homotopy H on a p-cochain $\underline{\alpha}$ for the covering $\mathcal{U}$:

$$H(\underline{\alpha})_{j_0,\ldots,j_{p-1}} = \sum_{k=0}^{p-1} (-1)^k \; res \; (\alpha_{f(j_0),\ldots,f(j_k),f'(j_k),\ldots,f'(j_{p-1})}),$$

where res denotes the operation of restriction from $U_{f(j_0),\ldots,f(j_k),f'(j_k),\ldots,f'(j_{p-1})}$ to $V_{j_0,\ldots,j_{p-1}}$. ∎

Lemma 1.3.8 implies that the induced map $f_* : \check{H}^p(\mathcal{U}, A) \to \check{H}^p(\mathcal{V}, A)$ is independent of f. Clearly we have an inductive system $\mathcal{U} \mapsto \check{H}^p(\mathcal{U}, A)$ for a sheaf A.

1.3.9. Definition. *Let A be a sheaf of abelian groups on a space X. The Čech cohomology groups $\check{H}^p(X, A)$ are defined as*

$$\check{H}^p(X, A) = \varinjlim_{\mathcal{U}} \check{H}^p(\mathcal{U}, A), \qquad (1-20)$$

where the direct limit is taken over the ordered set of open coverings $\mathcal{U}$ of X.

The problem of getting a long exact sequence for Čech cohomology from a short exact sequence may be formulated as follows: We introduce the notion of exact sequence of presheaves. The sequence $0 \to A \to B \to C \to 0$ of presheaves on X is called exact if for every open set U the sequence $0 \to A(U) \to B(U) \to C(U) \to 0$ is exact. Note that if these presheaves are all sheaves, a short exact sequence of sheaves does not in general give rise to an exact sequence of presheaves.

1.3.10. Lemma. *Let $0 \to A \to B \to C \to 0$ be an exact sequence of presheaves of abelian groups over X. Then we have a long exact sequence of groups*

$$0 \to \check{H}^0(X, A) \to \check{H}^0(X, B) \to \check{H}^0(X, C) \to \check{H}^1(X, A) \to \cdots \qquad (1-21)$$

Proof. For any open covering $\mathcal{U}$, we have an exact sequence of complexes

$$0 \to C^\bullet(\mathcal{U}, A) \to C^\bullet(\mathcal{U}, B) \to C^\bullet(\mathcal{U}, C) \to 0.$$

Then we have a long exact sequence of Čech cohomology groups for this covering. The direct limit of these exact sequences is still exact. ∎

Similarly one defines the Čech hypercohomology groups $\check{H}^p(X, K^\bullet)$ with coefficients in a bounded below complex of presheaves $K^\bullet$. We now study the relation of Čech cohomology (and hypercohomology) with sheaf cohomology (and hypercohomology). They will be isomorphic for a paracompact space, so we will now introduce this notion. An open covering (U_i) of X is said to be *locally finite* if every x in X has a neighborhood which meets only finitely many U_i's.

1.3.11. Definition. *A space X is said to be <u>paracompact</u> if X is Hausdorff and for every open covering $(U_i)_{i \in I}$, there exists a refinement $(V_j)_{j \in J}$ which is a locally finite open covering.*

A closed subset of a paracompact space is paracompact. Here are some basic properties of paracompact spaces. They are proved in [**Mun**].

1.3.12. Proposition. (1) *Let Z be a closed subset of a paracompact space X. Then every neighborhood of Z contains a closed neighborhood.*

(2) *Let $(U_i)_{i \in I}$ be an open covering of a paracompact space X. There exists an open covering $(V_i)_{i \in I}$ such that $\overline{V_i} \subseteq U_i$ for all i.*

(3) *Any metric space is paracompact.*

The infinite-dimensional manifolds considered in this book are paracompact (although not locally compact). This will be discussed in §4.

1.3.13. Theorem. [**Go**] (1) *There are natural group homomorphisms $\check{H}^p(X, K^\bullet) \to H^p(X, K^\bullet)$.*

(2) *Assume that for any presheaf A of abelian groups on X, such that the associated sheaf A is 0, the Čech cohomology groups $\check{H}^p(X, A)$ are all 0. Then the above homomorphism $\check{H}^p(X, K^\bullet) \to H^p(X, K^\bullet)$ is an isomorphism for any bounded below complex of sheaves $K^\bullet$.*

(3) *If X is paracompact, the condition of (2) is always satisfied. Hence for X paracompact, Čech hypercohomology is canonically isomorphic to sheaf hypercohomology.*

Proof. Consider a bounded below complex of sheaves $K^\bullet$ and an injective resolution $I^\bullet$ of $K^\bullet$. Let $\mathcal{U} = (U_i)_{i \in I}$ and $\mathcal{V} = (V_j)_{j \in J}$ be two open coverings of X, such that $\mathcal{U} \prec \mathcal{V}$. Choose a map $f : J \to I$ such that $V_j \subseteq U_{f(j)}$, and let $f_* : C^\bullet(\mathcal{U}, K^\bullet) \to C^\bullet(\mathcal{V}, K^\bullet)$ be the corresponding morphism of double complexes. Then we can construct a diagram of resolutions of $K^\bullet$, which is commutative up to homotopy:

$$C^\bullet(\mathcal{U}, \mathcal{K}^\bullet) \xrightarrow{\ f_\bullet\ } C^\bullet(\mathcal{V}, \mathcal{K}^\bullet)$$
$$\searrow \qquad \swarrow$$
$$I^\bullet$$

So on hypercohomology groups we get a commutative diagram

$$\check{H}^p(\mathcal{U}, \mathcal{K}^\bullet) \longrightarrow \check{H}^p(\mathcal{V}, \mathcal{K}^\bullet)$$
$$\searrow \qquad \swarrow$$
$$H^p(X, K^\bullet)$$

Therefore we get homomorphisms

$$can : \check{H}^p(X, K^\bullet) = \varinjlim_{\mathcal{U}} \check{H}^p(\mathcal{U}, K^\bullet) \to H^p(X, K^\bullet).$$

This proves (1).

Under the assumption of (2), we first prove that $can : \check{H}^p(X, A) \to H^p(X, A)$ is an isomorphism for all $p \geq 0$ and all sheaves A. For $p = 0$, we already have an isomorphism $\check{H}^0(\mathcal{U}, A) \xrightarrow{\sim} H^0(X, A)$ for any open covering. Now the assumption implies that for F any presheaf of abelian groups and for $\tilde{F}$ the associated sheaf, we have isomorphisms $\check{H}^p(X, F) \xrightarrow{\sim} \check{H}^p(X, \tilde{F})$. This implies that for every exact sequence of sheaves $0 \to A \to B \to C \to 0$, we have a corresponding long exact sequence of Čech cohomology groups. Indeed, the sheaf C is the associated sheaf to the presheaf $F(U) = B(U)/A(U)$. Lemma 1.3.10 gives a long exact sequence

$$\cdots \to \check{H}^p(X, A) \to \check{H}^p(X, B) \to \check{H}^p(X, F) = \check{H}^p(X, C) \to \cdots$$

This exact sequence maps to the exact sequence of sheaf cohomology groups. Assume we know that $can : \check{H}^i(X, A) \to H^i(X, A)$ is an isomorphism for all $i \leq p$ and all sheaves A on X. Given a sheaf A, let $A \hookrightarrow I$ be a monomorphism, where I has the property of Lemma 1.3.5. We have a commutative diagram of exact sequences:

$$\begin{array}{ccccccc}
\check{H}^p(X, I) & \to & \check{H}^p(X, I/A) & \xrightarrow{\ \delta\ } & \check{H}^{p+1}(X, A) & \to & 0 \\
\downarrow{\scriptstyle can} & & \downarrow{\scriptstyle can} & & \downarrow{\scriptstyle can} & & \\
H^p(X, I) & \to & H^p(X, I/A) & \to & H^{p+1}(X, A) & \to & 0
\end{array}$$

since $\check{H}^{p+1}(X, I) = 0$ by Lemma 1.3.5. Since the first two vertical maps are isomorphisms, so is the third. So $can : \check{H}^p(X, A) \xrightarrow{\sim} H^p(X, A)$. In the case of bounded below complexes of sheaves, one establishes a convergent spectral sequence

$$E_2^{p,q} = \check{H}^p(X, \underline{H}^q(K^\bullet)),$$

with E_∞-term the graded quotients of some filtration of the Čech hypercohomology groups of $K^\bullet$. This spectral sequence maps to the corresponding spectral sequence for sheaf hypercohomology. Since we have an isomorphism of the E_2 terms, we have an isomorphism of Čech hypercohomology with sheaf hypercohomology. This proves (2).

Now assume X is paracompact, and let A be a presheaf of abelian groups such that the associated sheaf $\tilde{A}$ is 0. It is not difficult to prove that every class in $\check{H}^p(X, A) = \varinjlim_{\mathcal{U}} \check{H}^p(\mathcal{U}, A)$ is represented by a Čech p-cocycle α of some locally finite open covering $\mathcal{U} = (U_i)_{i \in I}$ of X, with coefficients in A. According to Proposition 1.3.12, there exists an open covering $(V_i)_{i \in I}$ of X such that $\overline{V_i} \subseteq U_i$ for all i. We will find a covering $\mathcal{W}$ refining $\mathcal{U}$, such that α induces the 0-cochain of $\mathcal{W}$. This will establish (3). We choose for any $x \in X$ an open neighborhood W_x of x with the following conditions:

(a) W_x meets only finitely many U_i's.

(b) whenever $x \in V_i$, we have $W_x \subseteq V_i$.

(c) W_x does not meet V_i unless $x \in U_i$.

Now we use the assumption that the associated sheaf $\tilde{A}$ is 0. Given an open set V, $s \in A(V)$ and $x \in V$, there exists an open neighborhood Z of x such that s has zero image in $A(V)$. This is because the stalk of A at x is 0. Given x, there are only finitely many components $\alpha_{i_0,...,i_p}$ where each U_{i_j} contains x. Replacing W_x by a smaller neighborhood, we may assume that α has zero restriction to W_x, giving the property

(d) If $x \in U_{i_0,...,i_p}$, then $\alpha_{i_0,...,i_p}$ has zero restriction to $W_x \cap U_{i_0,...,i_p}$.

We have the open covering $\mathcal{W} = (W_x)_{x \in X}$. Now we can choose a mapping $f : X \to I$ such that $x \in V_{f(x)}$, hence $W_x \subseteq V_{f(x)} \subset U_{f(x)}$ by (b). This induces a homomorphism $f_* : C^p(\mathcal{U}, A) \to C^p(\mathcal{W}, A)$, and we claim that the image of α is 0. Indeed, let $x_0, \ldots, x_p \in X$ be such that $W_{x_0} \cap \cdots \cap W_{x_p} \neq \emptyset$; then we have $W_{x_k} \subseteq V_{f(x_k)}$. Put $i_k = f(x_k)$. For $0 \leq k \leq p$, W_{x_0} meets V_{i_k}, hence is contained in V_{i_k}. By (c), this implies that $x_0 \in U_{i_0,...,i_p}$. By condition (d), $\alpha_{i_0,...,i_p}$ has zero restriction to $W_{x_0} \cap U_{i_0,...,i_p}$, hence to $W_{x_0} \cap \cdots \cap W_{x_p}$. ∎

1.4. de Rham cohomology

We have seen in §2 that given a sheaf A of abelian groups on a space X and a resolution $K^\bullet$ of A by acyclic sheaves, the sheaf cohomology groups $H^p(X, A)$ are the cohomology groups of the complex $\Gamma(X, K^0) \to \Gamma(X, K^1) \to \cdots$. It will therefore be useful to have general results which say that certain sheaves are acyclic. For a smooth manifold M, there is a natural resolution of the constant sheaf $\mathbb{R}_M$, namely the de Rham complex.

We first recall the notion of smooth manifolds modeled on (possibly infinite-dimensional) topological vector spaces. While our treatment will be rather short, a very nice survey of this subject appears in [**Ham**]. We consider a topological vector space E, which we will assume to be locally convex and Hausdorff. A mapping f from an open set U of E to another topological vector space F is said to be continuously differentiable (or of class C^1) if for every $x \in U$ and $v \in E$, the limit $df(x, v) = \lim_{t \to 0} \frac{f(x+tv)-f(x)}{t}$ exists and is continuous as a function of (x, v). Similarly one has the notion of functions of class C^2, and so on. The n-th derivative $d^{(n)}f(x; v_1, v_2, \ldots, v_n)$ is supposed to be a continuous function in $U \times E^n$, which is of course multilinear in the last n variables. We will call a mapping f smooth if it is of class C^k for all $k \in \mathbb{N}$, i.e., $f \in C^\infty$. A differential form ω of degree n (or simply n-form) on the open set U of E is a smooth function $\omega : U \times E^n \to \mathbb{R}$ which is multilinear and alternating in the last n variables. The exterior differential $d\omega$ of a n-form is the $(n+1)$-form

$$d\omega(x; v_1, v_2, \ldots, v_{n+1}) = \sum_{j=1}^{n+1} (-1)^{j+1} \, D\omega(-; v_1, \ldots, \hat{v}_j, \ldots, v_{n+1})(x, v_j).$$

$$(1-22)$$

Here $D\omega(-; v_1, \ldots, \hat{v}_j, \ldots, v_{n+1})$ is a smooth function on U, and we take its partial derivative in the direction of v_j. The differential d satisfies $dd = 0$, hence we have the *de Rham complex*

$$\cdots \xrightarrow{d} A^p(U) \xrightarrow{d} A^{p+1}(U) \xrightarrow{d} \cdots, \qquad (1-23)$$

where $A^p(U)$ denotes the vector space of p-forms on U. This de Rham complex is denoted by $A^\bullet(U)$.

1.4.1. Lemma. *(Poincaré lemma) Let U be a convex open subset of the topological vector space E. Then the de Rham complex (1-23) has $H^p(A^\bullet(M)) = 0$ for $p > 0$, and $H^0(A^\bullet(M)) = \mathbb{R}$.*

Proof. The usual proof (see e.g., [**B-T**]) applies without change. ∎

A smooth manifold modeled on E is a Hausdorff space M equipped with an open covering $(U_i)_{i \in I}$ and with homeomorphisms $\phi_i : U_i \xrightarrow{\sim} V_i$, for V_i an open set of E. It is required that the transition functions

$\phi_j \circ \phi_i^{-1} : \phi_i(U_{ij}) \xrightarrow{\sim} \phi_j(U_{ij})$ be smooth maps between open subsets of E. Given a smooth manifold M modeled on E, one has the notion of smooth functions, of p-forms, of the de Rham complex, and so on.

For U an open set of M, we define the space $A^p(U)$ of degree p differential forms on M, to consist of families $\omega_i \in A^p(\phi_i(U_i \cap U))$ satisfying the glueing condition $(\phi_j \circ \phi_i^{-1})^* \omega_j = \omega_i$ on $\phi_i(U_{ij} \cap U)$. We then have the sheaf $\underline{A}_M^p$ of p-forms, such that $\underline{A}_M^p(U) = A^p(U)$. We have the complex of sheaves

$$\underline{A}_M^0 \xrightarrow{d} \underline{A}_M^1 \xrightarrow{d} \cdots, \qquad (1-24)$$

which is called the de Rham complex of sheaves and is denoted by $\underline{A}_M^\bullet$.

1.4.2. Definition. *The <u>de Rham cohomology</u> groups $H_{DR}^p(M)$ of a smooth manifold are the cohomology groups of the de Rham complex $A^\bullet(M)$.*

Our main purpose in this section is to compare the de Rham cohomology of a manifold M with its Čech cohomology $H^p(M, \mathbb{R}_M)$.

1.4.3. Proposition. *Let M be a smooth manifold. The complex of sheaves $\underline{A}_M^\bullet$ is a resolution of the constant sheaf $\mathbb{R}_M$.*

Proof. It is enough to show that for any $x \in M$, the stalk complex $\underline{A}_{M,x}^\bullet$ is a resolution of $\mathbb{R}$. Since M is a smooth manifold, we may as well assume that M is an open subset of E. Since E is locally convex, there exists a fundamental system of open neighborhoods U of x which are convex. Then we have $\underline{A}_{M,x}^p = \varinjlim_{U \text{ convex}} A^p(U)$. The complex $\mathbb{R} \to A^0(U) \to A^1(U) \to \cdots$ is acyclic by Lemma 1.4.1. The result follows by using the fact that a direct limit of exact sequences is exact. ∎

We wish to show that for a large class of manifolds, the sheaves $\underline{A}_M^p$ are acyclic. For this purpose we will use the theory of soft sheaves. It will turn out that soft sheaves are acyclic.

First we need to define the inverse image of a sheaf A on Y under a continuous map $f : X \to Y$. We will start with a rather abstract definition, and then give a more concrete description. The abstract definition proceeds in two steps. First we construct a presheaf B on X by

$$B(U) = \varinjlim_{\substack{V \subseteq Y \text{ open} \\ f(U) \subseteq V}} A(V).$$

Then the inverse image $f^{-1}A$ is the sheaf associated to the presheaf B. This *inverse image* $f^{-1}(A)$ will have stalk at $x \in X$ equal to the stalk $A_{f(x)}$.

Here is the more concrete description. A section $x \in U \mapsto s_x \in A_{f(x)}$ over an open set U of X will be called continuous, if for any $x \in U$, there exists a neighborhood V of $f(x)$ in Y and a section σ of A over V, such that $s_z = \sigma_{f(z)} \in A_{f(z)}$ for all $z \in U \cap f^{-1}(V)$. Then $f^{-1}(A)(U)$ is the set of such continuous sections $x \in U \mapsto s_x$. Both definitions of the sheaf $f^{-1}(A)$ can be found in the literature.

For A a sheaf on X and for $Z \xrightarrow{i} X$ a closed subspace, we set $\Gamma(Z, A) = \Gamma(Z, i^{-1}A)$.

1.4.4. Lemma. *Let X be a paracompact space, and let $i : Z \hookrightarrow X$ be the inclusion of a closed subset. Then for any sheaf A of sets on X we have $\Gamma(Z, A) = \varinjlim_U \Gamma(U, A)$, where the direct limit is taken over all open neighborhoods of Z.*

Proof. There is a natural map $\varinjlim_U \Gamma(U, A) \to \Gamma(Z, A)$. The crucial point is to show this map is surjective. Let $s \in \Gamma(Z, A)$. For all $x \in Z$, let $s_x \in i^{-1}(A)_x = A_x$ be the germ of s at x. For each $x \in Z$, there exists a neighborhood U'_x in x, and a section σ_x of A over U'_x, such that $(\sigma_x)_y = s_y$ for all $y \in U'_x$ (this is a consequence of the definition of $i^{-1}(A)$). Then since X is paracompact there exists a locally finite family $(U_i)_{i \in I}$ of open sets of X, such that $Z \subseteq \cup_i U_i$ and each U_i is contained in some U'_x. We can then find another such family $(V_i)_{i \in I}$ such that $\overline{V}_i \subseteq U_i$. For each $i \in I$, we have a section s_i of A over U_i, such that s_i coincides with s over $U_i \cap Z$. For each $x \in X$, let Y_x be an open neighborhood of x in X such that the set T_x of i, for which $Y_x \cap V_i$ is non-empty, is a finite set. By shrinking Y_x if necessary, we may assume that $x \in U_i$ for $i \in T_x$. For each $i, j \in T_x$, the sections s_i and s_j agree over some neighborhood N_{ij} of x. Let W_x be the intersection of Y_x with all N_{ij} (for $i, j \in T_x$); then over W_x, we have a well-defined section u_x of A, which at every y is equal to s_i for every i such that $y \in W_x \cap U_i$. For $x, y \in Z$, the sections u_x, u_y obviously have the same restriction to $W_x \cap W_y$. Since A is a sheaf, we obtain a section u of A over $\cup_{x \in Z} W_x$, whose restriction to W_x equals u_x. Hence $s \in \Gamma(Z, A)$ is the restriction to Z of $u \in \Gamma(\cup_{x \in Z} W_x, A)$. ∎

1.4.5. Definition. *A sheaf A of sets on a space X is called <u>soft</u> if for every closed subset Z of X, the restriction map $\Gamma(X, A) \to \Gamma(Z, A)$ is surjective.*

Note that if A is a soft sheaf on X, and Z is a closed subset of X, then $A_{/Z}$ is a soft sheaf on Z.

1.4.6. Theorem. *Let X be a paracompact space. Then*

(1) *Any injective sheaf of abelian groups on X is soft.*

(2) *Any soft sheaf of abelian groups on X is acyclic.*

Proof. We will show that an injective sheaf A on any space X is *flabby*, i.e., for every open set U of X, the restriction map $\Gamma(X, A) \to \Gamma(U, A)$ is surjective. Now Lemma 1.4.4 implies that a flabby sheaf on a paracompact space is soft. In §1, we constructed an embedding $A \hookrightarrow I$, where $I(U) = \prod_{x \in U} A_x$. This sheaf I is both injective and flabby. Since A is injective, A is a direct summand of the sheaf I. But a direct summand of a flabby sheaf is also flabby. This proves (1).

We will now show that given an exact sequence $0 \to F \to G \to H \to 0$ in $\mathcal{AB}(X)$ with F soft, the map $\Gamma(X, G) \to \Gamma(X, H)$ is surjective. Let $s \in \Gamma(X, H)$. As X is paracompact, there exists a locally finite open covering $(U_i)_{i \in I}$ of X and a section t_i of G over U_i such that t_i projects to s. Next we have an open covering (V_i) such that $\overline{V_i} \subseteq U_i$. Set $Z_i = \overline{V_i}$. Consider the set of pairs (J, t_J) consisting of a subset J of I and a section t_J of G over the closed subset $Z_J := \cup_{i \in J} Z_i$ which projects to s. This is ordered by $(J, t_J) < (K, t_K)$ if and only if $J \subseteq K$ and $t_J = (t_K)_{/Z_J}$. Any increasing chain of elements in this set has a maximal element. Hence by Zorn's lemma, there exists a maximal element (J, t_J). Assume that there exists $i \in I \setminus J$. We have the section $t_J - t_i$ of F over $Z_J \cap Z_i$. Since F is soft, we can extend it to a section u of F over X. Then $t_J \in \Gamma(Z_J, G)$ and $t_i + u \in \Gamma(Z_i, G)$ have the same restriction to $Z_J \cap Z_i$. Hence they glue together to give a section $t_{J \cup \{i\}}$ of G over $Z_{J \cup \{i\}}$ which projects to s. This contradicts the fact that (J, t_J) was a maximal element, showing that $\Gamma(X, G) \to \Gamma(X, H)$ is surjective.

This implies easily that given an exact sequence of sheaves $0 \to F \to G \to H \to 0$ with F and G soft, H is also soft. We now prove by induction on $n \geq 1$ that for any soft sheaf F, we have $H^n(X, F) = 0$. To prove the case $n = 1$, choose a monomorphism $F \to I$ with I injective. We have the exact sequence

$$0 \to \Gamma(X, F) \to \Gamma(X, I) \to \Gamma(X, I/F) \to H^1(X, F) \to 0,$$

since $H^1(X, I) = 0$. Since F is soft, the map $\Gamma(X, I) \to \Gamma(X, I/F)$ is onto, hence $H^1(X, F) = 0$. Now let $n \geq 1$, and assume that $H^n(X, F) = 0$ for any soft sheaf F. Construct as above an exact sequence $0 \to F \to I \to I/F \to 0$, with I injective, hence soft. I/F is also soft. The exact sequence

$$H^n(X, I/F) = 0 \to H^{n+1}(X, F) \to H^{n+1}(X, I) = 0$$

shows $H^{n+1}(X, F) = 0$, which completes the inductive step. ∎

1.4.7. Corollary. *Let M be a smooth paracompact manifold such that the sheaves $\underline{A}^p_M$ are soft. Then the sheaf cohomology $H^p(M, \mathbb{R}_M)$ is canonically isomorphic to the de Rham cohomology $H^p_{DR}(M)$.*

Since soft sheaves are acyclic, it follows from Proposition 1.2.12 that one can use a resolution by soft sheaves to compute the cohomology of a sheaf of abelian groups on a paracompact space X. We need a criterion for a sheaf to be soft. An endomorphism of a sheaf A of abelian groups is a morphism $A \to A$ in $\mathcal{AB}(X)$. There is a sheaf of rings $\underline{End}(A)$ with global sections equal to $End_X(A)$.

If s is a global section of a sheaf A of groups over X, the *support $Supp(s)$* of s is the complement of the set of x such that s has zero restriction to a neighborhood of x.

1.4.8. Definition. *Let A be a sheaf of abelian groups on a space X, and let $\mathcal{U} = (U_i)_{i \in I}$ be an open covering of X. A <u>partition of unity</u> for the sheaf A subordinate to the covering $\mathcal{U}$ is a family f_i of endomorphisms of the sheaf A which satisfy the following properties:*

(1) For any open set V and any section s of A over V, the section $f_i(s)$ has support contained in $V \cap U_i$.

(2) Every point x in X has an open neighborhood V such that only finitely many s_i are non-zero in V.

(3) We have $\sum_{i \in I} s_i = Id$ as endomorphisms of A. Note that the $\sum_{i \in I} s_i$ makes sense because by (2), the sum is locally finite.

To relate this abstract notion of partition of unity with the classical one, let M be a smooth manifold. Say f_i is a smooth partition of unity subordinate to the open covering (U_i). This means that f_i is a smooth function on M with support in U_i, that the sum $\sum_{i \in I} f_i$ is locally finite and equal to 1. Then f_i induces an endomorphism of the sheaf $\underline{A}^p_M$ of p-forms, namely multiplication by f_i. The endomorphism of $\underline{A}^p_M$ will be denoted by f_i. Conditions (1), (2) and (3) of Definition 1.4.8 are satisfied, so that (f_i) is a partition of unity of the sheaf $\underline{A}^p_M$.

1.4.9. Proposition. *Let A be a sheaf of abelian groups on a paracompact space. Assume that for any open covering $\mathcal{U} = (U_i)_{i \in I}$, there is a partition of unity of A subordinate to the covering. Then the sheaf A is soft.*

Proof. Let $Z \xrightarrow{i} X$ be a closed subset of X, and let $s \in \Gamma(Z, i^{-1}(A))$ be a section of $i^{-1}(A)$ over Z. According to Lemma 1.4.3, there exists an open set U containing Z, and a section s' of A over U, which restricts to s. Then $(U, X \backslash Z)$ is an open covering of X. Let (f_1, f_2) be a corresponding partition of unity. $f_1 = Id - f_2$ is equal to Id in a neighborhood of Z. There is a global section σ of A such that $\sigma = f_1(s')$ over U and $\sigma = 0$ over $X \backslash Supp(f_1)$. Indeed U and $X \backslash Supp(f_1)$ are two open sets covering X, and the restriction of σ to the intersection $U \backslash Supp(f_1)$ is equal to 0. The restriction of σ to a neighborhood of Z is equal to s', so σ maps to $s \in \Gamma(Z, i^{-1}(A))$. ∎

It is a classical theorem of Dieudonné (see [**Lan**]) that a paracompact finite-dimensional manifold has partitions of unity. The case of infinite-dimensional manifolds, which we will need later in this book, is more difficult; in particular, certain Banach manifolds do not have partitions of unity. The case of Hilbert manifolds is treated in the literature [**A-M-R**, §5.5] [**Lan**]. However we will need more general manifolds. We say the topological vector space E is an *ILH space* if $E = \lim_n H_n$ is an inverse limit of separable Hilbert spaces H_n. The topology on E is the inverse limit topology. This is the coarsest topology which makes all the projection maps $p_n : E \to H_n$ continuous. We require that for any open ball B in H_n, we have

$$p_n^{-1}(\overline{B}) = \overline{p_n^{-1}(B)} \qquad (1-25)$$

(implying that the image of p_n is dense). Examples of topological vector spaces of type ILH include the space $C^\infty(S^1)$ of smooth functions $S^1 \to \mathbb{R}$, with its natural Fréchet topology. Indeed $C^\infty(S^1)$ identifies with an inverse limit of Sobolev spaces (see Lemma 3.1.1 for details). An open set U will be called *round* if it is of the form $U = p_n^{-1}(V)$, for some n and some open ball V in H_n. The round open subsets form a basis of the topology of E. An open set U of E is said to be *scalloped* if there exist some round open subsets $V, V_1, \ldots, V_q$ such that $U = V \cap^c \overline{V_1} \cdots \cap^c \overline{V_q}$.

1.4.10. Lemma. *Let $E = \lim_n H_n$ be an ILH space satisfying condition (1-25). Then for every scalloped open set U of E, there exists a smooth function f on E which has support contained in $\overline{U}$, and such that $f(x) > 0$ for $x \in U$.*

Proof. Let $U = V \cap^c \overline{V_1} \cap \cdots \cap^c \overline{V_q}$. Say $V = p_n^{-1}(W)$, for W an open ball in H_n. There is a smooth function g on H_n, such that $g(x) = 0$ for $x \in H_n \backslash \overline{W}$ and $g(x) > 0$ for $x \in W$. Indeed one can take for g a suitable smooth function of the distance to the center of the ball. Then $\phi = g \circ p_n$ is a smooth function on E such that $\phi(y) = 0$ for $y \notin p_n^{-1}(\overline{W}) = \overline{V}$ and

$\phi(y) > 0$ for $y \in p_n^{-1}(W)$. For $1 \le j \le q$, let $V_j = p_{n_j}^{-1}(W_j)$, for W_j an open ball in the Hilbert space H_{n_j}. There exists a smooth function g_j on H_{n_j}, such that

$$0 \le g_j(x) < 1 \ for \ x \notin \overline{W_j}$$
$$g_j(x) = 1 \ for \ x \in \overline{W_j}$$

Let $\phi_j = \phi \circ p_{n_j}$. We have

$$0 \le \phi_j(x) < 1 \ for \ x \notin \overline{V_j}$$
$$\phi_j(x) = 1 \ for \ x \in \overline{V_j}$$

Hence $f(x) = \phi(x) \prod_{j=1}^q (1 - \phi_j(x))$ satisfies the requirements. ∎

Before stating the next lemma, we note that every open subset U of E has the property that it is second countable (i.e., its topology has a countable basis). Hence U is *Lindelöf*, i.e., for every open covering $(U_i)_{i \in I}$, there exists a countable subset J of I such that $(U_i)_{i \in J}$ is a covering. Also U is metrizable.

1.4.11. Lemma. *With E as in Lemma 1.4.10, let W be an open subset of E, and let $(Z_j)_{j \in \mathbb{N}}$ be a covering of W by round open sets contained in W. There is an open covering $(V_j)_{j \in \mathbb{N}}$ such that*

(1) $V_j \subseteq Z_j$ *for all j;*

(2) *each V_j is scalloped;*

(3) *the closed covering $(\overline{V}_j)$ is locally finite.*

Proof. Each Z_j is of the form $Z_j = p_{n_j}^{-1}(B_{a_j}(x_j))$, where $B_{a_j}(x_j)$ is the open ball of center x_j and radius a_j in H_{n_j}. Define V_j inductively. Set $V_1 = Z_1$. Having defined V_{j-1}, set

$$V_j = Z_j \cap^c \overline{p_{n_1}^{-1} B_{r_{1j}}(x_1)} \cap \cdots \cap^c \overline{p_{n_{j-1}}^{-1} B_{r_{j-1,j}}(x_{j-1})}$$

where $r_{1j} = a_1 - \frac{1}{j}, \ldots, r_{j-1,j} = a_{j-1} - \frac{1}{j}$. Each V_j is a scalloped open set contained in Z_j. We claim that the V_j cover W. Let $x \in W$, and let j be the smallest integer such that $x \in Z_j$. Then $x \in V_j$, otherwise for some $k \in \{1, \ldots, j-1\}$, x would belong to $\overline{p_{n_k}^{-1} B_{r_{k,j}}(x_k)} \subseteq Z_k$, a contradiction. We will show that the covering $(\overline{V}_j)$ is locally finite. Let $x \in W$, say $x \in Z_j$. Put $y = p_{n_j}(x) \in B_{a_j}(x_j)$. Let s be a > 0 number such that the ball $B_s(y)$ is contained in $B_{a_j}(x_j)$. Let $t = s/2$. For all k large enough, the ball $B_t(y)$ is contained in $B_{r_{jk}}(x_j)$, and therefore $p_{n_j}^{-1}(B_t(y))$ does not meet $\overline{V}_k$. Hence the open neighborhood $p_{n_j}^{-1}(B_t(y))$ of x intersects only finitely many $\overline{V}_k$. ∎

Note that Lemma 1.4.11 clearly implies that W is paracompact.

1.4.12. Proposition. *Let E be an ILH space, satisfying condition (1-25). Then for any open covering $\mathcal{U} = (U_i)_{i \in I}$ of E, there exists a smooth partition of unity subordinate to $\mathcal{U}$.*

Proof. As E is paracompact, we may assume that (U_i) is locally finite. Let (Z_i) be an open covering of E such that $\overline{Z}_i \subset U_i$ for all i. If we can find smooth functions $f_i \geq 0$ with support in $\overline{Z}_i$, such that $\sum_i f_i = 1$, we will have a partition of unity for (U_i). Replacing the covering (Z_i) by a refinement, we may as well assume that each Z_i is a round open set. Since E is Lindelöf, we may replace I by a countable subset and still get a covering. Hence we may assume that $\mathbb{N}$ is the indexing set. We denote our covering by $(Z_j)_{j \in \mathbb{N}}$ and construct scalloped open sets $V_j \subset Z_j$ as in Lemma 1.4.10. By Lemma 1.4.11, there exists for each $j \in \mathbb{N}$ a smooth function $g_j \geq 0$, with support in $\overline{V}_j$, such that $g_j(x) > 0$ for $x \in V_j$. The function $g = \sum_j g_j$ on E is smooth since the covering $(\overline{V}_j)$ is locally finite, hence the sum is locally finite. Since every $x \in E$ belongs to some V_j, we have $g(x) \geq g_j(x) > 0$. So the functions $f_j = \frac{g_j}{g}$ are smooth ≥ 0, with $Supp(f_j) \in \overline{V}_j \subset \overline{Z}_j$, and $\sum_j f_j = 1$. This gives our partition of unity for the covering (U_i). ∎

1.4.13. Corollary. *Let E be an ILH space. Let A, B be non-empty disjoint closed subsets of E. Then there exists a smooth function g on E such that $g(x) = 0$ for $x \in A$, $g(x) = 1$ for $x \in B$, and $0 \leq g(x) \leq 1$ for all x.*

Proof. $(E \backslash B, E \backslash A)$ is an open covering of E. Let (f, g) be a corresponding partition of unity. We have $0 \leq g(x) = 1 - f(x) \leq 1$. For $x \in A$, we have $g(x) = 1 - f(x) = 0$; for $x \in B$, we have $g(x) = 1 - f(x) = 1$. ∎

1.4.14. Theorem. *Let M be a paracompact manifold, modeled on a ILH space E satisfying (1-25). Then for any open covering $\mathcal{U} = (U_i)_{i \in I}$ of M, there exists a smooth partition of unity subordinate to $\mathcal{U}$.*

Proof. We may as well assume that each U_i is diffeomorphic to a round open set in E, and that the covering is locally finite. There exists an open covering (V_i) such that $\overline{V}_i \subset U_i$. Since U_i is diffeomorphic to E, we see from Corollary 1.4.11 that there exists a smooth function g_i on U_i, with support in $\overline{V}_i$, such that $g_i(x) > 0$ for all $x \in V_i$. Setting $g_i(x) = 0$ for $x \notin U_i$, we obtain a smooth function g_i on M, with support in $\overline{V}_i$, which is positive on V_i. Thus $g = \sum_i g_i$ is a positive smooth function, and thus the functions $f_i = \frac{g_i}{g}$ comprise a partition of unity subordinate to (U_i). ∎

We summarize the results we have obtained for manifolds modeled on ILH spaces.

1.4.15. Theorem. *Let M be a paracompact manifold, modeled on an ILH space E satisfying (1-25). Then the sheaves $\underline{A}_M^p$ are soft, and we have canonical isomorphisms*

$$\check{H}^p(M, \mathbb{R}) \widetilde{\to} H^p(M, \mathbb{R}) \widetilde{\to} H_{DR}^p(M).$$

If M is a convex open set of E, these groups are 0 for $p > 0$.

We wish to write down an explicit isomorphism between Čech cohomology groups and de Rham cohomology groups. Such an isomorphism was first given in [**Weil1**]. Observe that the isomorphism $H^p(M, \mathbb{R}_M) \to H^p(M, \underline{A}_M^\bullet)$ is induced by the quasi-isomorphism $\mathbb{R}_M \to \underline{A}_M^\bullet$. As regards the isomorphism $H_{DR}^p(M) \to H^p(M, \underline{A}_M^\bullet)$, it is very natural from the point of view of Čech hypercohomology. Namely we map $A^\bullet(M)$ to the first column $\prod_i A^\bullet(U_i)$ by the restriction maps from M to each U_i.

$$
\begin{array}{ccccccc}
A^2(M) & \to & \prod_i A^2(U_i) & \xrightarrow{\delta} & \prod_{i,j} A^2(U_{ij}) & \xrightarrow{\delta} & \cdots \\
\Big\uparrow{\scriptstyle d} & & \Big\uparrow{\scriptstyle d} & & \Big\uparrow{\scriptstyle d} & & \\
A^1(M) & \xrightarrow{\delta} & \prod_i A^1(U_i) & \xrightarrow{\delta} & \prod_{i,j} A^1(U_{ij}) & \xrightarrow{\delta} & \cdots \\
\Big\uparrow{\scriptstyle d} & & \Big\uparrow{\scriptstyle d} & & \Big\uparrow{\scriptstyle d} & & \\
A^0(M) & \to & \prod_i A^0(U_i) & \xrightarrow{\delta} & \prod_{i,j} A^0(U_{ij}) & \xrightarrow{\delta} & \cdots
\end{array}
$$

Let $\mathcal{U} = (U_i)$ be an open covering of M. We note that the cohomology class of the Čech p-cocycle $\underline{c}$ of the covering $\mathcal{U}$ with coefficients in $\mathbb{R}_M$ will correspond to the cohomology class of the closed p-form ω under this isomorphism if we can find a degree $(p - 1)$ cochain $(\underline{\alpha}_q)_{0 \leq q \leq p-1}$ in the Čech double complex $C^\bullet(\underline{A}_M^\bullet)$ with the following properties:

(1) Each $\underline{\alpha}_q$ is a Čech $(p - 1 - q)$-cochain with values in $\underline{A}_M^q$, and we have

$$(\delta\underline{\alpha}_q)_{i_0,\ldots,i_{p-q}} = (-1)^{p-q-1}d((\underline{\alpha}_{q-1})_{i_0,\ldots,i_{p-q}})$$

for $1 \leq q \leq p - 1$.

(2) $(d\underline{\alpha}_{p-1})_i$ is the restriction of ω to U_i.

(3) $\delta\underline{\alpha}_0 = -\underline{c}$.

$$
\begin{array}{c}
\omega \\
\uparrow d \\
\underline{\alpha}_{p-1} \;\xrightarrow{\;\delta\;}\; * \\
\qquad\qquad \uparrow -d \\
\qquad\qquad \underline{\alpha}_{p-2} \;\xrightarrow{\;\delta\;}\; * \\
\qquad\qquad\qquad\qquad \uparrow d \\
\qquad\qquad\qquad\qquad \underline{\alpha}_{p-3} \\
\qquad\qquad\qquad\qquad\qquad\qquad \ddots \\
\qquad\qquad\qquad\qquad\qquad\qquad\qquad \underline{\alpha}_0 \;\xrightarrow{\;\delta\;}\; -c
\end{array}
$$

Indeed, these conditions mean that the total boundary $\delta + (-1)^p\, d$ (where p is the first degree) of the Čech cochain $\sum_{q=0}^{p-1} \underline{\alpha}_q$ is equal to the difference $\omega - c$, where ω is viewed as a 0-cochain with values in $\underline{A}_M^p$, and c as a p-cochain with values in $\underline{A}_M^0$. Starting from c, we will construct $\underline{\alpha}_1$, $\underline{\alpha}_2$, and so on, inductively. For this, we choose a partition of unity (f_i) subordinate to $\mathcal{U}$.

1.4.16. Lemma. *Let* $H : C^{\bullet}(\mathcal{U}, \underline{A}_M^p) \to C^{\bullet-1}(\mathcal{U}, \underline{A}_M^p)$ *be given by*

$$
H(\underline{\alpha})_{i_1,\ldots,i_p} = \sum_i f_i \alpha_{i,i_1,\ldots,i_p}
$$

Then for each $p \geq 0$, H *gives a homotopy of the Čech complex for the sheaf* $\underline{A}_M^p$.

Proof. We have

$$
\delta(H(\underline{\alpha}))_{i_0,\ldots,i_p} = \sum_{j=0}^p (-1)^j\, H(\underline{\alpha})_{i_0,\ldots,\hat{i}_j,\ldots,i_p} = \sum_{j=0}^p \sum_i (-1)^j f_i \alpha_{i,i_0,\ldots,\hat{i}_j,\ldots,i_p}.
$$

On the other hand we have

$$
H(\delta(\underline{\alpha}))_{i_0,\ldots,i_p} = \sum_i f_i \delta(\underline{\alpha})_{i,i_0,\ldots,i_p}
$$

$$
= \sum_i f_i \alpha_{i_0,\ldots,i_p} - \sum_i f_i \sum_{j=0}^p (-1)^j \alpha_{i,i_0,\ldots,\hat{i}_j,\ldots,i_p}
$$

$$
= \alpha_{i_0,\ldots,i_p} - \delta(H(\underline{\alpha}))_{i_0,\ldots,i_p} \qquad \blacksquare
$$

We now have a means of finding $\underline{\alpha}_q$ by induction on q, starting with $q = 0$. First we take $\underline{\alpha}_0 = -H(c)$, since $\delta(H(c)) = c$ by Lemma 1.4.16. This gives

$$(\underline{\alpha}_0)_{i_1,\ldots,i_p} = -\sum_i f_i c_{i,i_0,\ldots,i_p}.$$

For $1 \le q \le p-1$, we want $\delta(\underline{\alpha}_q) = (-1)^{p-q-1} d\underline{\alpha}_{q-1}$. But $d\underline{\alpha}_{q-1}$ is in the kernel of δ, since $\delta d\underline{\alpha}_{q-1} = d\delta\underline{\alpha}_{q-1} = (-1)^{p-q} dd\underline{\alpha}_{q-2}$ using the inductive hypothesis. Therefore we have $d\underline{\alpha}_{q-1} = \delta H(d\underline{\alpha}_{q-1})$, so we set $\underline{\alpha}_q = (-1)^{p-q-1} H(d\underline{\alpha}_{q-1})$, yielding

$$(\underline{\alpha}_1)_{i_2,\ldots,i_p} = -(-1)^{p-2} \sum_{i_0,i_1} c_{i_0,i_1,\ldots,i_p} f_{i_0} df_{i_1},$$

and so on. Finally, the last term is

$$(\underline{\alpha}_{p-1})_i = -(-1)^{(p-1)(p-2)/2} \sum_{i_1,\ldots,i_p} c_{i_1,\ldots,i_p,i}\, f_{i_1} df_{i_2} \wedge \cdots \wedge df_{i_p}$$

The closed p-forms $\omega_i = (d\underline{\alpha}_{p-1})_i$ on U_i have the same restriction to U_{ij}, since $\delta d\underline{\alpha}_{p-1} = d\delta\underline{\alpha}_{p-1} = dd\underline{\alpha}_{p-2} = 0$. Hence we have proved the following.

1.4.17. Proposition. *(Weil [Weil1]) Let M be a smooth manifold on which the sheaves $\underline{A}_M^p$ are soft (e.g., M is a paracompact smooth manifold modeled on an ILH space satisfying (1-25)). Let $\mathcal{U}$ be an open covering of M, and let $(f_i)_{i \in I}$ be a partition of unity subordinate to $\mathcal{U}$. Then for $\underline{c}$ a degree p Čech cocycle with coefficients in the constant sheaf $\mathbb{R}_M$, let ω be the closed p-form on M such that*

$$(\omega)_{/U_i} = (-1)^{\frac{p(p+1)}{2}} \sum_{i_1,\ldots,i_p} c_{i_1,\ldots,i_p,i}\, df_{i_1} \wedge \cdots \wedge df_{i_p} \qquad (1-26).$$

Then the cohomology class $[\underline{c}] \in H^p(M, \mathbb{R}_M)$ and the cohomology class $[\omega] \in H_{DR}^p(M)$ correspond to each other under the isomorphism $H^p(M, \mathbb{R}_M) \tilde{\to} H_{DR}^p(M)$.

We note that this coincides exactly with the formula in [Weil1] if c is skew-symmetric.

It is possible, but more difficult, to write down explicitly a degree p Čech cocycle corresponding to a closed p-form. We refer to [B-T] for such a formula.

1.5. Deligne and Cheeger-Simons cohomologies

Deligne cohomology was invented by Deligne around 1972 for the purpose of having one cohomology theory for algebraic varieties which includes ordinary singular (or Čech) cohomology and the intermediate jacobians of Griffiths. We will mostly deal with a smooth analog of this theory, which we call smooth Deligne cohomology. It is called Cheeger-Simons cohomology in [B-K]. We will see the relation with the Cheeger-Simons groups of differential characters later on.

For M a smooth paracompact manifold which is finite-dimensional (or, more generally, modeled on an ILH space), we saw in §4 that the Čech cohomology groups $\check{H}^p(M, \mathbb{R}_M)$ are canonically isomorphic to the de Rham cohomology groups. We gave an explicit isomorphism in Proposition 1.4.17. Essentially, the construction of this isomorphism amounted to moving upwards on a "staircase" oriented southeast to northwest. This staircase was inside the Čech double complex for the complex of sheaves

$$0 \to \mathbb{R} \to \underline{A}^0_M \to \underline{A}^1_M \cdots \to \underline{A}^{p-1}_M \qquad (1-27)$$

where $\mathbb{R}$ is put in degree 0. We truncate the de Rham complex at $\underline{A}^{p-1}_M$ if we are interested in degree p cohomology classes.

The complex of sheaves (1-27) is essentially the smooth Deligne complex of index p, except for some twist which we will now try to motivate. The motivation comes from the theory of characteristic classes, which produces cohomology classes in $H^{2p}(M, \mathbb{Z})$ associated to a bundle over M. However, these classes only become integral after some natural expression in the curvature of the bundle is divided by a factor $(2\pi\sqrt{-1})^p$. Except for this arbitrary factor, the expression is purely algebraic over $\mathbb{Q}$, of the type $P(\Omega)$, where Ω is the curvature and P is some invariant polynomial with rational coefficients. Deligne's viewpoint is to eliminate this unnatural division by a power of $2\pi\sqrt{-1}$. So he introduces the cyclic subgroup $\mathbb{Z}(p) = (2\pi\sqrt{-1})^p \cdot \mathbb{Z}$ of $\mathbb{C}$ and views a characteristic class as lying in $H^{2p}(M, \mathbb{Z}(p))$ instead of $H^{2p}(M, \mathbb{Z})$. Similarly, for a subring B of $\mathbb{R}$ (the most interesting examples being $B = \mathbb{Z}, \mathbb{Q}$ and $\mathbb{R}$), he introduces the free cyclic B-module $B(p) = (2\pi\sqrt{-1})^p \cdot B \subset \mathbb{C}$. Note that product in $\mathbb{C}$ induces an isomorphism $B(p) \otimes_B B(q) \xrightarrow{\sim} B(p+q)$ of B-modules. In particular, $\mathbb{R}(p)$ is equal to $\mathbb{R}$ for p even and to $\sqrt{-1} \cdot \mathbb{R}$ for p odd.

Recall that $A^p(M)_{\mathbb{C}} = A^p(M) \otimes \mathbb{C}$ denotes the space of complex-valued p-forms on M and $\underline{A}^p_{M,\mathbb{C}}$ the sheaf of complex-valued differential forms. Let $i : B(p)_M \to \underline{A}^0_{M,\mathbb{C}}$ denote the inclusion of constant into smooth functions.

1.5.1. Definition. *Let M be a smooth manifold. For B a subring of $\mathbb{R}$*

and for $p \geq 0$, the _smooth Deligne complex_ $B(p)_D^\infty$ is the complex of sheaves:

$$B(p)_M \xrightarrow{i} \underline{A}^0_{M,\mathbb{C}} \xrightarrow{d} \underline{A}^1_{M,\mathbb{C}} \xrightarrow{d} \cdots \xrightarrow{d} \underline{A}^{p-1}_{M,\mathbb{C}} \qquad (1-28)$$

The hypercohomology groups $H^q(M, B(p)_D^\infty)$ are called the _smooth Deligne cohomology groups_ of M, and are sometimes denoted by $H_D^q(M, B(p)^\infty)$.

We now compute Deligne cohomology groups and show that Deligne cohomology is a refinement of ordinary cohomology. First we see that $B(0)_D^\infty$ is simply equal to B_M, so that $H^q(M, B(0)_D^\infty)$ is just sheaf cohomology $H^q(M, B_M)$. The complex $\mathbb{Z}(1)_D^\infty$ is much more interesting. We have

$$\mathbb{Z}(1)_D^\infty = \quad [\mathbb{Z}(1) \to \underline{\mathbb{C}}_M] \qquad (1-29)$$

We will need the notion of translation of a complex of groups (or of sheaves). For $K^\bullet$ a complex and $p \in \mathbb{Z}$, the complex $K^\bullet[p]$ is the complex obtained by translating $K^\bullet$ by p steps to the left. Thus $(K^\bullet[p])^q = K^{p+q}$, and $H^q(K^\bullet[p]) = H^{p+q}(K^\bullet)$.

1.5.2. Proposition. _The complex of sheaves $\mathbb{Z}(1)_D^\infty$ is quasi-isomorphic to $\underline{\mathbb{C}}_M^*[-1]$. Hence we have $H^q(M, \mathbb{Z}(1)_D^\infty) = H^{q-1}(M, \underline{\mathbb{C}}_M^*)$ for all q._

Proof. Consider the following morphism of complexes of sheaves exp : $\mathbb{Z}(1)_D^\infty \to \underline{\mathbb{C}}_M^*[-1]$.

$$\begin{array}{ccc} \mathbb{Z}(1) & \to & \underline{\mathbb{C}}_M \\ \downarrow 0 & & \downarrow exp \\ 0 & \to & \underline{\mathbb{C}}_M^* \end{array}$$

ϕ is a quasi-isomorphism of complex of sheaves, since the sequence

$$0 \to \mathbb{Z}(1) \to \underline{\mathbb{C}}_M \xrightarrow{exp} \underline{\mathbb{C}}_M^* \to 0$$

is exact by Corollary 1.1.9. ∎

Note that this result already gives some justification for the decision we made to start this complex of sheaves with $\mathbb{Z}(1)$ instead of $\mathbb{Z}$. One can generalize Proposition 1.5.2 to any subring B of $\mathbb{R}$, but we will not get into this here, except to say that $H^1(M, B(1)_D^\infty)$ has to do with multi-valued smooth functions f for which two branches of f at some point differ by a constant in $B(1) \subset \mathbb{C}$.

We now study smooth Deligne cohomology in general. There is a natural homomorphism $\kappa : H^q(M, B(p)_D^\infty) \to H^q(M, B(p))$ since the complex $B(p)_D^\infty$ projects to the constant sheaf $B(p)_M$. We have an exact sequence of complexes of sheaves

$$0 \to \sigma_{\leq p-1}(\underline{A}_{M,\mathbb{C}}^\bullet)[-1] \to B(p)_D^\infty \to B(p) \to 0, \qquad (1-30)$$

where $\sigma_{\leq p-1}(\underline{A}_{M,\mathbb{C}}^\bullet)$ denotes the complex $\underline{A}_{M,\mathbb{C}}^0 \to \cdots \to \underline{A}_{M,\mathbb{C}}^{p-1}$ obtained by chopping off the part of the complex $\underline{A}_{M,\mathbb{C}}^\bullet$ in degrees $\geq p$, giving a long exact sequence of hypercohomology groups. We assume now that the sheaves $\underline{A}_M^p$ are soft, hence so are their complexifications $\underline{A}_{M,\mathbb{C}}^p$. Then the hypercohomology groups of the complex of sheaves $\sigma_{\leq p-1}\underline{A}_{M,\mathbb{C}}^p$ are simply

$$H^q(M, \sigma_{\leq p-1}\underline{A}_{M,\mathbb{C}}^p) = \left\{ \begin{array}{l} H_{DR}^q(M) \otimes \mathbb{C} \text{ if } q \leq p-2 \\ A_{DR}^{p-1}(M)_\mathbb{C}/d\ A^{p-2}(M)_\mathbb{C} \text{ if } q = p-1 \\ 0 \text{ if } q \geq p \end{array} \right\}.$$
$$(1-31)$$

1.5.3. Theorem. *Let M be a smooth paracompact manifold such that the sheaves $\underline{A}_M^p$ are soft (for example, M is finite-dimensional or satisfies the assumptions of Theorem 1.4.15). The smooth Deligne cohomology groups $H^\bullet(M, B(p)_D^\infty)$ are as follows:*

(1) *For $q \leq p-1$, the group $H^q(M, B(p)_D^\infty)$ fits in the exact sequence*

$$0 \to H^{q-1}(M, B(p)) \to$$
$$\to H^{q-1}(M, \mathbb{C}) \to H^q(M, B(p)_D^\infty) \xrightarrow{\kappa} \text{Tors } H^q(M, B(p)) \to 0,$$
$$(1-32)$$

where Tors $H^q(M, B(p))$ *denotes the torsion subgroup of $H^q(M, B(p))$.*

(2) *The group $H^p(M, B(p)_D^\infty)$ fits in the exact sequence*

$$0 \to A^{p-1}(M)_\mathbb{C}/A^{p-1}(M)_{\mathbb{C},0} \to H^p(M, B(p)_D^\infty) \xrightarrow{\kappa} H^p(M, B(p)) \to 0,$$
$$(1-33)$$

where $A^{p-1}(M)_{\mathbb{C},0}$ denotes the group of closed complex-valued $(p-1)$-forms on M whose cohomology class belongs to the image of $H^{p-1}(M, B(p)) \to H^{p-1}(M, \mathbb{C})$.

(3) *For $q \geq p+1$, we have $H^q(M, B(p)_D^\infty) \xrightarrow{\sim} H^q(M, B(p))$.*

Proof. We use the long exact sequence associated to the exact sequence

(1-30) of complex of sheaves. This gives an exact sequence

$$H^{q-1}(M, B(p)) \xrightarrow{f_{q-1}} H^{q-1}(\sigma_{\leq p-1} A^\bullet(M)_\mathbb{C}) \to H^q(M, B(p)^\infty_D) \to$$
$$\to H^q(M, B(p)) \xrightarrow{f_q} H^q(\sigma_{\leq p-1} A^\bullet(M)_\mathbb{C}).$$

The groups $H^\bullet(\sigma_{\leq p-1} A^\bullet(M)_\mathbb{C})$ were determined in (1-31). The map $f_q : H^q(M, B(p)) \to H^q(\sigma_{\leq p-1} A^\bullet(M)_\mathbb{C})$ is the composite of the map $H^q(M, B(p)) \to H^q(M, \mathbb{C})$, of the isomorphism

$$H^q(M, \mathbb{C}) \simeq H^q_{DR}(M) \otimes \mathbb{C},$$

and of the projection map

$$H^q_{DR}(M) \otimes \mathbb{C} = H^q(A^\bullet(M)_\mathbb{C}) \to H^q(\sigma_{\leq p-1} A^\bullet(M)_\mathbb{C}).$$

For $q \leq p - 1$, we have $H^{q-1}(\sigma_{\leq p-1} A^\bullet(M)_\mathbb{C}) = H^{q-1}_{DR}(M) \otimes \mathbb{C} = H^{q-1}(M, \mathbb{C})$. For these values of q, the map $H^q(M, \mathbb{C}) \to H^q(\sigma_{\leq p-1} A^\bullet(M)_\mathbb{C})$ is injective, so the kernel of f_q is the torsion subgroup of $H^q(M, B(p))$. This proves (1). For $q = p$, the map f_{p-1} is the map from $H^{p-1}(M, B(p))$ to $A^{p-1}(M)_\mathbb{C}/dA^{p-2}(M)_\mathbb{C}$. Its image consists of those degree $p - 1$ complex-valued de Rham cohomology classes which belong to the image of $H^{p-1}(M, B(p))$ in $H^{p-1}(M, \mathbb{C})$. Hence the cokernel of f_{p-1} is equal to $A^{p-1}(M)_\mathbb{C}/A^{p-1}(M)_{\mathbb{C},0}$. On the other hand, $H^p(\sigma_{\leq p-1} A^\bullet(M)_\mathbb{C}) = 0$. This proves (2).

(3) is clear, since both $H^{q-1}(\sigma_{\leq p-1} A^\bullet(M)_\mathbb{C})$ and $H^q(\sigma_{\leq p-1} A^\bullet(M)_\mathbb{C})$ are 0 for $q \geq p$. ∎

The theorem shows that the Deligne cohomology group is most interesting for $q = p$, when it maps onto $H^p(M, B(p))$ with a large kernel, consisting of all $(p-1)$-forms modulo those which are closed and have periods in $B(p)$. There is a morphism d of complexes

$$B(p)_M \xrightarrow{i} \underline{A}^0_{M,\mathbb{C}} \xrightarrow{d} \cdots \xrightarrow{d} \underline{A}^{p-1}_{M,\mathbb{C}}$$
$$\downarrow d \qquad\qquad (1-34)$$
$$\underline{A}^p_{M,\mathbb{C}}$$

This induces a homomorphism $d : H^p(M, B(p)^\infty_D) \to A^p(M)_\mathbb{C}{}^{cl}$, where $A^p(M)_\mathbb{C}{}^{cl}$ is the space of closed complex-valued p-forms. The following result shows that classes in smooth Deligne cohomology can be used to compare classes in Čech cohomology with classes in de Rham cohomology. This will be used in Chapter 2; see [Br-ML1] for applications to formulas for characteristic classes.

1.5.4. Proposition. *Let $a \in H^p(M, \mathbb{R}(p)_D^\infty)$. Then the Čech cohomology class $\kappa(a) \in H^p(M, \mathbb{R}(p))$ corresponds to the de Rham cohomology class of $(-1)^p \cdot da$ in $H^p(A^\bullet(M)_\mathbb{C})$ under the isomorphism between Čech and de Rham cohomologies.*

Proof. Let $(a_0, \ldots, a_p)$ be a representative for a in the Čech double complex of some open covering with coefficients in the complex of sheaves $\mathbb{R}(p)_D^\infty$. So a_0 is a degree p Čech cocycle with coefficients in $\mathbb{R}(p)$, which represents $\kappa(a)$. For $i \geq 1$, a_i is a Čech $(p - i)$-cocycle with coefficients in $\underline{A}_{M,\mathbb{C}}^{i-1}$. We have $\delta(a_1) = (-1)^{p-1} i(a_0)$ in $C^p(\mathcal{U}, \underline{A}_{M,\mathbb{C}}^0)$. It follows that the total boundary of $(a_1, \ldots, a_p)$ in the Čech double complex is equal to $(-1)^{p-1} i(a_0) + da_p$. This means that the cohomology classes of $(-1)^p \cdot a_0$ and of the p-form da_p correspond to each other under the Čech-de Rham isomorphism of §4. ∎

One can define characteristic classes with values in this Deligne cohomology group; for instance, the k-th Chern class of a complex vector bundle equipped with a connection may be defined as an element of $H^{2k}(M, \mathbb{Z}(k)_D)$. An explicit formula for such a class is given in [**Br-ML1**]. In case the usual Chern class is 0, one obtains a complex-valued $2k - 1$-form defined modulo $A_{\mathbb{C},0}^{2k-1}$, which can be identified with the differential form of Chern and Simons [**Cher-S**].

In fact, the cohomology group $H^p(M, \mathbb{Z}(p)_D^\infty)$ is very closely related to the group of *differential characters* in the sense of Cheeger and Simons [**Chee-S**]. The definition we give here is slightly different from [**Chee-S**], in that we consider cochains with complex (as opposed to real) coefficients. We let $S_p(M)$ be the group of smooth singular chains on M, and $Z_p(M)$ the group of smooth singular p-cycles. We have the boundary map $\partial : S_p(M) \to S_{p-1}(M)$. For $S_p(M)^*$ the dual group of smooth singular cocycles, we have the transpose coboundary map $\partial^* : S_p(M)^* \to S_{p+1}(M)^*$.

1.5.5. Definition. [**Chee-S**] *Let M be a smooth finite-dimensional manifold. A differential character of degree p on M is a homomorphism $c : Z_{p-1}(M) \to \mathbb{C}/\mathbb{Z}(p)$ such that there exists a complex-valued p-form α for which*

$$c(\partial \gamma) = \int_\gamma \alpha \mod \mathbb{Z}(p) \tag{1-35}$$

for any smooth singular p-chain γ, with boundary $\partial \gamma$. The group of degree p differential characters is denoted by $\hat{H}^p(M, \mathbb{Z}(p))$.

Note that our notation for the group of differential characters is different from the notation in [**Chee-S**].

1.5.6. Lemma. [**Chee-S**] *We have an exact sequence*

$$0 \to A^{p-1}(M)_{\mathbf{C}}/A^{p-1}(M)_{\mathbf{C},0} \to \hat{H}^p(M, \mathbf{Z}(p)) \to H^p(M, \mathbf{Z}(p)) \to 0.$$

Proof. Let $q : \mathbf{C} \to \mathbf{C}/\mathbf{Z}(p)$ be the projection. One defines a map $\phi :$ $\hat{H}^p(M, \mathbf{Z}(p)) \to H^p(M, \mathbf{Z}(p))$ as follows. Let c be a differential character, and α a corresponding p-form satisfying (1-34). Since $\mathbf{C}/\mathbf{Z}(p)$ is a divisible group (hence an injective abelian group), and $S_{p-1}(M)$ is a free abelian group, there exists a group homomorphism $c' : S_{p-1}(M) \to \mathbf{C}$ such that $q(c'(\gamma)) = c(\gamma)$ for $\gamma \in Z_{p-1}(M)$. The difference $\sigma = \alpha - \partial^*(c')$ is a $\mathbf{Z}(p)$-valued p-cochain, according to (1-34). We have $\partial^* \sigma = \partial^* \alpha$, hence $\partial^* \sigma = d\alpha$ is a differential form which gives rise to a cochain with values in $\mathbf{Z}(p)$. This implies $d\alpha = 0$, hence $\partial^* \sigma = 0$. Thus σ is a smooth p-cocycle with values in $\mathbf{Z}(p)$. We claim that the cohomology class of σ is independent of the choices. If we change c' to $c' + a$, where a takes values in $\mathbf{Z}(p)$, we change σ to $\sigma - \partial^*(a)$, which has the same cohomology class. If we change c' to $c' + b$, where b vanishes on $p - 1$-cycles, then σ does not change. This proves that the cohomology class of σ is well-defined. We define $\phi(c) = [\sigma]$.

We show next that ϕ is onto. Let $u \in H^p(M, \mathbf{Z}(p))$, and let α be a closed p-form representing u. Let σ be a $\mathbf{Z}$-valued smooth p-cocycle in the class of u. Then the difference cochain $\alpha - \sigma$ vanishes on $Z_p(M)$, so it is of the form $\partial^* c'$ or some smooth $p - 1$-cochain c'. Let c be the restriction to $Z_{p-1}(M)$ of $q \circ c'$. Now c and α satisfy (1-35), and we see that $\phi(c) = u$.

Let $c \in \hat{H}^p(M, \mathbf{Z}(p))$ be in the kernel of ϕ. This means that one can choose the extension c' of c to all $(p - 1)$-chains so that $\partial^* c' = \alpha$. Since singular cohomology is isomorphic to de Rham cohomology, we also have $\alpha = d\beta$ for some $(p - 1)$-form β. $c' - \beta$ is a cocycle, so its restriction to $Z_{p-1}(M)$ is represented by a closed $(p - 1)$-form β_1. c' and $\beta + \beta_1$ agree on $Z_{p-1}(M)$. Thus we have shown that c is given by some $(p - 1)$-form. This $(p - 1)$-form is determined precisely up to the ambiguity of a $(p - 1)$-form ω such that $\int_\gamma \omega \in \mathbf{Z}(p)$ for all $(p - 1)$-cycles γ, i.e., up to an element of $A^p(M)_{\mathbf{C},0}$. ∎

Given this result, the following proposition is not surprising.

1.5.7. Proposition. *The group $\hat{H}^p(M, \mathbf{Z}(p))$ is canonically isomorphic to the smooth Deligne cohomology group $H^p(M, \mathbf{Z}(p)_D^\infty)$.*

We refer to [**E**] [**D-H-Z**] for a construction of this isomorphism (see also [**Br-ML1**]).

The notion of differential character is very concrete and well-suited to

differential-geometric applications. One advantage of Deligne cohomology is
a very pleasant description of products

$$\cup : B(p)_D^\infty \otimes B(q)_D^\infty \rightarrow B(p+q)_D^\infty,$$

due to Deligne and Beilinson [Be]. To avoid confusion, we will use $\tilde{d}$ to
denote the differential in a Deligne complex, which does not always coincide
with exterior differentiation d. For $x \in B(p)_D^\infty$, we denote by $deg(x)$ its
degree in this complex of sheaves. We then put

$$x \cup y = \left\{ \begin{array}{ll} x \cdot y & \text{if deg } (x) = 0 \\ x \wedge dy & \text{if deg(x)} > 0 \text{ and deg(y)} = q \\ 0 & \text{otherwise} \end{array} \right\} \qquad (1-36)$$

1.5.8. Proposition. (1) *Each $\cup$ is a morphism of complexes.*

(2) *The multiplication $\cup$ is associative and is commutative up to homotopy.*

(3) *The class 1 is a left identity for $\cup$ (hence a right identity up to homotopy).*

Proof. We have

$$\tilde{d}(x \cup y) = \left\{ \begin{array}{ll} x \cdot dy & \text{if deg(x)} = 0, deg(y) \leq q \\ dx \wedge dy & \text{if deg(x)} > 0, deg(y) = q \\ 0 & \text{otherwise} \end{array} \right\} = \tilde{d}x \cup y + (-1)^{deg(x)} x \cup \tilde{d}y,$$

which proves (1).

The associativity of $\cup$ is easily checked. For commutativity of $\cup$ up to
homotopy, we want to compare the morphisms of complexes $\cup$ and $\cup \circ s$,
where $s : B(p)_D^\infty \otimes B(q)_D^\infty \xrightarrow{\sim} B(q)_D^\infty \otimes B(p)_D^\infty$ is the transposition $s(x \otimes y) =$
$(-1)^{deg(x)deg(y)} y \otimes x$. There is a homotopy between these two morphisms,
given by

$$H(x \otimes y) = \left\{ \begin{array}{l} 0 \text{ if deg(x)} = 0 \text{ or deg(y)} = 0 \\ (-1)^{deg(x)} x \wedge y \text{ otherwise} \end{array} \right.$$

(3) is obvious. ■

An illustration of this cup-product will be given in Chapter 2, in connection with a method used to construct line bundles with connection.

So far we have studied smooth Deligne cohomology. Deligne cohomology itself was invented by Deligne for complex manifolds.

1.5.9. Definition. *Let X be a complex manifold. For a subring B of $\mathbb{R}$, the __Deligne complex__ $B(p)_D$ is the complex of sheaves*

$$B(p)_M \xrightarrow{i} \Omega^0_M \xrightarrow{d} \Omega^1_M \xrightarrow{d} \cdots \xrightarrow{d} \Omega^{p-1}_M \qquad (1-37)$$

The hypercohomology groups $H^q(M, B(p)_D)$ are called the __Deligne cohomology groups__, and are sometimes denoted by $H^q_D(M, B(p))$.

In analogy with Proposition 1.5.2, we have

1.5.10. Proposition. *The complex of sheaves $\mathbb{Z}(1)_D$ is quasi-isomorphic to $\mathcal{O}^*_X [-1]$, hence we have $H^q(X, \mathbb{Z}(1)_D) = H^{q-1}(X, \mathcal{O}^*_X)$.*

Deligne's motivation for introducing this cohomology theory was to have a rich theory of cohomology classes associated to algebraic subvarieties of a projective complex algebraic manifold X. To explain this, we note the exact sequence of complexes of sheaves

$$0 \to \Omega^\bullet_X / F^p \Omega^\bullet_X [-1] \to \mathbb{Z}(p)_D \to \mathbb{Z}(p) \to 0, \qquad (1-38)$$

where

$$F^p \Omega^\bullet_X = [\quad \Omega^p_X \xrightarrow{d} \cdots \xrightarrow{d} \Omega^n_X]$$
$$\underset{deg\ p}{|} \qquad\qquad\qquad \underset{deg\ n}{|}$$

is the so-called truncated de Rham complex. This truncation gives a decreasing filtration of the holomorphic de Rham complex $\Omega^\bullet_X$, as well as a filtration of any injective resolution $I^{\bullet\bullet}$ of $\Omega^\bullet_X$. The corresponding spectral sequence is the second spectral sequence of the double complex $\Gamma(X, I^{\bullet\bullet})$. This is called the Hodge to de Rham spectral sequence, or the Fröhlicher spectral sequence. Hodge theory says that this spectral sequence degenerates at E_1 and that the corresponding filtration on the abutment $H^\bullet(X, \mathbb{C})$ is the Hodge filtration F^p (see [**De1**]). It follows that $H^q(X, \Omega^\bullet_X / F^p \Omega^\bullet_X) = H^q(X, \mathbb{C}) / F^p\ H^q(X, \mathbb{C})$.

Therefore the exact sequence of complexes of sheaves (1-38) induces the long exact sequence of cohomology groups

$$\cdots \to H^{q-1}(X, \mathbb{Z}(p)) \to H^{q-1}(X, \mathbb{C}) / F^p\ H^{q-1}(X, \mathbb{C}) \to$$
$$\to H^q(X, \mathbb{Z}(p)_D) \to H^q(X, \mathbb{Z}(p)) \to \cdots \qquad (1-39)$$

For algebraic cycles, the crucial Deligne cohomology group is $H^{2p}(X, \mathbb{Z}(p)_D)$. Let $\mathrm{Hdg}^p(X)$ the group of Hodge cohomology classes, i.e., the group of classes γ in $H^{2p}(X, \mathbb{Z}(p))$ such that the image of γ in $H^{2p}(X, \mathbb{C})$ belongs to $F^p H^{2p}(X, \mathbb{C})$.

1.5.11. Theorem. *(Deligne)* (1) *There is an exact sequence*

$$0 \to J^p \to H^{2p}(X, \mathbb{Z}(p)_D) \to \mathrm{Hdg}^p(X) \to 0, \qquad (1-40)$$

where

$$J^p = H^{2p-1}(X, \mathbb{C})/F^p H^{2p-1}(X, \mathbb{C}) + H^{2p-1}(X, \mathbb{Z}(p)) \qquad (1-41)$$

is the <u>Griffiths intermediate jacobian</u> of [**Gri1**].

(2) *Every algebraic subvariety Z of X of pure codimension p has a cohomology class in $H^{2p}(X, \mathbb{Z}(p)_D)$. The image in $H^{2p}(X, \mathbb{Z}(p)_D)$ of this class is the class of Z in Čech cohomology.*

We refer to [**R-S-S**] or [**E-V**] for a proof and for discussions of the geometric significance of this theorem, as well as for the relation with Griffiths' work.

The product $B(p)_D \otimes B(q)_D \to B(p+q)_D$ is constructed in exactly the same way as in the smooth case, so we will not need to repeat the description.

Beilinson [**Be1**] [**Be2**] has developed Deligne cohomology much further and has introduced a version with growth conditions, which is much better suited to non-compact algebraic manifolds. The theory of Beilinson is called *Deligne-Beilinson cohomology*. Beilinson has found regulator maps from algebraic K-theory of algebraic manifolds to their Beilinson-Deligne cohomology groups, and has given conjectures for varieties over $\mathbb{Q}$, expressing values of their Hasse-Weil L-functions in terms of these regulators. This is far from the topic of this book, and so for further information we will refer the reader to the aforementioned articles of Beilinson, the book [**R-S-S**] and the survey article [**Sou**].

1.6. The Leray spectral sequence

In this section we study the spectral sequence invented by J. Leray during the Second World War [**Le**]. Its purpose is to analyze the cohomology of a space Y, with the help of a mapping $f : Y \to X$, in terms of the cohomology of X and that of the fibers $Y_x = f^{-1}(x)$. The spectral sequence is mostly used for fibrations, but not exclusively.

Let A be a sheaf of abelian groups on Y. There is a corresponding direct image sheaf $f_*(A)$ on X. We define it by $f_*(A)(U) = A(f^{-1}(U))$ for every open set U of X. If $V \subset U$, then the restriction map $A(f^{-1}(U)) \to A(f^{-1}(V))$ induces the restriction map $f_*(A)(U) \to f_*(A)(V)$.

1.6.1. Proposition. (1) *Given an exact sequence* $0 \to A \to B \to C \to 0$ *in* $\mathcal{AB}(Y)$, *the sequence* $0 \to f_*(A) \to f_*(B) \to f_*(C)$ *is exact in* $\mathcal{AB}(X)$. *That is, the functor* f_* *is left-exact.*

(2) *We have* $\Gamma(X, f_*(A)) \overset{\sim}{\to} \Gamma(Y, A)$ *for any sheaf* A *on* Y.

Proof. According to Proposition 1.1.7, it is enough to check that we have an exact sequence on stalks. Hence it is enough to check that the sequence $0 \to \Gamma(U, f_*(A)) \to \Gamma(U, f_*(B)) \to \Gamma(U, f_*(C))$ is exact for every open set U in Y. But this is true since the sequence $0 \to \Gamma(f^{-1}(U), A) \to \Gamma(f^{-1}(U), B) \to \Gamma(f^{-1}(U), C)$ is exact by Lemma 1.1.10. This proves (1); (2) follows from the definition of f_*. ∎

Note that for A a sheaf on X, we have $\Gamma(X, A) = p_*(A)$, where $p : X \to *$ is the projection of X to a point.

Since the functor f_* is left-exact, but not exact in general, it is appropriate to define higher direct images. We follow the method of §1 and §2, based on injective resolutions.

1.6.2. Proposition and Definition. Let $f : Y \to X$ be a continuous map. For $K^\bullet$ a bounded below complex of sheaves on Y, we define the *higher direct image sheaf* $R^q f_*(K^\bullet)$ to be the q-th cohomology sheaf of the total complex of the double complex of sheaves $f_*(I^{\bullet\bullet})$, where $I^{\bullet\bullet}$ is an injective resolution of $K^\bullet$. The sheaf $R^q f_*(K^\bullet)$ on X is independent of the injective resolution $I^{\bullet\bullet}$. For any exact sequence $0 \to A^\bullet \to B^\bullet \to C^\bullet \to 0$ of bounded below complexes of sheaves on Y, we have a long exact sequence of sheaves of abelian groups on X:

$$\cdots R^{q-1} f_*(C^\bullet) \overset{\delta}{\longrightarrow} R^q f_*(A^\bullet) \to R^q f_*(B^\bullet) \to R^q f_*(C^\bullet) \overset{\delta}{\longrightarrow} \cdots \quad (1-42)$$

Proof. The proof is completely similar to those of Proposition 1.2.7 and Corollary 1.2.9, and uses the same ingredients, notably Proposition 1.2.6 and Proposition 1.2.8. ∎

We now are ready to set up the Leray spectral sequence. For a bounded below complex of sheaves $K^\bullet$ on Y, we let $I^{\bullet\bullet}$ be an injective resolution of $K^\bullet$, and $I^\bullet$ the total complex of $I^{\bullet\bullet}$. We now consider the complex of sheaves $f_*(I^\bullet)$ on X, which has $R^p f_*(K^\bullet)$ as cohomology sheaves. Taking an injective resolution $J^{\bullet\bullet}$ of $f_*(I^\bullet)$ which satisfies all the conditions of Proposition 1.2.6, we look at the double complex $\Gamma(X, J^{\bullet\bullet})$. We claim that the total cohomology of this double complex is equal to the hypercohomology $H^n(Y, K^\bullet)$. Indeed we note the following

1.6.3. Lemma. *Let $f : Y \to X$ be a continuous map. Then for any injective sheaf I of abelian groups on Y, the sheaf $f_*(I)$ on X is injective.*

Proof. We note that for a sheaf A of abelian groups on Y and a sheaf B of abelian groups on X, we have a canonical isomorphism

$$Hom_X(B, f_*(A)) \xrightarrow{\sim} Hom_Y(f^{-1}(B), A) \qquad (1-43)$$

Let $A \hookrightarrow C$ be a monomorphism in $\mathcal{AB}(Y)$. We have to show that the map $Hom_Y(C, f_*(I)) \to Hom_Y(A, f_*(I))$ is surjective. This amounts to showing that $Hom_X(f^{-1}(C), I) \to Hom_X(f^{-1}(A), I)$ is surjective. This holds true because I is an injective sheaf and $f^{-1}(A) \to f^{-1}(C)$ is a monomorphism. ∎

The cohomology of the row $\Gamma(X, J^{\bullet,p})$ is the cohomology $H^q(X, f_*(I^p))$, which is 0 for $q > 0$. Hence the spectral sequence for the second filtration degenerates, and the cohomology of the double complex $\Gamma(X, J^{\bullet\bullet})$ identifies with the cohomology of the complex $\Gamma(X, f_*(I^p)) = \Gamma(Y, I^p)$, i.e., with $H^p(Y, I^\bullet) = H^p(Y, K^\bullet)$.

Now we look at the first filtration of $\Gamma(X, J^{\bullet\bullet})$. The spectral sequence of Leray will in fact be the spectral sequence for the first filtration. The E_1-term is the vertical cohomology, i.e., we have

$$E_1^{p,q} = Ker(d : \Gamma(X, J^{p,q}) \to \Gamma(X, J^{p,q+1}))/d\Gamma(X, J^{p,q-1}).$$

Recall the notations of the proof of Proposition 1.2.6, in particular

(1) $S^{\bullet,q} = Ker(d : J^{\bullet,q} \to J^{\bullet,q+1})$, which gives an injective resolution of $Ker(d_K : f_*I^q \to f_*I^{q+1})$.

(2) $R^{\bullet,q} = J^{\bullet,q-1}/S^{\bullet,q-1}$, which gives an injective resolution of $d(f_*I^{q-1})$.

(3) $H^{\bullet,q} = S^{\bullet,q}/R^{\bullet,q}$, which gives an injective resolution of $\underline{H}^q(f_*I^\bullet) = R^q f_*(K^\bullet)$.

We therefore get

$$E_1^{p,q} = \Gamma(X, S^{p,q})/\Gamma(X, R^{p,q}) = \Gamma(X, H^{p,q}) \qquad (1-44)$$

The d_1-differential $\Gamma(X, H^{p,q}) \to \Gamma(X, H^{p+1,q})$ is induced by the differential

$\delta : H^{p,q} \to H^{p+1,q}$ in the injective resolution $H^{\bullet,q}$ of $R^q f_*(K^\bullet)$. Hence we have $E_2^{p,q} = H^p(X, R^q f_*(K^\bullet))$. So we have established

1.6.4. Theorem. *For a bounded below complex of sheaves $K^\bullet$ on Y, there is a convergent spectral sequence in the category $\mathcal{AB}(X)$, with E_2-term*

$$E_2^{p,q} = H^p(X, R^q f_*(K^\bullet)), \qquad (1-45)$$

and with E_∞-term the graded quotients of some filtration of $H^n(Y, K^\bullet)$. This spectral sequence is called the _Leray spectral sequence_ for the complex of sheaves $K^\bullet$ on Y and the mapping $f : Y \to X$.

We also note the spectral sequence

$$E_2^{p,q} = R^p f_*(\underline{H}^q(K^\bullet)) \qquad (1-46)$$

in the category of sheaves of abelian groups over X, with E_∞-term the graded quotients of some filtration of $R^n f_*(K^\bullet)$. This is derived by a similar method. From this spectral sequence one obtains

1.6.5. Proposition. *Let $\phi : K^\bullet \to L^\bullet$ be a quasi-isomorphism of bounded below complexes of sheaves over X. Then ϕ induces an isomorphism of the spectral sequence for $K^\bullet$ with the spectral sequence for $L^\bullet$.*

Proof. We will use Proposition 1.2.4 twice. First, the morphism of spectral sequences of type (1-45) induced by ϕ is an isomorphism at the E_2-level, hence also at the E_∞-level. So we have $R^p f_*(K^\bullet) \to R^p f_*(L^\bullet)$; hence we have an isomorphism at the E_2-level of the Leray spectral sequences, so also at the E_r and E_∞-levels. ■

To exploit the Leray spectral sequence for a complex of sheaves $K^\bullet$ on Y, one needs a good understanding of the higher direct image sheaves $R^q f_*(K^\bullet)$.

1.6.6. Proposition. *Let $K^\bullet$ be a bounded below complex of sheaves on Y, and let $f : Y \to X$ be a continuous map. The sheaf $R^q f_*(K^\bullet)$ identifies with the sheaf on X associated with the presheaf $U \mapsto H^q(U, K^\bullet)$.*

Proof. Let $I^\bullet$ be the total complex associated with an injective resolution $I^{\bullet\bullet}$ of $K^\bullet$, which has the property that the restriction of each I^q to any open set of Y is an injective sheaf. Then the sheaf $R^q f_*(K^\bullet)$ on X is the

q-th cohomology sheaf of the complex of sheaves $f_*(I^\bullet)$. This cohomology sheaf is the sheaf associated to the presheaf $U \mapsto H^q(\Gamma(U, f_*(I^\bullet)))$. This cohomology group is equal to $H^q(f^{-1}(U), K^\bullet)$. ∎

The Leray spectral sequence is of special interest for the constant sheaf A_Y on Y, where A is some abelian group. Suppose that Y, X and F are smooth manifolds satisfying the hypothesis of Corollary 1.4.7 and suppose that $f : Y \to X$ is a locally trivial smooth fibration with fiber F. We observe the following homotopy invariance for Čech cohomology.

1.6.7. Lemma. *For Z a convex open set in an ILH space satisfying the hypothesis of Corollary 1.4.7, we have $H^q(Z \times F, A) \xrightarrow{\sim} H^q(F, A)$.*

Proof. We will give the proof for $A = \mathbb{R}$. Using the Leray spectral sequence for the second projection $p_2 : Z \times F \to F$, one easily checks, using Theorem 1.4.15, that $R^i p_2 * (\mathbb{R}_{Z \times F}) = 0$. The Leray spectral sequence degenerates, which gives the isomorphism of the statement. ∎

Since every point x of X has a fundamental system of open neighborhoods which are diffeomorphic to a convex open set in an ILH space, we see that the stalk of $R^q f_*(A_Y)$ at x is equal to $H^q(F, A)$. So the direct image sheaf $R^q f_*(A_Y)$ is *locally constant*, with stalk equal to the cohomology group $H^q(F, A)$. In fact, $R^q f_*(A_Y)$ is constant over any contractible open set of X. For simplicity suppose that X is connected. Then the locally constant sheaf $R^q f_*(A_Y)$ is described by a *monodromy representation* $\rho : \pi_1(X, x) \to \mathrm{Aut}\, H^q(F, A)$. Observe that a representation $\rho : \pi_1(X, x) \to \mathrm{Aut}(V)$, where V is an abelian group, defines a locally constant sheaf, say $\tilde{V}$ on X. To describe this sheaf $\tilde{V}$, let $\pi : \tilde{X} \to X$ be a universal covering space, and view $\pi_1(X, x)$ as the group of deck transformations of π. One puts

$$\tilde{V}(U) = \{s \in \Gamma(\pi^{-1}(U), V) : g^*(s) = \rho(g)^{-1} \cdot s, \ \forall g \in \pi_1(X, x)\}.$$

We denote by $\underline{H}^q(F, A)$ the locally constant sheaf associated to the representation $\rho : \pi_1(X, x) \to \mathrm{Aut}\, H^q(F, A)$.

In case $A = \mathbb{R}$, one can use the de Rham complex $\underline{A}_Y^\bullet$ instead of the constant sheaf $\mathbb{R}_Y$, since by Proposition 1.6.5 the Leray spectral sequence for $\mathbb{R}_Y$ identifies with the Leray spectral sequence for $\underline{A}_Y^\bullet$. There is a very concrete way to construct the Leray spectral sequence for $\underline{A}_Y^\bullet$, due to H. Cartan. The cohomology of Y is that of the global de Rham complex $A^\bullet(Y)$, and one can filter it by subcomplexes $\cdots \supset F^p \supset F^{p+1} \supset \cdots$. Recall that a vector field ξ on Y is called *vertical* if $df(\xi) = 0$.

1.6.8. Definition. *Let $f : Y \to X$ be a smooth fibration between smooth manifolds satisfying the assumptions of Theorem 1.4.14. We define F^p as the subcomplex of $A^\bullet(Y)$ such that $F^p(A^n(Y))$ consists of those n-forms ω which satisfy*

$$i(\xi_1) \cdots i(\xi_{n-p+1}) \cdot \omega = 0 \qquad (1-47)$$

for any vertical vector fields $\xi_1, \ldots, \xi_{n-p+1}$.

Equivalently, $\omega \in F^p$ means that for $(x_1, \ldots, x_k, z_1, \ldots, z_m)$ local coordinates on Y such that $(x_1, \ldots, x_k)$ are local coordinates on X, ω has an expression

$$\omega = \sum_{q \geq p} \sum_{j_1, \ldots, j_q, k_1, \ldots, k_{n-q}} f_{j_1, \ldots, k_{n-q}} \, dx_{j_1} \wedge \cdots \wedge dx_{j_q} \wedge dz_{k_1} \wedge \cdots \wedge dz_{j_{n-q}}$$

$$(1-48)$$

In other words, every term in ω contains at least p differentials dx_j coming from X.

The spectral sequence for the complex $A^\bullet(Y)$, filtered by the subcomplexes F^p, is convergent, since $F^{n+1}(A^n(Y)) = 0$ and $A^n(Y) \subset F^0$. Let us denote this spectral sequence by $(E_C)_r^{p,q}$. Before we compute the $(E_C)_1$-term, recall that one defines the space $A^q(Y/X)$ of *relative differential forms* of degree q as $A^q(Y)/f^{-1} A^1(X) \wedge A^{q-1}(Y)$. There is an exterior differential $d : A^q(Y/X) \to A^{q+1}(Y/X)$. The E_1-term of the spectral sequence E_C is then

$$(E_C)_1^{p,q} = A^p(X) \otimes_{C^\infty(X)} A^{n-p}(Y/X) \qquad (1-49).$$

The tensor product is taken over the ring $C^\infty(X)$ of smooth functions on X. The differential is $d_1 = Id \otimes d$, where $d : A^{n-p}(Y/X) \to A^{n-p+1}(Y/X)$ is the exterior differential.

We note that we have a complex of sheaves $\underline{A}^\bullet(Y/X)$ on Y. In the next statement, $\underline{C}_X^\infty$ denotes the sheaf of C^∞-functions on X.

1.6.9. Lemma. *The q-th cohomology sheaf of the complex of sheaves $f_*(A^\bullet(Y/X))$ on Y identifies with $\underline{C}_X^\infty \otimes \underline{H}^q(F, \mathbb{R})$.*

Proof. There is a relative Poincaré lemma which says that $\underline{A}^\bullet(Y/X)$ is a resolution of the inverse image sheaf $f^{-1}\underline{C}_X^\infty$. As the sheaf $\underline{A}^p(Y/X)$ restricts to a soft sheaf on any open set of Y, we can use this resolution to compute the higher direct image $R^q f_*(f^{-1}\underline{C}_X^\infty)$. We claim that the natural map from $\underline{C}_X^\infty \otimes R^q f_*(\mathbb{R}_Y)$ to $R^q f_*(f^{-1} C_X^\infty)$ is an isomorphism. It is enough to check that we have an isomorphism of sections over a contractible open set V of

X, such that $f^{-1}(V) = V \times F$. Then we take a good covering $\mathcal{U} = (U_i)$ of F, and we note that $\Gamma(V, \underline{C}_X^\infty \otimes R^q f_*(\mathbb{R}_Y))$ is the degree q cohomology of the complex

$$\to \prod_{i_0,\ldots,i_q} C^\infty(V) \otimes \Gamma(U_{i_0,\ldots,i_q}, \mathbb{R}) \to \prod_{i_0,\ldots,i_{q+1}} C^\infty(V) \otimes \Gamma(U_{i_0,\ldots,i_{q+1}}, \mathbb{R}) \to .$$

Since $\Gamma(V, R^q f_*(f^{-1} C_X^\infty))$ is given in the same way, the result follows. ∎

From Lemma 1.6.9, we get an easier description of $(E_C)_1^{p,q}$, namely that it is the space of p-forms on X with values in the local system $\underline{H}^q(F, \mathbb{R})$. Then the d_1-differential is just the exterior differential for such twisted differential forms. So the E_2-term is

$$(E_C)_2^{p,q} = H^p(X, \underline{H}^q(F, \mathbb{R})). \qquad (1-50)$$

1.6.10.   Theorem. *The spectral sequence of Cartan, obtained from the filtration F^p of $A^\bullet(Y)$, identifies (for $r \geq 2$) with the Leray spectral sequence for the sheaf $\mathbb{R}_Y$.*

Proof. We will find a third spectral sequence which maps isomorphically to both spectral sequences. The Leray spectral sequence is the same for $\mathbb{R}_Y$ and for the de Rham complex $\underline{A}_Y^\bullet$. We have the resolution $\underline{A}_X^q \otimes f_*(\underline{A}_Y^p)$ of $f_*(\underline{A}_Y^\bullet)$, where both vertical and horizontal differentials are induced by exterior differentiation. This is a resolution by a complex of soft sheaves on X, and furthermore the images of the vertical differentials (resp. their kernels, resp. the vertical cohomology) form a soft resolution of the appropriate image, kernel or cohomology sheaves. Repeating the analysis carried out previously for an injective resolution $J^{\bullet\bullet}$ of $f_*(\underline{A}_Y^\bullet)$, we find that the second spectral sequence (E') of the double complex $\Gamma(X, \underline{A}_X^q \otimes f_*(\underline{A}_Y^p))$ has E_2-term

$$(E')_2^{p,q} = H^q(X, \underline{H}^q(F, \mathbb{R})).$$

Hence the morphism of double complexes of sheaves from $\underline{A}_X^q \otimes f_*(\underline{A}_Y^p)$ to $J^{p,q}$ gives (for the double complexes of global sections) an isomorphism $E_2^{p,q} \xrightarrow{\sim} (E'_2)^{p,q}$. Since both spectral sequences are convergent, this is an isomorphism of spectral sequences.

Comparing the total complex of the double complex $\Gamma(X, \underline{A}_X^q \otimes f_*(\underline{A}_Y^p))$ with the de Rham complex $A^\bullet(Y)$, we observe that the first complex maps to the second one by wedge product of differential forms, and the second filtration maps to F^p. Hence we get a morphism of spectral sequences of filtered complexes, which is an isomorphism at the E_2-level. ∎

The Cartan-Leray spectral sequence is a very convenient way to view the Leray spectral sequence of a smooth fibration. We will have occasion to use it in Chapters 4 and 5. In fact, we will give a geometric interpretation for some of the differentials in this spectral sequence.

As the Cartan-Leray spectral sequence is a first-quadrant spectral sequence, according to §2. we have edge homomorphisms $E_2^{n,0} = H^n(X,\mathbb{R}) \to H^n(Y,\mathbb{R})$; this is just the pull-back map f^* on cohomology. Assume now that the fiber F of f is a compact connected manifold of dimension k, and that the fibration $f : Y \to X$ is oriented. We then have, for any abelian group A: $R^n f_* A_Y = 0$ for $n > k$, and $R^k f_* A_Y \xrightarrow{\sim} A_X$. We thus obtain a natural homomorphism $H^n(Y,A) \to H^{n-k}(X, R^k f_* A) \xrightarrow{\sim} H^{n-k}(X,A)$. This homomorphism is called *fiber integration* and is denoted by $\int_f$. For $A = \mathbb{R}$, $\int_f$ is induced by the natural morphism of complexes $\int_f : A^\bullet(Y) \to A^{\bullet-k}(X)$, which is also called fiber integration.

An interesting special case occurs when the fiber F is the sphere S^k. Then the spectral sequence is spherical in the sense of §2, and the exact sequence (1-10) becomes

$$\cdots \to H^{n-k-1}(X,A) \xrightarrow{\cup Eu} H^n(X,A) \xrightarrow{f^*} H^n(Y,A) \xrightarrow{\int_f} H^{n-k}(X,A) \to \cdots$$
$$(1-51)$$

where Eu denotes the Euler class $Eu(f) \in H^{k+1}(X,\mathbb{Z})$ of the fibration. This is called the *Gysin exact sequence* of the S^k-fibration. In Chapter 3, we will use the case $k = 1$ of the Gysin sequence.

We conclude with some comments on <u>basic</u> differential forms. A p-form ω on Y is called basic if it satisfies the two conditions:

(B1) For any vertical vector field ξ on Y, we have $i(\xi)\omega = 0$.

(B2) For any vertical vector field ξ on Y, we have $i(\xi)d\omega = 0$.

In terms of the Cartan filtration, (R1) means that $\omega \in F^p$ and (R2) means that $d\omega \in F^{p+1}$.

The following is well known.

1.6.11. Lemma. *Assume that $f : Y \to X$ is a smooth fibration, such that its fiber F is connected. Then f^* identifies $A^p(X)$ with the space of basic p-forms on Y.*

Chapter 2

Line Bundles and Central Extensions

We discuss line bundles, connection and curvature, and the group of isomorphism classes of line bundles L equipped with a connection ∇ (§2.1 and §2.2). This group turns out to be isomorphic to a Deligne cohomology group (§2.2). In fact Deligne cohomology provides a tool for constructing line bundles with connections. If the infinitesimal action of some Lie algebra on the underlying manifold preserves the isomorphism class of (L, ∇), then a central extension of this Lie algebra acts on sections of the line bundle L. The action is written down explicitly using hamiltonian vector fields (prequantization à la Kostant-Souriau) (§2.3). Similarly, if a Lie group action preserves (L, ∇) up to isomorphism, a central extension of the Lie group acts on sections of the line bundle (Kostant). A similar central extension exists in a holomorphic context (Mumford). In §2.5, we discuss results of Weinstein, which give a similar central extension when there is no line bundle (because the given 2-form is not integral).

2.1. Classification of line bundles

The contents of this section are very classical [**Weil2**] [**Ko1**]. We describe line bundles over a manifold up to isomorphism, both in the smooth and in the holomorphic case. Connections and curvature will be taken up in §2.2.

We will work over a smooth manifold M, which we will assume to be paracompact, since this implies that sheaf cohomology is isomorphic to Čech cohomology (cf. §1.3). We will also assume that M satisfies the hypothesis of Theorem 1.4.15, so that de Rham cohomology $H^p_{DR}(M)$ is isomorphic to Čech cohomology $\check{H}^p(M, \mathbb{R})$.

2.1.1. Definition. *A line bundle $p : L \to M$ is a (locally trivial) complex vector bundle of rank 1. For any $x \in M$, the fiber $L_x := p^{-1}(x)$ is a one-dimensional vector space over $\mathbb{C}$.*

If L_1 and L_2 are line bundles over M, an isomorphism of line bundles is a diffeomorphism $\phi : L_1 \xrightarrow{\sim} L_2$ which satisfies the following two conditions:

(a) ϕ is compatible with the projections to M, that is, $p_1 = p_2 \circ \phi$;

(b) *for any $x \in M$, the induced diffeomorphism $(L_1)_x \xrightarrow{\sim} (L_2)_x$ is $\mathbb{C}$-linear.*

We make some comments on this definition. The local triviality condition for $p : L \to M$ means that for any $x \in M$, there is a neighborhood U of x and a bundle isomorphism between the trivial line bundle $\mathbb{C} \times U$ and $L_{/U} = p^{-1}(U)$. Such an isomorphism amounts to a smooth section $s : U \to L$ which is nowhere vanishing on U; given s, define $\phi : \mathbb{C} \times U \xrightarrow{\sim} L_{/U} = p^{-1}(U)$ by $\phi(y) = (\frac{y}{s(p(y))}, p(y))$. Here the $\mathbb{C}^*$-valued function $\frac{y}{s(p(y))}$ is characterized by $\frac{y}{s(p(y))} \cdot s(p(y)) = y$; this is well-defined because $p(y) = p(s(p(y)))$.

Now we give the construction of the $\mathbb{C}^*$-bundle associated to a line bundle. If $L \to M$ is a line bundle, we have the zero section $0 : M \hookrightarrow L$; its image is called the zero-section of L. We denote by L^+ the complement in L of the zero-section. If y_1, y_2 are points of L lying in the same fiber, there is a well-defined number $\frac{y_1}{y_2}$ such that $y_1 = \frac{y_1}{y_2} \cdot y_2$. The mapping $L^+ \mapsto M$ is a locally trivial principal bundle with structure group $\mathbb{C}^*$, where $\mathbb{C}^*$ acts by dilations on the fibers.

One can recover L from L^+ as an associated bundle, namely, $L = (L^+ \times \mathbb{C})/\mathbb{C}^*$, where $\lambda \in \mathbb{C}^*$ acts on $L^+ \times \mathbb{C}$ by $\lambda \cdot (y, u) = (\lambda^{-1} \cdot y, \lambda \cdot u)$. This construction $(L^+ \times \mathbb{C})/\mathbb{C}^*$ is denoted by $L^+ \times^{\mathbb{C}^*} \mathbb{C}$ and is called the *contracted product* of the $\mathbb{C}^*$-manifolds L^+ and $\mathbb{C}$. In concrete terms, a line bundle over M amounts to the same thing as a principal bundle with structure group $\mathbb{C}^*$ (in short, a principal $\mathbb{C}^*$-bundle). This is a stronger statement than the fact that the set of isomorphism classes of line bundles is in bijection with the set of isomorphism class of $\mathbb{C}^*$-bundles. Indeed, much more is true: if L is a line bundle and L^+ the corresponding $\mathbb{C}^*$-bundle, the group of automorphisms of the line bundle L is equal to the group of automorphisms of the $\mathbb{C}^*$-bundle; in fact, both groups are isomorphic to the group $\Gamma(M, \underline{\mathbb{C}}^*)$ of smooth $\mathbb{C}^*$-valued functions.

We state this more formally using categories. For a fixed manifold M, let C_1 denote the category whose objects are line bundles over M, and whose morphisms are line bundle isomorphisms. Similarly, let C_2 be the category of principal $\mathbb{C}^*$-bundles $Q \to M$, whose morphisms are principal bundle isomorphisms. Observe that the association $L \mapsto L^+$ defines a functor $F : C_1 \to C_2$; so that for every isomorphism $\phi : L_1 \to L_2$ between line bundles, one has a corresponding isomorphism $F(\phi) : F(L_1) = L_1^+ \xrightarrow{\sim} F(L_2) = L_2^+$. One says that F is an *equivalence of categories* (see [McL1]) if the following two conditions are fulfilled:

(1) For L_1, L_2 objects of C_1, the map $Hom_{C_1}(L_1, L_2) \to Hom_{C_2}(F(L_1), F(L_2))$ is bijective;

(2) Every object of C_2 is isomorphic to an object of the form $F(L)$, for some object L of C_1.

2.1.2. Proposition. *Let C_1 be the category of line bundles over M, C_2 the category of $\mathbb{C}^*$-principal bundles over M, $F : C_1 \to C_2$ the functor $F(L) = L^+$. Then F is an equivalence of categories between C_1 and C_2.*

Proof. Property (1) has already been observed above in the case $L_1 = L_2$. The case where L_1 and L_2 are isomorphic then follows immediately. Now we claim that if L_1 and L_2 are not isomorphic as line bundles, L_1^+ and L_2^+ are not isomorphic as $\mathbb{C}^*$-bundles. In fact, such an isomorphism would imply that the associated line bundles $L_1^+ \times^{\mathbb{C}^*} \mathbb{C}$ is isomorphic to $L_2^+ \times^{\mathbb{C}^*} \mathbb{C}$, hence that L_1 is isomorphic to L_2.

(2) follows from the construction of the line bundle associated to a $\mathbb{C}^*$-bundle. ∎

We note that a functor $F : C_1 \to C_2$ is an equivalence of categories if and only if there exists a *quasi-inverse* functor $G : C_2 \to C_1$, for which there are invertible natural transformations $\phi_1 : G \circ F \to Id_{C_1}$ and $\phi_2 : F \circ G \to Id_{C_2}$. In our case $G(P) = P \times^{\mathbb{C}^*} \mathbb{C} \to M$.

If $p : L \to M$ is a line bundle, and if $f : N \to M$ is a smooth map between manifolds, the *pull-back* f^*L of L is the line bundle $L \times_M N \xrightarrow{p_2} N$. For s a section of L over an open subset U of M, we denote by f^*s the corresponding section of f^*L over $f^{-1}(U)$. If $f : U \hookrightarrow M$ is the inclusion of an open set, the pull-back f^*L is often denoted $L_{/U}$, and called the restriction of L to the open set U. The corresponding notion for $\mathbb{C}^*$-bundles is the pull-back $P \times_M N$ of a $\mathbb{C}^*$-bundle $P \to M$.

Let us give the important example of the "tautological" line bundle over a projective space and of the corresponding $\mathbb{C}^*$-bundle. Let E be a finite-dimensional vector space over $\mathbb{C}$, and let $\mathbb{P} = (E \setminus 0)/\mathbb{C}^*$ be the projective space of lines in E. Recall that $\mathbb{P}$ is a manifold, and that the projection map $p : E \setminus 0 \to \mathbb{P}$ is a $\mathbb{C}^*$-fibration. For definiteness, let $E = \mathbb{C}^n$, with coordinates $(z_1, \ldots, z_n)$; in this case $\mathbb{P} = \mathbb{C}\mathbb{P}_{n-1}$. For $x \in \mathbb{C}\mathbb{P}_{n-1}$, a point $(z_1, \ldots, z_n)$ in $p^{-1}(x)$ is called a set of *homogeneous coordinates* for x. Note that $E \setminus 0$ is covered by the open sets $V_i = \{(z_1, \ldots, z_n) : z_i \neq 0\}$ (where $1 \leq i \leq n$); then $\mathbb{C}\mathbb{P}_{n-1}$ itself is covered by the open sets $U_i = p(V_i)$. Over U_i, the projection p admits the smooth section s_i characterized by $z_i = 1$.

The corresponding line bundle $L \to \mathbb{P}$ may be described as the *blow-up* of the origin in E. In other words, $L \subset E \times \mathbb{P}$ is the submanifold consisting of pairs (v, l), where v is a point of E, l is a line in E, and $v \in l$. Note that the fiber of L over l is exactly the line l itself; this is why this line bundle is called "tautological."

For a separable Hilbert space E, the projective space $\mathbb{P}$ of E is a manifold modeled on $E/\mathbb{C}$. For (e_n) an orthonormal basis of E, $E \setminus 0$ is covered by the open sets $V_i = \{(\sum_n z_n e_n; z_i \neq 0\}$. $\mathbb{P}$ is covered by the open sets $U_i = p(V_i)$. The open set U_i identifies with the subspace $(\mathbb{C} \cdot e_i)^{\perp}$ of E; to $v \in (\mathbb{C} \cdot e_i)^{\perp}$ corresponds the line spanned by $v + e_i$. Again $p : E \setminus 0 \to \mathbb{P}$ is a principal $\mathbb{C}^*$-bundle.

The meaning of Proposition 2.1.2 is that we can go back and forth between line bundles and $\mathbb{C}^*$-bundles without losing any information. In some instances, the language of line bundles is more convenient. This is certainly the case if one is interested in the space $\Gamma(M, L)$ of smooth sections of a line bundle L; this space of sections is a module over the ring $C^{\infty}(M)_{\mathbb{C}}$ of smooth complex-valued functions. In other cases, the language of $\mathbb{C}^*$-bundles is more convenient. For instance, the tensor product $L_1 \otimes L_2$ of line bundles corresponds to the so-called *contracted product* of $\mathbb{C}^*$-bundles. The contracted product $Q_1 \times^{\mathbb{C}^*} Q_2$ of two $\mathbb{C}^*$-bundles is the quotient of the fiber product $Q_1 \times_M Q_2$ by the following action of $\mathbb{C}^*$:

$$\lambda \cdot (y_1, y_2) = (\lambda^{-1} \cdot y_1, \lambda \cdot y_2).$$

Just like tensor product, the operation of contracted product is associative and commutative (up to natural isomorphisms). For s_1 and s_2 sections of Q_1 and Q_2, we obtain a section $s_1 \times s_2$ of $Q_1 \times^{\mathbb{C}^*} Q_2$. Similarly, for s_1 and s_2 sections of the line bundles L_1 and L_2, we obtain a section $s_1 \otimes s_2$ of $L_1 \otimes L_2$. The inverse of a line bundle corresponds to the operation which transforms the $\mathbb{C}^*$-bundle $Q \to M$ into the bundle $\bar{Q} \to M$, where $\bar{Q} = Q$, but the action of $\lambda \in \mathbb{C}^*$ on $\bar{Q}$ is the action of λ^{-1} on Q. For line bundles, the inverse is given by the construction of the dual L^* of L; a section of L^* is the same thing as a function on L which is $\mathbb{C}^*$-homogeneous of degree 1.

In this way one obtains a commutative group structure on the set of isomorphism classes of line bundles over M. This group is a smooth version of the Picard group of a complex manifold, and for this reason we will denote it by $Pic^{\infty}(M)$. We wish to describe this commutative group in topological terms. The idea is that a line bundle L is classified by a certain obstruction to finding a global non-vanishing section of L — or, in other words, a global section of the $\mathbb{C}^*$-bundle L^+. Locally one can find such a section. It follows that there exists an open covering $(U_i)_{i \in I}$ of M such that $L^+_{/U_i}$ has a section s_i. Then over the intersection $U_{ij} := U_i \cap U_j$ of two open sets of the covering, we have the smooth invertible function $g_{ij} = \frac{s_i}{s_j}$. Over the intersection U_{ijk} of three open sets, we have the cocycle relation

$$g_{ik} = g_{ij} \cdot g_{jk}.$$

Hence the g_{ij} form a degree 1 Čech cocycle of the covering (U_i) with coefficients in the sheaf $\underline{\mathbb{C}}_M^*$ of smooth invertible complex-valued functions, which was studied in Chapter 1.

2.1.3. Theorem. (*Weil* [Weil2], *Kostant* [Ko1]) (1) *The cohomology class $c_1(L) \in H^1(M, \underline{\mathbb{C}}_M^*)$ of the Čech cocycle g_{ij}^{-1} is independent of the open covering and of the choice of the sections s_i. It is called the first Chern class of L.*

(2) *The assignment $L \mapsto c_1(L)$ gives an isomorphism between the group $Pic^\infty(M)$ of isomorphism classes of line bundles over M and the sheaf cohomology group $H^1(M, \underline{\mathbb{C}}_M^*)$.*

Proof. We first show that for a given open covering $\mathcal{U} = (U_i)$, the Čech cohomology class of g_{ij} is independent of the section s_i of L^+ over U_i. Any other such section is of the form $s_i' = f_i \cdot s_i$, where f_i is an invertible $\mathbb{C}^*$-valued function over U_i, i.e., a section of $\underline{\mathbb{C}}_M^*$ over U_i. Then we have, for the Čech 1-cocycle g_{ij}' corresponding to s_i',

$$g_{ij}' = \frac{s_i'}{s_j'} = \frac{f_i}{f_j} \cdot \frac{s_i}{s_j}$$
$$= \frac{f_i}{f_j} \cdot g_{ij}.$$

Let $\underline{f}$ be the Čech 0-cochain with coefficients in $\underline{\mathbb{C}}_M^*$ which takes the value f_i on U_i. The Čech coboundary of $\underline{f}$ is equal to the difference between the cocycles g_{ij} and g_{ij}'.

Next we have to show the independence of the open covering. We may assume that the open covering $\mathcal{V} = (V_j)_{j \in J}$ is a refinement of $\mathcal{U} = (U_i)_{i \in I}$, which means (cf. §1.3) that there is a map $f : J \to I$ such that $V_j \subseteq U_{f(j)}$ for any $j \in J$. Then, for $s_i \in \Gamma(U_i, L^+)$, we obtain a section t_j of L^+ over V_j, as the restriction of $s_{f(j)}$ to $V_j \subseteq U_{f(j)}$. The Čech 1-cocycle for the covering V_j obtained from these t_j is clearly the image of the 1-cocycle under the induced map $f_* : C^1(\mathcal{U}, \underline{\mathbb{C}}_M^*) \to C^1(\mathcal{V}, \underline{\mathbb{C}}_M^*)$. This proves (1).

The proof of (2) proceeds as follows. First it is easy to see that $[L] \mapsto c_1(L)$ defines a group homomorphism c_1 from $Pic^\infty(M)$ to $H^1(M, \underline{\mathbb{C}}_M^*)$. c_1 is injective, since the triviality of the cohomology class of g_{ij} means that there exists a section of $\underline{\mathbb{C}}_M^*$ over U_i such that $g_{ij} = \frac{f_i}{f_j}$; but then $s_i' = \frac{s_i}{f_i}$ is a section of L^+ over U_i, and we have $s_i' = s_j'$ over $U_i \cap U_j$. So the s_i' give a global section of L^+ over M, which shows that L^+ (hence also L) is the trivial bundle.

The surjectivity of ϕ is proved by constructing a $\mathbb{C}^*$-bundle $Q \to M$ associated to a given 1-cocycle g_{ij} of (U_i) with coefficients in $\underline{\mathbb{C}}_M^*$. We start with a space $W = \coprod_{i \in I} U_i \times \mathbb{C}^*$, and we construct a quotient space by making the following identifications: for $x \in U_{ij}$ and $\lambda \in \mathbb{C}^*$, we identify $(x, \lambda) \in U_i \times \mathbb{C}^*$ with $(x, \lambda \cdot g_{ij}(x)) \in U_j \times \mathbb{C}^*$. This quotient space maps to M admits a natural action of $\mathbb{C}^*$, and is in fact a principal $\mathbb{C}^*$-bundle over M. This last fact depends on the cocycle relation for g_{ij}. There is a section s_i of Q over U_i, such that $s_i(x)$ is the class of $(x, 1) \in U_i \times \mathbb{C}^*$. By construction, we have $\frac{s_i}{s_j} = g_{ij}$ over U_{ij}, concluding the proof. ∎

Let $f : N \to M$ be a smooth mapping between manifolds. The operation of pulling back line bundles from M to N induces a group homomorphism $f^* : H^1(M, \underline{\mathbb{C}}_M^*) \to H^1(N, \underline{\mathbb{C}}_N^*)$ which is the composite of the inverse image map $H^1(M, \underline{\mathbb{C}}_M^*) \to H^1(N, f^{-1}\underline{\mathbb{C}}_M^*)$ with the map induced by the morphism of sheaves $f^{-1}\underline{\mathbb{C}}_M^* \to \underline{\mathbb{C}}_N^*$.

As an example, we return to the $\mathbb{C}^*$-bundle $\mathbb{C}^n \setminus 0 \xrightarrow{p} \mathbb{CP}_{n-1}$. Using the covering of $\mathbb{CP}_{n-1}$ by the U_i (defined by the condition $z_i \neq 0$), we have $g_{ij} = \frac{z_j}{z_i}$. This is the 1-cocycle for the tautological line bundle.

Now we have the exponential exact sequence of sheaves

$$0 \to \mathbb{Z}(1) = 2\pi\sqrt{-1} \cdot \mathbb{Z} \to \underline{\mathbb{C}}_M \xrightarrow{exp} \underline{\mathbb{C}}_M^* \to 0$$

which gives a long exact sequence of cohomology groups. The cohomology groups $H^i(M, \underline{\mathbb{C}}_M)$ are 0 for $i > 0$, from our assumptions on M, the fact that $\underline{\mathbb{C}}_M$ is a soft sheaf (Theorem 1.4.15) and the fact that a soft sheaf on a paracompact space is acyclic (Theorem 1.4.6). So we have an isomorphism of cohomology groups $H^1(M, \underline{\mathbb{C}}_M^*) \xrightarrow{\sim} H^2(M, \mathbb{Z}(1))$. This implies

2.1.4. Corollary. *The group $Pic^\infty(M)$ is canonically isomorphic to the Čech cohomology group $H^2(M, \mathbb{Z}(1))$.*

Let us write down a Čech 2-cocycle $\underline{\mu}$ with values in $\mathbb{Z}(1)$ representing $c_1(L)$. If g_{ij} are the transition cocycles for the line bundle L, and if one has a branch $Log(g_{ij})$ of the logarithm of g_{ij} over U_{ij}, then $\mu_{ijk} = -Log(g_{jk}) + Log(g_{ik}) - Log(g_{ij})$ is a $\mathbb{Z}(1)$-valued Čech 2-cocycle which represents $c_1(L)$.

It is often important to consider metrics on line bundles.

2.1.5. Definition. *Let $L \to M$ be a line bundle. A hermitian metric on L is a smooth function $h : L \to \mathbb{R}_+ \cup \{0\}$ such that $h(\lambda \cdot x) = |\lambda|^2 \cdot h(x)$ for $x \in L$ and $\lambda \in \mathbb{C}$, and $h(x) > 0$ if $x \in L^+$. A hermitian line bundle is a line bundle equipped with a hermitian metric. An isomorphism of hermitian line bundles is an isomorphism of line bundles which is compatible with the metrics.*

Notice that if h is a hermitian metric on L, then for two sections s_1 and s_2 of L over some open set, one may define a complex-valued function (s_1, s_2) by $(s_1, s_2) = \frac{s_1}{s} \cdot (\overline{\frac{s_2}{s}}) \cdot h(s)$ for any non-vanishing section s. The pairing $(\ ,\ )$ is hermitian. One has $(s, s) = h(s)$.

To a hermitian line bundle $L \to M$ with metric h, one associates a circle bundle $L^1 \to M$, where $L^1 = \{x \in L : h(x) = 1\}$. This is indeed a principal bundle with the circle group $\mathbb{T}$ as structure group. Conversely, let $P \to M$ be a principal $\mathbb{T}$-bundle; we construct the line bundle $L := P \times^{\mathbb{T}} \mathbb{C} \to M$, where $\mathbb{T}$ acts on $\mathbb{C}$ by scalar multiplication. The hermitian metric h is then defined by $h(x, \lambda) = |\lambda|^2$, for $x \in P$ and $\lambda \in \mathbb{C}$. Now we have the following analog of Proposition 2.1.2.

2.1.6. Proposition. *The functors $(L, h) \mapsto L^1$ and $P \mapsto P \times^{\mathbb{T}} \mathbb{C}$ are quasi-inverse functors between the category of hermitian line bundles over M, and the category of $\mathbb{T}$-bundles over M.*

Using partitions of unity, one easily proves

2.1.7. Proposition. *Every line bundle over a paracompact manifold satisfying the assumptions of Theorem 1.4.15 admits a hermitian metric.*

The analog of Theorem 2.1.3 and Corollary 2.1.4 for hermitian line bundles is

2.1.8. Theorem. *The group of isomorphism classes of hermitian line bundles over M is isomorphic to the sheaf cohomology group $H^1(M, \mathbb{T}_M)$, hence also to $H^2(M, \mathbb{Z}(1))$.*

This gives another proof of the existence of a hermitian metric on any line bundle. We now turn to holomorphic line bundles over complex manifolds.

2.1.9. Definition. *Let M be a complex manifold. A <u>holomorphic line bundle</u> $L \mapsto M$ is a holomorphic vector bundle of rank 1. For any $x \in M$, the fiber $L_x := p^{-1}(x)$ is a one-dimensional vector space over $\mathbb{C}$.*

If L_1 and L_2 are holomorphic line bundles over M, an isomorphism of holomorphic line bundles is a complex-analytic diffeomorphism $\phi : L_1 \xrightarrow{\sim} L_2$ which satisfies the following two conditions:

(a) *ϕ is compatible with the projections to M, that is, $p_1 = p_2 \circ \phi$;*

(b) *for any $x \in M$, the induced diffeomorphism $(L_1)_x \xrightarrow{\sim} (L_2)_x$ is $\mathbb{C}$-linear.*

For a holomorphic line bundle $L \to M$, the complement L^+ of the zero section gives a locally trivial holomorphic $\mathbb{C}^*$-bundle $L^+ \to M$. The holomorphic analog of Proposition 2.1.2 holds, i.e., the category of holomorphic line bundles over M is equivalent to the category of holomorphic $\mathbb{C}^*$-bundles. The analog of Theorem 2.1.3 is phrased in terms of the transition cocycles g_{ij} associated to an open covering (U_i) and to a holomorphic section s_i of L^+ over U_i; then $g_{ij} = \frac{s_i}{s_j}$ is an invertible holomorphic function over U_{ij}, i.e., a section of the sheaf $\mathcal{O}_M^*$ of germs of invertible holomorphic functions.

2.1.10. Theorem. (1) *The cohomology class in $H^1(M, \mathcal{O}_M^*)$ of the Čech cocycle g_{ij} is independent of the open covering and of the choice of the sections s_i.*

(2) *The* Picard group *$Pic(M)$ of isomorphism classes of holomorphic line bundles over M is isomorphic to the sheaf cohomology group $H^1(M, \mathcal{O}_M^*)$.*

To analyze the Picard group, one uses the exponential exact sequence:

$$0 \to \mathbb{Z}(1) \to \mathcal{O}_M \xrightarrow{exp} \mathcal{O}_M^* \to 0. \qquad (2-1)$$

This gives the exact sequence of groups

$$\cdots \to H^1(M, \mathbb{Z}(1)) \to H^1(M, \mathcal{O}_M) \to H^1(M, \mathcal{O}_M^*)$$
$$\to H^2(M, \mathbb{Z}(1)) \to H^2(M, \mathcal{O}_M) \to \cdots$$

The group $Pic(M) = H^1(M, \mathcal{O}_M^*)$ appears as an extension of the discrete group $Ker\ (H^2(M, \mathbb{Z}(1)) \to H^2(M, \mathcal{O}_M))$ by the connected Lie group $H^1(M, \mathcal{O}_M)/Im\ (H^1(M, \mathbb{Z}(1)))$. The important case of a complex torus is studied in detail in [Weil2]. Line bundles on tori are related to the theory of theta-functions.

We note that the tautological line bundle over the projective space $\mathbb{P}$ of a vector space E is a holomorphic line bundle. The transition functions $g_{ij} = \frac{z_i}{z_j}$ are therefore also holomorphic. The Picard group $Pic(\mathbb{P})$ is equal to $H^2(\mathbb{P}, \mathbb{Z}(1)) = \mathbb{Z}$, because $H^j(\mathbb{P}, \mathcal{O}_{\mathbb{P}}) = 0$ for $j = 1$ or 2.

An important notion is that of a *flat line bundle*. A *local system* of complex vector spaces of rank r is a sheaf of vector spaces which is locally isomorphic to the constant sheaf $\mathbb{C}^r$.

2.1.11. Definition. *A* flat line bundle *over M consists of a line bundle $L \to M$, together with a subsheaf E consisting of germs of sections of L, such that*

(1) *E is a local system of $\mathbb{C}$-vector spaces of rank 1;*

(2) *For any $x \in M$, there exists an element $s \in E_x$ (the stalk of E at x), which, as a germ of section of L, does not vanish at x.*

There is a natural notion of pull-back for flat line bundles.

2.1.12. Proposition. (1) *A flat line bundle L amounts to the same thing as a local system of complex vector spaces of rank 1.*

(2) *The group of isomorphism classes of flat line bundles over M is naturally isomorphic to $H^1(M, \mathbb{C}^*)$.*

Proof. We will show how a local system E of complex vector spaces of rank 1 defines a line bundle L over M. We will define this line bundle by describing its space of sections over an open set U: It is the space of functions $x \in U \mapsto \lambda_x$, where $\lambda_x \in E_x$ satisfy the following condition. For any section s of the dual local system E^* over an open set $V \in U$, the function $x \in V \mapsto \langle \lambda_x, s_x \rangle \in \mathbb{C}$ must be smooth. The rest of (1) is left to the reader.

To prove (2), one proceeds as in Theorem 2.1.3, but one chooses the s_i to belong to E, so that the g_{ij} are locally constant functions. One may alternatively use the description of a flat line bundle as given by a character of the fundamental group. ∎

2.2. Line bundles with connection

Let M be a smooth paracompact manifold, and let $p : L \to M$ be a line bundle over M. We want to be able to differentiate a section s over an open set U of M; the "differential" of s will be a 1-form over U with values in L. We denote by $(A^1 \otimes L)(U)$ the space of 1-forms with values in L, defined over U. It is the space of sections over U of the tensor product vector bundle $T^*M \otimes L$, where T^*M is the cotangent bundle.

2.2.1. Definition. *A connection ∇ on a line bundle L is a rule which to every open set U and to every section s of L over U, associates an element $\nabla(s)$ of $(A^1 \otimes L)(U)$, i.e., an L-valued differential form, so that the following properties hold:*

(1) *The assignment $s \mapsto \nabla(s)$ is compatible with restrictions to smaller open sets.*

(2) *For given U, the map $s \mapsto \nabla(s)$ is $\mathbb{C}$-linear.*

(3) *The Leibniz rule holds, that is,*

$$\nabla(f \cdot s) = f \cdot \nabla(s) + df \otimes s.$$

Here f is any smooth complex-valued function, s a section of L over U, and $df \otimes s$, the tensor product of a degree 1 differential form with a section of L is an element of $(A^1 \otimes L)(U)$.

For example, let $L \to \mathbb{CP}_{n-1}$ be the tautological line bundle. For an open set U of $\mathbb{CP}_{n-1}$, a section s of L over U assigns to a point p of U a set $(z_1(p), \ldots, z_n(p))$ of homogeneous coordinates representing p. We define the differential form $\nabla(s)$ with values in L as the 1-form obtained by projecting the vector-valued 1-form $(dz_1, \ldots, dz_n)$ onto the complex line spanned by $(z_1, \ldots, z_n)$. This means precisely that

$$\nabla(s) = \frac{\bar{z}_1 dz_1 + \cdots + \bar{z}_n dz_n}{|z_1|^2 + \cdots + |z_n|^2} \otimes (z_1, \ldots, z_n). \qquad (2-2)$$

All the required properties for a connection are easily verified.

If $p : L \to M$ is a line bundle, and $f : N \to M$ is a smooth mapping between manifolds, then any connection ∇ on L induces a connection $f^*\nabla$ on the pull-back line bundle f^*L, which is characterized by the property that $(f^*\nabla) \cdot (f^*s) = f^*\nabla(s)$ for any local section s of L. The connection $f^*\nabla$ is called the pull-back connection on f^*L.

Note that if ∇ is a connection, then for any complex-valued 1-form α on M we have a new connection $\nabla + \alpha$ defined by $(\nabla + \alpha)(s) = \nabla(s) + \alpha \otimes s \in A^1 \otimes L$.

2.2.2. Proposition. *Given a connection ∇ on L, the space $A^1(M) \otimes \mathbb{C}$ of complex-valued 1-forms identifies with the set $Co(L)$ of connections on L, via the map $\alpha \in A^1(M) \otimes \mathbb{C} \mapsto \nabla + \alpha$.*

Proof. The point is that if ∇' is another connection on L, then $\nabla' - \nabla$ is given by a 1-form. To show this, one subtracts the Leibniz rules for ∇' and for ∇ and one sees that $\alpha(s) := \nabla'(s) - \nabla(s)$ is linear over smooth complex-valued functions. This means that α is given by a 1-form. ∎

2.2.3. Corollary. *For any line bundle L, the space $Co(L)$ of connections is an affine space under the translation action of the vector space $A^1(M) \otimes \mathbb{C}$. In particular, L admits a connection.*

Proof. Let $(U_i)_{i \in I}$ be a locally finite open covering of M such that $L_{/U_i}$ has a nowhere vanishing section s_i. Over U_i, there is a connection ∇_i defined by

$\nabla_i(f \cdot s_i) = df \otimes s_i$. Let h_i be a partition of unity subordinate to (U_i). Then $\nabla = \sum_{i \in I} h_i \cdot \nabla_i$ is a connection on L. (The point of this proof is that in an affine space, a finite linear combination $\sum_i \lambda_i x_i$ of points x_i makes sense, as long as $\sum_i \lambda_i = 1$). ∎

If ∇ is a connection, then for any local vector field v, we have an operator ∇_v defined on local sections of L, given by $\nabla_v(s) = \langle \nabla(s), v \rangle$. Indeed, the contraction of the L-valued 1-form $\nabla(s)$ with the vector field v produces a section of L.

Another way of viewing connections is in terms of a 1-form on the total space L^+ of the associated $\mathbb{C}^*$-bundle. Note that the Lie algebra of $\mathbb{C}^*$ is canonically identified with $\mathbb{C}$. Since $\mathbb{C}^*$ acts on L^+, the element z of $\mathbb{C}$ induces a vector field on L^+, denoted by $z \cdot \xi$; the action of $z \cdot \xi$ on functions is given by

$$(z \cdot \xi) \cdot f(y) = \frac{d}{dt} \, f(y \cdot exp(tz))_{t=0}.$$

As usual, the action of $\mathbb{C}^*$ on L^+ is written on the right. Let $p_+ : L^+ \to M$ denote the projection map.

2.2.4. Definition. *A _connection 1-form_ on L^+ is a complex-valued 1-form A on L^+ which satisfies the following conditions:*

(1) *A is $\mathbb{C}^*$-invariant;*

(2) *For any $z \in \mathbb{C}$, we have $\langle A, z \cdot \xi \rangle = z$.*

This is E. Cartan's definition of a connection, specialized to $\mathbb{C}^*$-bundles (see [K-N]). Because of condition (2), A is determined by the *horizontal tangent bundle* of L^+, which is defined as $Ker(A) \subset TL^+$. This horizontal tangent bundle, as a subbundle of the tangent bundle, is invariant under $\mathbb{C}^*$ by condition (1).

The precise relation of connection 1-forms on L^+ with connections on L is given by the following

2.2.5. Proposition. (1) *For a connection 1-form A on L^+, there is a unique connection ∇ on the line bundle L which satisfies the condition:*

(C) for any section s of L^+ over an open set, we have the equality of 1-forms with values in L

$$\nabla(s) = s^* A \otimes s.$$

Here $s^* A$ is the pull-back of the 1-form A under $s : M \to L^+$.

(2) *For every connection ∇ on L, there is a unique connection 1-form A on L^+ such that (C) holds for every section of L^+ over some open set.*

Proof. Given A, we have to define a connection ∇ on L. Take a section s of L over some open set. Locally, we may write $s = f_1 \cdot s_1$, where f_1 is a complex-valued function, and s_1 is a nowhere vanishing section. Condition (C) then forces the definition

$$\nabla(s) = f_1 \cdot s_1^* A \otimes s_1 + df_1 \otimes s_1.$$

We must show that this is independent of the choice of the expression $s = f_1 \cdot s_1$. Assume $s = f_2 \cdot s_2$ is another, similar expression. If we put $g = \frac{s_2}{s_1}$, we have $f_1 = g \cdot f_2$. From the properties of A, we see that $s_2^* A = \frac{dg}{g} + s_1^* A$. Hence we have

$$f_2 \cdot s_2^* A \otimes s_2 + df_2 \otimes s_2 = f_2 \cdot \frac{dg}{g} \otimes gs_1 + f_2 \cdot s_1^* A \otimes gs_1$$
$$+ g^{-1} df_1 \otimes gs_1 - g^{-2} f_1 dg \otimes gs_1$$

and this simplifies to

$$f_1 \cdot s_1^* A \otimes s_1 + df_1 \otimes s_1,$$

showing that $\nabla(s)$ is well-defined. It is clearly $\mathbb{C}$-linear, and we only have to show the Leibniz identity. If $s = f_1 \cdot s_1$, for any complex-valued function h, we have $h \cdot s = (fh) \cdot s_1$; hence we get

$$\nabla(h \cdot s) = hf_1 \cdot s_1^* A \otimes s_1 + d(hf_1) \otimes s_1$$
$$= h \cdot \nabla(s) + dh \otimes s$$

which finishes the proof of (1).

To prove (2), one first observes that A, if it exists, is uniquely characterized by condition (C); so it is enough to show the existence of A locally. Then we may assume $L = M \times \mathbb{C}$ is the trivial bundle, and $\nabla(f) = df + f \cdot B$ for some 1-form B on M. One sees that $A = B + \frac{dz}{z}$ works, where z is the coordinate on $\mathbb{C}$. ∎

2.2.6. Remark. From Definition 2.2.4, it is clear that the connection 1-forms comprise an affine space under the translation action of $A^1(M)_{\mathbb{C}}$ (this is an exercise for the reader). An alternative way of proving (2) above would be to say that the map $A \mapsto \nabla$ of affine spaces under $A^1(M)$ is necessarily bijective.

We now come to the *curvature* of a connection ∇ on L. This is the obstruction to finding a *horizontal* local section s of L^+, i.e., a section which satisfies $\nabla(s) = 0$. To uncover this obstruction, start from any local section s_1 of L^+, and try to modify it into a new section $s = fs_1$ which is horizontal. We want to solve the differential equation for f:

$$0 = \nabla(s) = f \cdot \nabla(s_1) + df \otimes s_1.$$

A more palatable way to write this down is: $\dfrac{df}{f} = -\dfrac{\nabla s_1}{s_1}$, where $\frac{\nabla s_1}{s_1}$ is simply a 1-form on M. Then f can exist only if $K = d(\frac{\nabla s_1}{s_1})$ is zero. Note that the 2-form we have written down is independent of s_1; in fact, if hs_1 is another section of L^+, we have $\frac{\nabla(hs_1)}{hs_1} = \frac{dh}{h} + \frac{\nabla s_1}{s_1}$, and $\frac{dh}{h}$ is closed.

2.2.7. Definition and Proposition, *Let ∇ be a connection on the line bundle L. The* <u>*curvature*</u> *$K = K(\nabla)$ of L is the unique 2-form such that for a local section s of L^+, we have the equality: $K = d(\frac{\nabla s}{s})$. K is a closed 2-form. For A the connection 1-form corresponding to ∇, we have $p_+^*(K) = dA$.*

Proof. We have yet to prove that $p_+^*(K) = dA$. This is a local question, so, as in the end of the proof of Proposition 2.2.5, we may assume that $L = M \times \mathbb{C}$; there is a 1-form B on M such that $\nabla(f) = df + f \cdot B$ and $A = B + \frac{dz}{z}$. Then we have $K = dB$ and $p_+^*(K) = p_+^*(dB) = dA$. ∎

2.2.8. Proposition. (1) *Let ∇ be a connection on the line bundle L, with curvature 0. Then the sheaf $L_{\nabla=0}$ of horizontal sections of L is a local system on M of rank 1, i.e., a flat line bundle on M. The corresponding line bundle identifies with L.*

(2) *Conversely, let E be a local system on M of rank 1, and let L be the corresponding line bundle. Then L has a unique connection for which any local section of E is horizontal. The curvature of this connection is 0.*

2.2.9. Lemma. *If K is the curvature of the connection ∇, then for any 1-form α, the curvature of the connection $\nabla + \alpha$ is equal to $K + d\alpha$.*

Recall that there is a tensor product operation on line bundles. Let L_1 and L_2 be line bundles with respective connections ∇_1 and ∇_2. There is a connection on $L_1 \otimes L_2$, denoted by $\nabla_1 + \nabla_2$, such that, if s_1, s_2 are sections of L_1 and L_2 over the same open set U, we have

$$(\nabla_1 + \nabla_2)(s_1 \otimes s_2) = s_1 \otimes \nabla_2(s_2) + \nabla_1(s_1) \otimes s_2. \qquad (2-3)$$

Here we use the fact that the tensor product of a section of L_1 with an L_2-valued 1-form is a 1-form with values in $L_1 \otimes L_2$. The same holds with L_1 and L_2 interchanged.

2.2.10. Lemma. *The curvature of $\nabla_1 + \nabla_2$ is equal to $K_1 + K_2$, for K_j the curvature of ∇_j.*

On the dual L^* of a line bundle L with connection ∇, there is a unique connection ∇^* such that the isomorphism $L \otimes L^* \xrightarrow{\sim} 1$ transforms the connection $\nabla + \nabla^*$ into the trivial connection on the trivial line bundle **1**. The curvature of ∇^* is the opposite of the curvature of ∇.

We will study pairs (L, ∇) of a line bundle L together with a connection ∇. An isomorphism between two such pairs (L_1, ∇_1) and (L_2, ∇_2) is an isomorphism $\phi : L_1 \xrightarrow{\sim} L_2$ which is compatible with the connections, i.e., $\phi(\nabla_1(s)) = \nabla_2(\phi(s))$, for s a local section of L_1. The tensor product operation on the set of isomorphism classes of line bundles with connection defines the structure of an abelian group. We will describe this group of isomorphism classes of pairs (L, ∇) of a line bundle with a connection. The description is due to Kostant [**Ko1**], except for the interpretation in terms of smooth Deligne cohomology, which is due to Deligne [**De2**] (see also [**Bl**] and [**Ma-Me**]).

Let $(U_i)_{i \in I}$ be an open covering of M such that there is a section s_i of L^+ over U_i. As in §2.1, we have the section $g_{ij} = \dfrac{s_i}{s_j}$ of the sheaf $\underline{\mathbb{C}}_M^*$ over U_{ij}, which gives a Čech 1-cocycle with values in $\underline{\mathbb{C}}_M^*$. Next we use the connection to get a 1-form $\alpha_i = \dfrac{\nabla(s_i)}{s_i}$ over U_i. We have $\alpha_i - \alpha_j = \dfrac{dg_{ij}}{g_{ij}}$ over U_{ij}. Let us call $\underline{\alpha}$ the Čech 0-cochain α_i with values in the sheaf $\underline{A}_{M,\,\mathbb{C}}^1$ of complex-valued 1-forms. Also let $\underline{g}$ denote the Čech 1-cocycle g_{ij} with values in $\underline{\mathbb{C}}_M^*$. Then we have $\delta(\alpha) = -d\,log(g)$, where $d\,log : \underline{\mathbb{C}}_M^* \to \underline{A}_{M,\mathbb{C}}^1$ is a homomorphism of abelian sheaves. Here δ denotes the Čech coboundary, with the notations of Chapter 1. This can best be phrased using the 2-term complex of sheaves

$$\underline{\mathbb{C}}_M^* \xrightarrow{d\,log} \underline{A}_{M,\,\mathbb{C}}^1 \qquad (2-4)$$

with $\underline{\mathbb{C}}_M^*$ placed in degree 0. Then $(g, -\alpha)$ gives a cocycle in the Čech double complex associated to this complex of sheaves.

We now have the following version of Theorem 2.1.3 for line bundles with connection.

2.2.11. Theorem.

(1) *The cohomology class of* $(\underline{g}, -\underline{\alpha})$ *in* $H^1(M, \underline{\mathbb{C}}_M^* \xrightarrow{d \ log} A_M^1, \mathbb{C})$ *is independent of the sections* s_i *and of the open covering. Hence it is an intrinsic invariant of the pair* (L, ∇).

(2) *In this manner, the group of isomorphism classes of line bundles with connection is identified with* $H^1(M, \underline{\mathbb{C}}_M^* \xrightarrow{d \ log} A_M^1, \mathbb{C})$.

(3) *The group of isomorphism classes of pairs* (L, ∇) *with curvature* 0 *is isomorphic to* $H^1(M, \mathbb{C}^*)$.

Proof. For a given open covering, we show that a change in the choice of the s_i has the effect of adding a coboundary to $(\underline{g}, -\underline{\alpha})$. In fact, let $s_i' = f_i \cdot s_i$; then g_{ij} is replaced by $g_{ij}' = \frac{f_i}{f_j} \cdot g_{ij}$, as in the proof of Theorem 2.1.3. And α_i is replaced by $\alpha_i' = \alpha_i + d \ log(f_i)$. Therefore we have $(\underline{g}', -\underline{\alpha}') - (\underline{g}, -\underline{\alpha}) = (\delta + d \ log)(-\underline{f})$, for $\underline{f}$ the 0-cochain f_i.

The independence of the open covering is proved by the same argument as in the proof of Theorem 2.1.3. We then have a group homomorphism from the group of isomorphism classes of pairs (L, ∇) to the cohomology group $H^1(M, \underline{\mathbb{C}}_M^* \xrightarrow{d \ log} A_M^1, \mathbb{C})$. We first show it is injective. If the cohomology class of $(\underline{g}, -\underline{\alpha})$ is trivial, by the above calculation one can change s_i so as to achieve $g_{ij} = 1$ and $\alpha_i = 0$. The first equality means that the s_i glue together to form a global section s of L^+. The second equality means that s is horizontal. This implies that (L, ∇) is isomorphic to the trivial class.

For the surjectivity, one starts with a Čech 1-cocycle $(\underline{g}, -\underline{\alpha})$ of the open covering (U_i) with values in $\underline{\mathbb{C}}_M^* \xrightarrow{d \ log} A_M^1, \mathbb{C}$. The proof of Theorem 2.1.3 shows that there exists a line bundle L on M, and sections s_i of L^+ over U_i, such that $s_i = g_{ij} \cdot s_j$ holds over U_{ij}. Let ∇_i be the unique connection of L over U_i such that $\frac{\nabla_i(s_i)}{s_i} = \alpha_i$. We claim that ∇_i and ∇_j agree over U_{ij}. It suffices to show that $\nabla_i(s_i) = \nabla_j(s_i)$ over U_{ij}. We have

$$\nabla_j(s_i) = \nabla_j(g_{ij}s_j) = g_{ij}\alpha_j \otimes s_j + dg_{ij} \otimes s_j$$

$$= (\alpha_j + \frac{dg_{ij}}{g_{ij}}) \otimes s_i = \alpha_i \otimes s_i$$

$$= \nabla_i(s_i).$$

Hence the ∇_i glue together into a global connection ∇. The Čech cocycle associated to the pair (L, ∇) and the sections s_i is then exactly $(\underline{g}, -\underline{\alpha})$. This finishes the proof of (2).

(3) follows easily from Proposition 2.2.8. ∎

Note that by the basic properties of hypercohomology, there is an exact sequence

$$H^0(M,\underline{\mathbb{C}}_M^*) \xrightarrow{d \, log} H^0(M,\underline{A}_{M, \, \mathbb{C}}^1) \to H^1(M,\underline{\mathbb{C}}_M^* \xrightarrow{d \, log} \underline{A}_{M, \, \mathbb{C}}^1)$$
$$\to H^1(M,\underline{\mathbb{C}}_M^*) \to 0$$

where we use the fact that $H^1(M,\underline{A}_{M, \, \mathbb{C}}^1) = 0$. This means that the group of isomorphism classes of line bundles with connection is an extension of $H^1(M,\underline{\mathbb{C}}_M^*)$ by the quotient group of complex-valued 1-forms by those which are globally of the form $\frac{df}{f}$. This latter group is the group of isomorphism classes of connections on the trivial bundle.

Note the quasi-isomorphism of complexes of sheaves

$$
\begin{array}{ccccc}
0 & \to & \mathbb{Z}(1) & \to & \underline{\mathbb{C}}_M & \xrightarrow{d} & \underline{A}_{M,\mathbb{C}}^1 \\
& & & & \downarrow{\scriptstyle exp} & & \downarrow{\scriptstyle Id} \\
& & & & \underline{\mathbb{C}}_M^* & \xrightarrow{d \, log} & \underline{A}_{M, \, \mathbb{C}}^1
\end{array}
\qquad (2-5)
$$

The complex on top is quasi-isomorphic to the smooth Deligne complex $\mathbb{Z}(2)_D^\infty$. Note that this quasi-isomorphism involves multiplication by $2\pi\sqrt{-1}$. So we may rephrase Theorem 2.2.10 in the following way:

2.2.12. Theorem. *The group of isomorphism classes of pairs (L, ∇) of a line bundle on M with connection is canonically isomorphic to $\mathbb{Z}(-1) \otimes H^2(M, \mathbb{Z}(2)_D^\infty)$, where $H^2(M, \mathbb{Z}(2)_D^\infty)$ is a Deligne cohomology group.*

Note the following cohomological interpretation of the curvature. One has a morphism of complexes of sheaves

$$
\begin{array}{ccc}
\underline{\mathbb{C}}_M^* & \xrightarrow{d \, log} & \underline{A}_{M, \, \mathbb{C}}^1 \\
\downarrow & & \downarrow{\scriptstyle d} \\
0 & \to & \underline{A}_{M, \, \mathbb{C}}^2
\end{array}
$$

Call this morphism of complexes d.

2.2.13. Proposition. *The curvature of (L, ∇) is the opposite of the element of $H^0(M, \underline{A}_{M, \, \mathbb{C}}^2)$ obtained by applying d to the class of (L, ∇) in $H^1(M, \underline{\mathbb{C}}_M^* \xrightarrow{d \, log} \underline{A}_{M, \, \mathbb{C}}^1)$.*

Proof. This is clear, as $d(\underline{g}, -\underline{\alpha})$ is the section of $A_{M, \, \mathbb{C}}^2$ which is equal to $-d\alpha_i$ on U_i. Since $\alpha_i = \frac{\nabla(s_i)}{s_i}$, we have $d\alpha_i = K$, for K the curvature. ∎

The proposition has several consequences.

2.2.14. Theorem. (*Weil* [Weil2], *Kostant* [Ko1]) *Let K be the curvature of a pair (L, ∇). Then K is a closed 2-form, and its cohomology class is the image of $-c_1(L) \in H^2(M, \mathbb{Z}(1))$ in $H^2(M, \mathbb{C})$.*

Proof. Let $\underline{g}$ be the 1-cocycle with values in $\underline{\mathbb{C}}^*_M$ given by the transition cocycles, as in the proof of Theorem 2.2.10. Choose a branch $Log(g_{ij})$ of the logarithm for g_{ij} over U_{ij}. Then a $\mathbb{Z}(1)$-valued Čech 2-cocycle $\underline{\mu}$ representing $c_1(L)$ is given by $\mu_{ijk} = -Log(g_{jk}) + Log(g_{ik}) - Log(g_{ij})$. Then $(-2\pi\sqrt{-1} \cdot \underline{\mu}, -2\pi\sqrt{-1} \cdot Log(\underline{g}), 2\pi\sqrt{-1} \cdot \underline{\alpha})$ is a Čech 2-cocycle with values in the smooth Deligne complex $\mathbb{R}(2)^\infty_D$. We have $K_{U_i} = d\alpha_i$. And, from Theorem 1.5.4, it follows that the Čech cohomology class $-c_1(L)$ of $\underline{\mu}$ corresponds to the de Rham cohomology class of K under the isomorphism between Čech and de Rham cohomology. ∎

It may be useful to explain our choice for the sign convention regarding $c_1(L)$, as different authors have different conventions. In the case of the tautological line bundle over $\mathbb{CP}_1$, we find that the curvature K has restriction to $\mathbb{C}$ equal to $\dfrac{\bar{z}dz}{|z|^2 + 1}$. With the canonical orientation of the complex manifold $\mathbb{CP}_1$, we then find $\int_{\mathbb{CP}_1} K = 2\pi\sqrt{-1}$. On the other hand, we want the dual line bundle L^* (which has non-zero holomorphic sections) to correspond to the positive generator of $H^2(\mathbb{CP}_1, \mathbb{Z})$. This leads us to defining the first Chern class as having $-K$ as representative.

The integrality result for the curvature of a connection admits the following converse, also due to Weil and Kostant.

2.2.15. Theorem. (1) *If K is a closed complex-valued 2-form on M whose cohomology class is in the image of $H^2(M, \mathbb{Z}(1)) \to H^2(M, \mathbb{C})$, then there exists a line bundle L with connection on M, with curvature equal to K.*

(2) *Assume that K satisfies the condition in (1). The set of isomorphism classes of pairs (L, ∇) with curvature K is a principal homogeneous space under the group $H^1(M, \mathbb{C}^*)$ of isomorphism classes of flat line bundles over M.*

Proof. Let $K^\bullet = \underline{\mathbb{C}}^*_M \xrightarrow{d \, log} \underline{A}^1_{M, \, \mathbb{C}}$. We have an exact sequence of complexes of sheaves

$$0 \to \mathbb{C}^* \to K^\bullet \to \underline{A}^{2}_M{}^{cl}[-1] \to 0,$$

where $\mathbb{C}^*$ is a constant sheaf, and $\underline{A}^2_M{}^{cl}$ is the sheaf of germs of closed 2-forms.

On the level of cohomology groups, we get the exact sequence

$$0 \to H^1(M, \mathbb{C}^*) \to H^1(M, K^\bullet) \to A^2(M)_\mathbb{C}^{\text{cl}} \to H^2(M, \mathbb{C}^*).$$

Therefore a closed 2-form K is the curvature of some (L, ∇) iff the cohomology class $[K] \in H^2(M, \mathbb{C})$ has zero image in $H^2(M, \mathbb{C}^*)$, i.e., comes from $H^2(M, \mathbb{Z}(1))$. This proves (1).

The same exact sequence shows that the set of preimages of K in $H^1(M, K^\bullet)$ forms a principal homogeneous space under $H^1(M, \mathbb{C}^*)$. According to Theorem 2.1.12, this is the group of isomorphism classes of flat line bundles on M. One can prove (2) more concretely as follows. Let (L, ∇) have curvature K. Let (L', ∇') be another line bundle with connection, also with curvature K. Then $(L' \otimes L^*, \nabla' + \nabla^*)$ has curvature 0. Hence the group of isomorphism classes of pairs (L, ∇) with curvature equal to K is a principal homogeneous space under $H^1(M, \mathbb{C}^*)$. ∎

Let us next look at connections on hermitian line bundles.

2.2.16. Definition. *Let (L, h) be a hermitian line bundle. A connection ∇ on L is said to be hermitian (or compatible with h) if it satisfies:*

(H) *For any local section s of L^+, one has*

$$d\,(h(s)) = 2h(s) \cdot \Re(\frac{\nabla(s)}{s}).$$

Here $\Re$ denotes the real part of a complex-valued 1-form.

Note that if (H) is true for a section s, it holds for $s_1 = f \cdot s$, since

$$d\,(h(s_1)) = |f|^2 \cdot d(h(s)) + 2h(s) \cdot \Re(\bar{f} \cdot df)$$
$$= 2|f|^2 \cdot h(s) \cdot (\Re(\frac{\nabla(s)}{s}) + \Re\frac{df}{f})$$
$$= 2h(s_1) \cdot \Re(\frac{\nabla(s_1)}{s_1})$$

Therefore (H) is equivalent to

(H') *If s is a local section of L of length 1 (i.e., $h(s) = 1$), then $\dfrac{\nabla(s)}{s}$ is a purely imaginary 1-form.*

One easily obtains the results below for hermitian connections. The proofs are omitted. For (L, h) a hermitian line bundle we denote by L^1 the corresponding $\mathbb{T}$-bundle, consisting of all $x \in L$ such that $h(x) = 1$ (see §1).

2.2.17. Proposition. (1) *A connection on (L, h) is hermitian if and only if the restriction to L^1 of the connection 1-form A is purely imaginary.*

(2) *Given a hermitian line bundle (L, h) on M, the set of hermitian connections on (L, h) is an affine space under the vector space $\sqrt{-1} \cdot A^1(M)$ of purely imaginary 1-forms on M.*

(3) *The curvature of a hermitian connection is a purely imaginary 2-form.*

From (1) it follows that a hermitian connection ∇ amounts to the same thing as a $\mathbb{R}(1)$-valued 1-form A on L_1 which satisfies the analogs of the conditions of Proposition 2.2.5.

Then one has the hermitian analogs of Theorem 2.2.11 and its corollaries.

2.2.18. Theorem. (*Kostant, [Ko1]*) (1) *The group of isomorphism classes of hermitian line bundles with hermitian connection (L, h, ∇) is canonically isomorphic to the hypercohomology group $H^1(M, \mathbb{T}_M \xrightarrow{d \log} \sqrt{-1} \cdot \underline{A}^1_M)$.*

(2) *A closed purely imaginary 2-form K is the curvature of a hermitian line bundle with connection if and only if its cohomology class belongs to the image of $H^2(M, \mathbb{Z}(1)) \to \sqrt{-1} \cdot H^2(M, \mathbb{R})$.*

Given a 2-form K as in 2.2.28 (2), a hermitian line bundle with curvature K is often called a *quantum line bundle*.

Next we study connections on a holomorphic line bundle L. There are two ways of defining such a connection.

2.2.19. Definition. *Let L be a holomorphic line bundle on M, ∇ a connection on L.*

(1) ∇ *is a <u>holomorphic connection</u> if $\nabla(s)$ is a holomorphic 1-form, for any local holomorphic section s of L.*

(2) ∇ *is said to be <u>compatible with the holomorphic structure</u> if, for any local holomorphic section s of L, $\nabla(s)$ is purely of type $(1, 0)$.*

Note that (1) is much stronger than (2). Concerning holomorphic line bundles with holomorphic connection, one may define an isomorphism $\phi : (L_1, \nabla_1) \xrightarrow{\sim} (L_2, \nabla_2)$ to be an isomorphism $\phi : L_1 \xrightarrow{\sim} L_2$ of holomorphic line bundles, which is compatible with the connections. We then have

2.2.20. Theorem. (*Deligne [De2] [Ma-Me] [Bl]*) *The group of isomorphism classes of holomorphic line bundles equipped with a holo-*

morphic connection, is canonically isomorphic to the cohomology group $H^1(M, \mathcal{O}_M^* \xrightarrow{d\ log} \Omega_M^1)$, *hence to*

$$\mathbb{Z}(-1) \otimes H^2(M, \mathbb{Z}(2)_D) = (2\pi\sqrt{-1})^{-1} \cdot H^2(M, \mathbb{Z}(2)_D).$$

This is the exact holomorphic analog of Theorem 2.2.11. Concerning connections which are compatible with the holomorphic structure, we have a similar result:

2.2.21. Theorem. *The group of isomorphism classes of holomorphic line bundles equipped with a connection compatible with the holomorphic structure is canonically isomorphic to* $H^1(M, \mathcal{O}_M^* \xrightarrow{d\ log} \underline{A}_M^{(1,0)})$.

The following result describes which holomorphic line bundles admit a holomorphic connection.

2.2.22. Theorem. (1) *Every holomorphic line bundle admits a connection compatible with the holomorphic structure. The set of such connections is an affine space under* $A^{(1,0)}(M)$.

(2) *A holomorphic line bundle* L *admits a holomorphic connection if and only if the class of* L *in* $H^1(M, \mathcal{O}_M^*)$ *has image 0 in* $H^1(M, \Omega_M^1)$ *under the map induced on* H^1 *by the morphism of sheaves* $d\ log : \mathcal{O}_M^* \to \Omega_M^1$.

Proof. The proof of (1) is similar to that of Corollary 2.2.3. One first sees locally that the statement is true, and then one uses a partition of unity to show that a global connection exists, which is compatible with the holomorphic structure. The other connections with this property are obtained by adding to it a 1-form of type $(1,0)$.

(2) follows from the exact sequence

$$H^1(M, \mathcal{O}_M^* \xrightarrow{d\ log} \Omega_M^1) \to H^1(M, \mathcal{O}_M^*) \to H^1(M, \Omega_M^1) \qquad (2-6)$$

and Theorem 2.2.21. ∎

A remarkable consequence of Theorem 2.2.21 is that it allows one to use cohomological methods for the construction of holomorphic line bundles with holomorphic connection. This was exploited by Deligne [**De2**] and Beilinson [**Be1**] in the following context. Let M be a complex manifold, and let f, g be invertible holomorphic functions on M. Then there is a line bundle with holomorphic connection on M, denoted by (f, g). Recall that f gives a class $[f] \in H^1(M, \mathbb{Z}(1)_D)$ in Deligne cohomology. Similarly, $[g] \in H^1(M, \mathbb{Z}(1)_D)$. The cup-product $[f] \cup [g]$ is a class in $H^2(M, \mathbb{Z}(2)_D)$, which, by Theorem

2.2.12, is the isomorphism class of a holomorphic line bundle with holomorphic connection. The line bundle can be described using the explicit construction of the cup-product in Čech cohomology. However, [De2] gives a more intrinsic description of the line bundle (f, g) with connection. This line bundle is characterized by the fact that any local branch $Log(f)$ of a logarithm for f defines a section of L, denoted by $\{Log(f), g\}$.

One imposes a relation between local sections defined by different branches of the logarithm:

$$\{Log(f) + 2\pi\sqrt{-1} \cdot n, g\} = g^n \cdot \{Log(f), g\}. \qquad (2-7)$$

In other words, one considers the trivial line bundle over the infinite cyclic covering $\tilde{M}$ of M, which is the fiber product $\tilde{M} = M \times_{\mathbb{C}^*} \tilde{\mathbb{C}}^*$, where $\tilde{\mathbb{C}}^* \to \mathbb{C}^*$ is the universal covering, and M maps to $\mathbb{C}^*$ via f. One descends the trivial line bundle to a line bundle on M via the "monodromy automorphism" which is multiplication by g.

The holomorphic connection ∇ is characterized by the equation

$$\nabla(\{Log(f), g\}) = \frac{1}{2\pi\sqrt{-1}} \, Log(f) \cdot \frac{dg}{g} \otimes \{Log(f), g\}, \qquad (2-8)$$

which is compatible with the rule (2-7).

Let us compute the Čech cohomology class in $\check{H}^1(M, \mathcal{O}_M^* \to \Omega_M^1)$ corresponding to this line bundle with connection. Let (U_i) be an open covering of M, with U_i connected, such that on each U_i there exists a branch $Log_i(f)$ of the logarithm of f. If $U_{ij} \neq \emptyset$, we have $Log_i(f) - Log_j(f) = 2\pi\sqrt{-1} \cdot n_{ij}$ on U_{ij}, for some $n_{ij} \in \mathbb{Z}$. We have the section $s_i = \{Log_i(f), g\}$ of (f, g) over U_i. The corresponding transition cocycle is

$$g_{ij} = \frac{s_i}{s_j} = g^{n_{ij}}$$

by rule (2-7). Next we have

$$\frac{\nabla(s_i)}{s_i} = \frac{1}{2\pi\sqrt{-1}} \, Log_i(f) \cdot \frac{dg}{g}.$$

So a Čech 1-cocycle with coefficients in the complex of sheaves $\mathcal{O}_M^* \to \Omega_M^1$ representing (L, ∇) is

$$\left(g^{n_{ij}}, \frac{-1}{2\pi\sqrt{-1}} \, Log_i(f) \cdot \frac{dg}{g}\right). \qquad (2-9)$$

On the other hand, the class of f is represented by the 1-cocycle $(-2\pi\sqrt{-1} \cdot n_{ij}, Log_i(f))$ with values in $\mathbb{Z}(1)_D$. Similarly the class of g is represented by the 1-cocycle $(-2\pi\sqrt{-1}\cdot m_{ij}, Log_i(g))$, with $2\pi\sqrt{-1}\cdot m_{ij} = Log_i(g) - Log_j(g)$. The cup-product of these two classes is computed using the product $\mathbb{Z}(1)_D \otimes \mathbb{Z}(1)_D \to \mathbb{Z}(2)_D$ in §1.5 (cf. (1-36)). It is the 2-cocycle

$$((2\pi\sqrt{-1})^2 n_{ij}m_{jk}, -2\pi\sqrt{-1}\cdot n_{ij}\cdot Log_j(g), Log_i(f)\cdot d\ (Log_j(g))).$$

Applying the quasi-isomorphism of complexes of sheaves (2-5) and multiplication by $2\pi\sqrt{-1}$, we obtain the 2-cocycle (2-9) with values in $\mathbb{Z}(2)_D$. So we have proved

2.2.23. Proposition. *The class of the line bundle (f,g) in $H^2(M,\mathbb{Z}(2)_D)$ is the opposite of the cup-product $[f]\cup[g]$, where $[f],[g]\in H^1(M,\mathbb{Z}(1)_D)$.*

We refer to [De] [Be1] [E-V] for the many wonderful properties of the line bundle (f,g), which appears in important cases of Beilinson's conjecture on values of L-functions. We note here simply that the holomorphic line bundle with holomorphic connection $(1-g,g)$ is trivial if g is a holomorphic function which avoids the values 0 and 1. One needs only to see this for $M = \mathbb{C}^* \setminus 0$, with g replaced by the coordinate function z. A horizontal section is then given by $s(z) = exp(Li_2(z))\cdot\{Log(1-z), z\}$, where $Li_2(z) = \int_0^z Log(1-z)dz$ is the (multi-valued) dilogarithm function; the branch of $Li_2(z)$ thus depends on the choice of $Log(1-z)$ but the resulting expression for $s(z)$ itself is well-defined, independent of choices. The triviality of the line bundle $(1-z,z)$ is an avatar of the famous Steinberg relation [St]. For more details, see [De2].

Now for connections which are merely compatible with the holomorphic structure, here is an important fact.

2.2.24. Theorem. (1) *Let L be a holomorphic line bundle, and let ∇ be a connection which is compatible with the holomorphic structure. Then the curvature of a connection only has components of type $(2,0)$ and $(1,1)$.*

(2) *Let L be a holomorphic line bundle on M, which is also equipped with a hermitian structure h. There is a <u>unique</u> connection ∇ which is compatible with both the holomorphic structure and the hermitian structure. It satisfies $\nabla(s) = d' Log\, h(s) \otimes s$ if s is a local holomorphic section of L^+. The curvature K of ∇ is purely of type $(1,1)$. If s is a local holomorphic section, we have $d''d' Log(h(s)) = K$.*

Proof. In the situation of (1), let us check that K has no component of type $(0,2)$. This is a local question, so we may assume there is a holomorphic section s of L^+. Then $\frac{\nabla(s)}{s}$ is of pure type $(1,0)$, so $K = d(\frac{\nabla(s)}{s})$ has no component of type $(0,2)$.

To prove (2), let $Co(L)$ be the set of all connections on L, $Co_{hol}(L)$ the subset of connections which are compatible with the holomorphic structure, $Co_{her}(L)$ the subset of $Co(L)$ consisting of connections which are compatible with the hermitian structure. Then $Co(L)$ is an affine space under the vector space $A^1(M) \otimes \mathbb{C}$, $Co_{hol}(L)$ is an affine subspace under the real vector subspace $A^1(M)$, and $Co_{her}(M)$ is an affine subspace under the real vector subspace $\sqrt{-1} \cdot A^1(M)$. But these two vector subspaces of $A^1(M) \otimes \mathbb{C}$ are complementary subspaces, hence our two affine subspaces $Co_{hol}(L)$ and $Co_{her}(L)$ which are parallel to them must meet at one and only one point. This point ∇ is the unique connection which is compatible with both the holomorphic and hermitian structures. Let then s be a local holomorphic section of L^+. Then $\dfrac{\nabla(s)}{s}$ must be a 1-form of type $(1,0)$. Also $\dfrac{\nabla(h(s)^{-1/2} \cdot s)}{h(s)^{-1/2}s}$ must be a purely imaginary 1-form. We have

$$\frac{\nabla(h(s)^{-1/2} \cdot s)}{h(s)^{-1/2}s} = \frac{\nabla(s)}{s} - \frac{1}{2} \cdot d\, Log(h(s))$$

and this can be purely imaginary only if $\frac{\nabla(s)}{s} = d'Log(h(s))$. The curvature is then equal to $dd'\, Log(h(s)) = d''d'\, (Log(h(s))$. Clearly it is purely of type $(1,1)$. ∎

For instance, let $L \to \mathbb{CP}_{n-1}$ be the tautological line bundle. The connection 2-2 is clearly compatible with the holomorphic structure. It is also compatible with the hermitian structure $h(z_1, \ldots, z_n) = |z_1|^2 + \cdots + |z_n|^2$. In fact, if the section s satisfies $h(s) = 1$, then

$$0 = \sum_{i=1}^{n} (z_i d\bar{z}_i + \bar{z}_i dz_i) = 2\Re(\bar{z}_i dz_i),$$

so that $\frac{\nabla(s)}{s}$ is purely imaginary. So ∇ is the connection of part (2) of Theorem 2.2.24.

2.2.25. Corollary. (1) *For a holomorphic line bundle L over a complex manifold M, the image of $c_1(L)$ in $H^2(M, \mathbb{C})$ is represented by a closed, purely imaginary 2-form of pure type $(1,1)$.*

(2) *Assume M is compact Kähler. If L admits a holomorphic connection, $c_1(L)$ is a torsion class.*

Proof. Choose a hermitian structure on L, and let ∇ be the connection compatible with both the holomorphic and hermitian structures. Its curvature K represents $c_1(L)$, and it is purely imaginary of type $(1,1)$.

We sketch the proof of (2). As M is compact Kaehler, we have the Hodge decomposition: $H^2(M, \mathbb{C}) = \sum_{p=0}^{2} H^{p,2-p}$ (see [Weil2]). The fact that $c_1(L)$ has a representative of type $(1,1)$ means that only the component in $H^{(1,1)}$ may be non-zero. We have $H^{(1,1)} = H^1(M, \Omega_M^1)$. One can then show that the cohomology class of K is the element of $H^1(M, \Omega_M^1)$ obtained by applying $d\ log$ to the class of L in $H^1(M, \mathcal{O}_M^*)$. The existence of a holomorphic connection implies that this cohomology class is zero. $c_1(L)$ becomes zero in complex cohomology, hence it is torsion. ∎

There is a very interesting converse to part (2) of Theorem 2.2.24, which is a special case of a theorem of Griffiths.

2.2.26. Theorem. (*Griffiths* [Gri2]) *Let M be a complex manifold. Let L be a C^∞-line bundle over M, equipped with a connection ∇ such that the curvature K is purely of type $(1,1)$. Then there is a unique holomorphic structure for which a local section s of L is holomorphic if and only if $\nabla(s)$ is of type $(1,0)$.*

Proof. This is a local question, so we may assume M is an open subset of $\mathbb{C}^n$ and L is trivial. Then $\nabla(f) = df + f \cdot A$ for some 1-form A. It is enough to show that there is a non-vanishing function f such that $\nabla(f)$ is of type $(1,0)$. This is equivalent to $\frac{\partial f}{\partial \bar{z}_i} + A_i f = 0$ for $1 \le i \le n$, where $A_i = \langle A, \frac{\partial}{\partial \bar{z}_i} \rangle$. We have $\frac{\partial A_i}{\partial \bar{z}_j} = \frac{\partial A_j}{\partial \bar{z}_i}$, because the component of K of type $(0,2)$ is zero. It follows that this system of P.D.E.'s locally has a non-vanishing solution (as is easily seen in solving explicitly for $Log(f)$). ∎

2.3. Central extension of the Lie algebra of hamiltonian vector fields

We will use the following notations: $i(v) \cdot \omega$ is the interior product of a vector field v and a differential form ω, and $\mathcal{L}(v) \cdot \omega$ is the Lie derivative. We recall the useful formulas:

$$\mathcal{L}(v) = di(v) + i(v)d \quad , \quad i([v,w]) = [\mathcal{L}(v), i(w)]$$

(cf. [K-N]).

We will consider a smooth manifold M (possibly infinite-dimensional), which satisfies the hypotheses of Theorem 1.4.15. We suppose M is equipped with a closed real 2-form ω. Let us examine the Lie algebra of hamiltonian vector fields, following Kostant [Ko1]. First we introduce the Lie algebra

$Symp(M)$ of *symplectic* vector fields, which consists of the vector fields v on M which preserve ω, i.e., satisfy $\mathcal{L}(v) \cdot \omega = 0$. Let $I : TM \to T^*M$ be the bundle map given by inner product with ω, i.e., $I(v) = i(v)\omega$. The map need not have constant rank. The pair (M, ω) is called a Dirac manifold; the theory of Dirac manifolds is developed in [**Co**].

2.3.1. Proposition. (1) *A vector field v is symplectic if and only if $I(v)$ is a closed 1-form.*

(2) *Let α be a germ of 1-form on M. If α is of the form $I(v)$ for some symplectic vector field v, then $\langle \alpha, w \rangle$ vanishes for any vector field w such that $I(w) = 0$. The converse holds if M is finite-dimensional and I has constant rank.*

Proof. We have

$$\mathcal{L}(v)(\omega) = i(v)d\omega + d\, i(v)\omega = d\, I(v),$$

which proves (1). For (2), we note that at every point x, the image of the homomorphism $I_x : TM_x \to T^*M_x$ is orthogonal to the kernel of I_x, because ω is skew-symmetric. If M is finite-dimensional and ω has constant rank, then the image of I and its kernel are vector bundles of the same rank; so they must be equal. ∎

This justifies the following definition.

2.3.2. Definition. *A vector field v is called <u>hamiltonian</u> if there exists a function f such that $I(v) = df$. Such a vector field is symplectic.*

A symplectic vector field is locally hamiltonian but need not be globally hamiltonian. Let $\mathcal{H}_M \subseteq Symp(M)$ denote the space of hamiltonian vector fields.

2.3.3. Proposition. (1) $\mathcal{H}_M$ *is a Lie subalgebra of $Symp(M)$. There is an exact sequence of Lie algebras*

$$0 \to \mathcal{H}_M \to Symp(M) \to H^1(M, \mathbb{R}). \qquad (2-10)$$

(2) *If furthermore M is finite-dimensional and ω is non-degenerate, then this exact sequence can be extended by 0 to the right.*

Proof. Let v be a symplectic vector field, and let $\alpha = I(v)$, which is a closed 1-form. We first show that the linear map $Symp(M) \to H^1(M, \mathbb{R})$

which assigns to v the cohomology class of α is a Lie algebra character. Let v_1, v_2 be two symplectic vector fields. Let $\alpha_j = I(v_j)$ $(j = 1, 2)$ be the corresponding closed 1-forms. Then

$$
\begin{aligned}
i([v_1, v_2]) \cdot \omega &= \mathcal{L}(v_1) \cdot i(v_2) \cdot \omega - i(v_2) \cdot \mathcal{L}(v_1) \cdot \omega \\
&= \mathcal{L}(v_1) \cdot i(v_2) \cdot \omega = \mathcal{L}(v_1) \cdot \alpha_2 \\
&= d(i(v_1)\alpha_2) + i(v_1)d\alpha_2 = d(i(v_1)\alpha_2)
\end{aligned}
$$

which is an exact 1-form. The cohomology class of α is trivial if and only if $\alpha = df$ globally. This proves (1).

If ω is non-degenerate and M is finite-dimensional, $I : TM \to T^*M$ is a bundle isomorphism. Any Čech cohomology class in $H^1(M, \mathbb{R})$ is represented by a closed 1-form α. Then $v = I^{-1}(\alpha)$ is a symplectic vector field which maps to the class of α. ∎

Since we do not assume ω is non-degenerate, we can not in general make sense of a hamiltonian vector field X_f for every function f. We introduce the kernel $Ker(I)$ of I, which is an (often) singular distribution of tangent spaces.

2.3.4. Lemma. *The distribution $K = Ker(I) = Ker(\omega)$ is <u>integrable</u>, i.e., if two local vector fields v_1, v_2 belong to $Ker(\omega)$, so does their bracket $[v_1, v_2]$.*

Proof. Let v_1, v_2 belong to $Ker(\omega)$, i.e., $i(v_j)\omega = 0$. We have

$$
i([v_1, v_2])\omega = \mathcal{L}(v_1) \cdot i(v_2)\omega - i(v_2) \cdot \mathcal{L}(v_1) \cdot \omega = \mathcal{L}(v_1) \cdot i(v_2)\omega = 0,
$$

hence $[v_1, v_2]$ belongs to $Ker(\omega)$ too. ∎

2.3.5. Definition. *A smooth local function f is said to be <u>of hamiltonian type</u>, or simply <u>hamiltonian</u>, if df belongs to the image of $I : Vect(M) \to A^1(M)$.*

We note that the condition that f is of hamiltonian type is a local condition (as is easily seen using a partition of unity). If the rank of the map I of vector bundles is constant over M, then f is hamiltonian if and only if, for every $x \in M$, df is orthogonal to K_x. This pointwise characterization does not hold in case the rank is not constant. For instance, if $M = \mathbb{R}^2$, $\omega = (x^2 + y^2) \cdot dx \wedge dy$, the function x^2 is not of hamiltonian type.

There may be no globally defined non-constant function of hamiltonian type, even if ω has constant rank. However, if ω is not identically zero, such functions exist locally. One could therefore introduce a sheaf of germs of functions of hamiltonian type.

2.3.6. Proposition. (1) *Let $C^\infty(M/K)$ denote the space of global hamiltonian functions. Then $C^\infty(M/K)$ is an algebra. It is also a Lie algebra under the* <u>*Poisson bracket*</u> *$\{f, g\} = v \cdot g$, where v is a vector field such that $df = I(v)$.*

(2) *Assume M is connected. Let $\mathcal{K}$ be the Lie algebra consisting of global vector fields which belong to the distribution K. Then we have an exact sequence of Lie algebras*

$$0 \to \mathcal{K} \to \mathcal{H}_M \to C^\infty(M/K)/\mathbb{R} \to 0. \qquad (2-11)$$

Proof. These are local statements, and they are obvious when $\omega = 0$. So we may as well assume that ω is nowhere vanishing. We first show that $\mathcal{K}$ is a Lie ideal in $\mathcal{H}_M$ and that $\{f, g\}$ coincides, modulo constants, with the bracket induced on the quotient Lie algebra, which identifies with $(C^\infty(M/K))/\mathbb{R}$. The first statement is proved by the same sort of computation used in 2.3.3 and 2.3.4, so we will skip it. Next assume that $v, w \in \mathcal{H}_M$ and let $I(v) = df$ for $f \in C^\infty(M/K)$; similarly, let $I(w) = dg$ for $g \in C^\infty(M/K)$. Then $I([v, w]) = d(i(v)dg) = d\{f, g\}$ by the proof of Proposition 2.3.3. This shows that the bracket $\{,\}$ defines a Lie algebra structure on $C^\infty(M/K)$ modulo the constants. We still have to show that the Jacobi identity holds exactly, not just modulo constants. For this, we may assume that $C^\infty(M/K)$ contains non-constant functions. Fix functions $g, h \in C^\infty(M/K)$. Then the cyclic symmetrization of $\{f, \{g, h\}\}$ is given by a differential operator acting on f. Since this expression is always a constant function (for any f in the algebra $C^\infty(M/K)$), it has to be identically zero. This fact relies on the following algebraic lemma. ∎

2.3.7. Lemma. *Let A be a commutative $\mathbb{R}$-algebra without zero divisors, which has infinite dimension as a vector space. Let P be an (algebraic) differential operator with coefficients in A. If $P(f)$ is a constant for any $f \in A$, then $P = 0$.*

Proof. Since A is without zero-divisors, we may as well assume A is a field. We can reduce to the case where A is an algebraic extension of $\mathbb{R}(x_1, \ldots, x_n)$ for $n \geq 1$. The differential operator $\frac{\partial}{\partial x_i} P$ is zero, for any i. As the algebra of differential operators on A is without zero-divisors, P itself must be 0. ∎

Recall that a differential operator on A is a $\mathbb{R}$-linear endomorphism P of A which is of finite order. The notion of operator of order $\leq k$ is defined by induction. Operators of order 0 are just multiplication operators M_a for $a \in A$. For $k \geq 1$, an operator P is of order $\leq k$ if and only if $[M_a, P]$ is of order $\leq k - 1$ for all $a \in A$.

We choose $x \in M$ such that there exists an element of $C^{\infty}(M/K)$ whose derivative at x is non-zero. We then apply Lemma 2.3.7 to the algebra A consisting of the jets of infinite order at x of all elements of $C^{\infty}(M/K)$. Since A is a subalgebra of a formal power series, it is without zero divisors, completing the proof of Proposition 2.3.6.

2.3.8. Remark. We note another formula for $\{f, g\}$. If $df = I(v)$ and $dg = I(w)$, then $\{f, g\} = -\omega(v, w)$. The proof is immediate.

The closed 2-form ω on M^{2n} is called a *symplectic form* if it is everywhere of maximal rank, or equivalently if the degree $2n$-form ω^{n} is nowhere vanishing. The pair (M, ω) is called a *symplectic manifold*. In the case of a symplectic manifold, one has the exact sequence

$$0 \to \mathbb{R} \to C^{\infty}(M) \to \mathcal{H}_{M} \to 0,$$

which can easily be generalized.

2.3.9. Proposition. (1) *Assume M is connected. Let $\tilde{\mathcal{H}}_{M}$ be the Lie algebra of pairs (v, f), where $v \in \mathcal{H}_{M}$ and $f \in C^{\infty}(M/K)$ satisfy $I(v) = df$. We have an exact sequence*

$$0 \to \mathbb{R} \to \tilde{\mathcal{H}}_{M} \to \mathcal{H}_{M} \to 0. \qquad (2-12)$$

(2) *If ω is non-degenerate, $\tilde{\mathcal{H}}_{M}$ identifies with the Lie algebra $C^{\infty}(M)$.*

When ω is non-degenerate, the Lie algebra homomorphism $C^{\infty}(M) \to \mathcal{H}_{M}$ associates to a function f the corresponding hamiltonian vector field X_{f}, which is characterized by $I(X_{f}) = df$.

In general, the exact sequence of Proposition 2.3.9 gives the *Kostant-Souriau* central extension of the Lie algebra $\mathcal{H}_{M}$ of hamiltonian vector fields. Notice that it is the pull-back of the central extension $0 \to \mathbb{R} \to C^{\infty}(M/K) \to (C^{\infty}(M/K))/\mathbb{R} \to 0$ by the Lie algebra homomorphism $\mathcal{H}_{M} \to (C^{\infty}(M/K))/\mathbb{R}$. We wish to describe a representative Lie algebra 2-cocycle for this central extension. For this, we will choose a linear section $s : \mathcal{H}_{M} \to \tilde{\mathcal{H}}_{M}$ of the canonical surjection. Then $c(v, w) = [s(v), s(w)] - s([v, w])$ is a 2-cocycle for the Lie algebra $\mathcal{H}_{M}$, i.e., it is a skew bilinear form satisfying the condition

$$c([u, v], w) + c([v, w], u) + c([w, u], v) = 0. \qquad (2-13)$$

2.3.10. Proposition. [P-S] *Let $x \in M$, and assume M connected. Then a representative 2-cocycle for the central extension of Proposition 2.3.9 is given by*

$$c(v, w) = -\omega_x(v, w). \qquad (2-14)$$

Proof. We define a section $s : \mathcal{H}_M \to \tilde{\mathcal{H}}_M$ by $s(v) = (v, f)$, where f is the unique function such that $df = I(v)$ and $f(x) = 0$. Let w be another element of $\mathcal{H}_M$, for which $s(w) = (w, g)$. We have to compute the value at x of the component in $C^\infty(M/K)$ of $[s(v), s(w)]$; this component is by definition equal to $\{f, g\} = -\omega(v, w)$, and its value at x is the difference $[s(v), s(w)] - s([v, w])$. ∎

We refer to [P-S] [Bry2] for a study of this 2-cocycle, and to [Bry2] for a direct proof that its Lie algebra cohomology class is independent of $x \in M$.

2.3.11. Corollary. *Let $\mathcal{H}_{M_x}$ be the Lie subalgebra of $\mathcal{H}_M$ consisting of those vector fields in $\mathcal{H}_M$ which vanish at x. Then the restriction to $\mathcal{H}_{M_x}$ of the central extension splits.*

We now want to give a geometric interpretation of this central extension of the Lie algebra $\mathcal{H}_M$, when a line bundle $p : L \to M$ is given, together with a connection with curvature $K = 2\pi\sqrt{-1} \cdot \omega$. We denote by A the connection 1-form so that $dA = p_+^*(K)$. We need another interpretation of the curvature which involves the *horizontal lift v_h* of a vector field v on M.

2.3.12. Definition. *The horizontal lift v_h of a vector field v is the unique vector field v_h on L^+ which satisfies*

(1) v_h *is p-related to v;*

(2) $\langle A, v_h \rangle = 0$.

If $\tilde{v}$ is a $\mathbb{C}^*$-invariant vector field on L^+, then $\tilde{v}$ has a unique expression $\tilde{v} = v_h + f \cdot \xi$, for a vector field v on M and a complex-valued function f on M. Recall from §2 that for $z \in \mathbb{C}$, $z \cdot \xi$ is the vector field on L^+ giving the infinitesimal action of $\mathbb{C}^*$. Therefore, for f a complex-valued function on M, $f \cdot \xi$ is a vector field on L^+. Of course, $\tilde{v}$ is p-related to v.

2.3.13. Proposition. *For v, w vector fields on M, one has*

$$[v_h, w_h] = [v, w]_h - K(v, w) \cdot \xi$$

as vector fields on L^+.

Proof. Let us compute

$$\langle A, [v_h, w_h] \rangle = \mathcal{L}(v_h)i(w_h)A - i(w_h)\mathcal{L}(v_h)A$$
$$= -i(w_h)i(v_h)dA - i(w_h)di(v_h)A$$
$$= -i(w_h)i(v_h)dA = -(p_+^* K)(v_h, w_h) = -K(v, w)$$

so that the vertical component of $[v_h, w_h]$ is $-K(v, w) \cdot \xi$. ∎

2.3.14. Definition. *A $\mathbb{C}^*$-invariant vector field $\tilde{v}$ on L^+ is said to preserve the connection if $\mathcal{L}(\tilde{v}) \cdot A = 0$. Let $\mathcal{V}_L$ be the Lie algebra of invariant connection-preserving vector fields.*

2.3.15. Proposition. (1) *A vector field $\tilde{v}$ in $\mathcal{V}_L$ is p-related to a hamiltonian vector field v. The assignment $\tilde{v} \mapsto v$ is a surjective Lie algebra homomorphism $\phi : \mathcal{V}_L \to \mathcal{H}_M$.*

(2) *The kernel of ϕ is equal to $\mathbb{C} \cdot \xi$.*

Proof. Let $\tilde{v} \in \mathcal{V}_L$. Let $\tilde{v} = v_h + f \cdot \xi$ be its decomposition into horizontal and vertical components. We have

$$i(v_h)(p_+^*(K)) = i(\tilde{v})(p_+^*(K)) = i(\tilde{v})(dA) =$$
$$= \mathcal{L}(\tilde{v})(A) - d(i(\tilde{v})A) = -d(i(\tilde{v})(A)) = -df$$

It follows that $i(v)(K) = -df$, or $i(v)\omega = d(\frac{\sqrt{-1} \cdot f}{2\pi}) = d(-1/2\pi \cdot \Im(f))$, where $\Im(f)$ is the imaginary part of f. This means that v is a hamiltonian vector field. The same method shows that any hamiltonian vector field v is obtained in this fashion from some $\tilde{v} \in \mathcal{V}_L$, namely, $\tilde{v} = v_h + f \cdot \xi$, where $df = -i(v)K$. This proves (1).

We have, for $\zeta \in \mathbb{C}$: $\mathcal{L}(\zeta \cdot \xi)A = d\zeta + i(\zeta \cdot \xi)dK = 0$, so $\mathbb{C} \cdot \xi \subset \mathcal{V}_L$. On the other hand, a vector field in $Ker(\phi)$ is a $\mathbb{C}^*$-invariant vector field on L^+, so it is of the form $f \cdot \xi$, for some $\mathbb{C}$-valued function f on M; the condition $\mathcal{L}(f \cdot \xi)(A) = 0$ means that $df = 0$. This proves (2). ∎

The significance of the Lie algebra $\mathcal{V}_L$ is shown by the fact that it acts on functions on L^+ (as a Lie algebra of vector fields on L^+) and in particular on sections of L, which are the $\mathbb{C}^*$-homogeneous functions on L^+ of degree

-1. We state this more precisely in the next proposition.

2.3.16. Proposition. *The Lie algebra $\mathcal{V}_L$ acts on germs of sections of L by the formula:*

$$\tilde{v} \cdot s = \nabla_v(s) - f \cdot s,$$

where $\tilde{v} = v_h + f \cdot \xi$ is the decomposition of $\tilde{v}$ into its horizontal and vertical components.

Proof. Let ρ be the degree -1 homogeneous function corresponding to s, i.e., $\rho(y) = \frac{s(p(y))}{y}$. Since the question is local, we may assume $L^+ = M \times \mathbb{C}^*$; the connection is $\nabla(f) = df + fB$ for some complex 1-form B on M, and $v_h = v - \langle B, v \rangle \cdot \xi$. Then $s(y)$ takes values in $\mathbb{C}$, so that $\rho(x, z) = \frac{s(x)}{z}$. We have

$$\tilde{v}(\rho) = \frac{v \cdot s}{z} + (\langle B, v \rangle - f) \cdot \rho$$
$$= \frac{\nabla_v(s) - f \cdot s}{z}$$

which corresponds to the section $\nabla_v(s) - f \cdot s$ of L. ∎

Therefore the problem of defining a Lie algebra action of $\mathcal{H}_M$ on sections of L is transformed into the problem of finding a Lie algebra section for the central extension $\mathcal{V}_L \twoheadrightarrow \mathcal{H}_M$. For instance, this problem has a solution on the Lie subalgebra $\mathcal{H}_{M_z}$, which therefore acts on sections of L.

The relation of this central extension of $\mathcal{H}_M$ by $\mathbb{C}$ with the central extension by $\mathbb{R}$ found in Proposition 2.3.9 is described in the next proposition.

2.3.17. Proposition. *There is a morphism of exact sequences of Lie algebras*

$$
\begin{array}{ccccccccc}
0 & \longrightarrow & \mathbb{R} & \longrightarrow & \tilde{\mathcal{H}}_M & \longrightarrow & \mathcal{H}_M & \longrightarrow & 0 \\
& & \downarrow{\scriptstyle \times -2\pi\sqrt{-1}} & & \downarrow{\scriptstyle \psi} & & \downarrow{\scriptstyle Id} & & \\
0 & \longrightarrow & \mathbb{C} & \longrightarrow & \mathcal{V}_L & \longrightarrow & \mathcal{H}_M & \longrightarrow & 0
\end{array}
$$

where $\psi(v, f) = v_h - 2\pi\sqrt{-1}f \cdot \xi$.

Proof. The only non-obvious statement is that ψ is a Lie algebra homomorphism. We have $[(v, f), (w, g)] = ([v, w], v \cdot g) = ([v, w], -\omega(v, w))$, and

$$\psi([(v, f), (w, g)]) = [v, w]_h - 2\pi\sqrt{-1}\{f, g\} \cdot \xi = [v, w]_h + K(v, w) \cdot \xi.$$

On the other hand, we have

$$
\begin{aligned}
[v_h - 2\pi\sqrt{-1}f \cdot \xi, w_h - 2\pi\sqrt{-1}g \cdot \xi] &= \\
&= [v_h, w_h] - 2\pi\sqrt{-1}(v \cdot g)\xi + 2\pi\sqrt{-1}(w \cdot f)\xi. \\
&= [v, w]_h + K(v, w)) \cdot \xi = \psi([(v, f), (w, g)])
\end{aligned}
$$

∎

This means that the central extension $\mathcal{V}_L$ of the Lie algebra $\mathcal{H}_M$ by $\mathbb{C}$ is really induced from the central extension $\tilde{\mathcal{H}}_M$ by $\mathbb{R}(1) = \sqrt{-1} \cdot \mathbb{R}$.

If ω is symplectic, then for any smooth function f there is a unique hamiltonian vector field X_f such that $I(X_f) = I(df)$. We have $[X_f, X_g] = X_{\{f,g\}}$. For M connected, we have the central extension

$$
0 \to \mathbb{R} \to C^\infty(M) \to \mathcal{H}_M \to 0. \tag{2-15}
$$

2.3.18. Corollary. (A) *Let* $\Gamma(M, L)$ *be the space of smooth sections of L. For* $(v, f) \in \tilde{\mathcal{H}}_M$, *let* $\rho(v, f)$ *be the endomorphism of* $\Gamma(M, L)$ *such that*

$$
\rho(v, f) \cdot s = \frac{1}{2\pi\sqrt{-1}} \cdot \nabla_v(s) + f \cdot s.
$$

Then we have

(1) ρ *is a Lie algebra homomorphism from* $\tilde{\mathcal{H}}_M$ *to* $End(\Gamma(M, L))$, *with bracket* $2\pi\sqrt{-1} \cdot (AB - BA)$.

(2) $\rho(0, 1) = 1$;

(3) *If* ω *is non-degenerate, L has a hermitian structure compatible with the connection* ∇, *and if M is compact,* $\rho(v, f)$ *is a* *hermitian* *operator when* $\Gamma(M, L)$ *is equipped with the scalar product*

$$
B(s_1, s_2) = \int_M (s_1, s_2) \frac{\omega^n}{n!}.
$$

For a symplectic manifold, this is the *Kostant-Souriau prequantization* of the Lie algebra $C^\infty(M)$. The program of geometric quantization consists of associating with the symplectic manifold (M, ω) a much smaller subspace than $\Gamma(M, L)$, namely, the space of sections which have horizontal restriction to the leaves of a lagrangian foliation. We refer to [**Ko1**] [**Ko2**] [**G-S**] for the theory of geometric quantization. Many references can be found in the survey article [**Ki1**].

2.4. Central extension of a group of symplectic diffeomorphisms

Let $p : L \to M$ be a line bundle over a connected manifold M. Let A be a connection 1-form on L^+. In the last section we studied the Lie algebra $\mathcal{V}_L$ of connection-preserving vector fields on L^+ and showed that it is a central extension by $\mathbb{C}$ of the Lie algebra $\mathcal{H}_M$ of hamiltonian vector fields on M. In this section we will describe a corresponding central extension of Lie groups, which is due to Kostant [Ko1].

Let K be the curvature of the connection. Set $\omega = \dfrac{K}{2\pi\sqrt{-1}}$. We introduce the group $Diff(M, \omega)$ of diffeomorphisms of M which preserve ω. For finite-dimensional manifolds M, $Diff(M, \omega)$ is equipped with the topology of local uniform convergence of all derivatives of all orders.

2.4.1. Proposition. [Les] [E-M] [Om] *For M a closed (finite-dimensional) manifold, $Diff(M, \omega)$ is a Lie group, with Lie algebra $Symp(M)$. The action $Diff(M, \omega) \times M \to M$ is smooth.*

Here a Lie group G is defined to be a smooth manifold, equipped with a group law $m : G \times G \to G$ such that m and the mapping $g \mapsto g^{-1}$ are smooth. This is a rather weak definition. If M is compact and $\eta \in Symp(M)$, the one-parameter subgroup $exp(t \cdot \eta)$ is simply the flow of η. Of course, if M is not compact, the flow of η may not exist for all t, and so the exponential map is usually not globally defined. We note that there is a much stronger result for M closed and ω symplectic, namely, that $Diff(M, \omega)$ is an ILH-group ("inverse limit of Hilberts") in the sense of Omori [**Om, Theorem 8.2.1**]. The same is shown in [**R-S**] for important subgroups of $Diff(M, \omega)$.

We will finesse the difficulty of finding a Lie group structure on $Diff(M, \omega)$ by focusing on a somewhat different situation. Say we are given a Lie group G, and a smooth action of G on M which preserves ω. We would now like to lift the action of G on M to a connection-preserving action on L^+. To clarify this, recall the notion of bundle automorphism of L.

2.4.2. Definition. *Let $p : L \to M$ be a line bundle. A <u>bundle automorphism</u> of L is a diffeomorphism $\tilde{g}$ of L such that*

(1) *There exists a diffeomorphism g of M such that $p \circ \tilde{g} = g \circ p$;*

(2) *$\tilde{g}$ induces a linear map $L_x \xrightarrow{\sim} L_{g(x)}$ on the fibers.*

One says that $\tilde{g}$ is a bundle automorphism covering g.

2.4.3. Lemma. *Given a diffeomorphism g of M and a line bundle L with connection ∇ over M, the following two conditions are equivalent:*

(1) *There exists a bundle automorphism $\tilde{g}$ which covers g and is connection-preserving;*

(2) *The pull-back $g^*(L, \nabla)$ is isomorphic to (L, ∇).*

For $g \in G$, we know that g preserves the curvature K of (L, ∇). But according to Theorem 2.2.15, the set of isomorphism classes of pairs (L, ∇) with curvature K is a principal homogeneous space under $H^1(M, \mathbf{C}^*)$. So $g^*(L, \nabla)$ need not be isomorphic to (L, ∇). This motivates the next definition.

2.4.4. Definition. *Let $H \subseteq G$ be the subgroup consisting of those $g \in G$ which fix the isomorphism class of (L, ∇).*

Note that $H = G$ if the fundamental group of M is perfect (i.e., equal to its own commutator subgroup). We would like to lift H to a group of bundle diffeomorphisms of L^+.

2.4.5. Proposition (Kostant [Ko1], see also [G-S]) *Assume M is connected. Let $\tilde{H}$ be the group of bundle diffeomorphisms of L^+ which cover an element of H and are connection-preserving. Then $\tilde{H}$ has a natural Lie group structure, and there is a central extension of Lie groups*

$$1 \to \mathbf{C}^* \to \tilde{H} \overset{q}{\longrightarrow} H \to 1,$$

where $q(\tilde{h}) = h$ if $\tilde{h}$ is a lift of h. In particular, the mapping q is a locally trivial $\mathbf{C}^*$-fibration.

An element of $\tilde{H}$ is called a *quantomorphism.*

The proof of Proposition 2.4.5 essentially consists in finding a class of local sections of q which differ from one another by a smooth $\mathbf{C}^*$-valued function. This requires developing some useful machinery about parallel transport in a connection.

2.4.6. Proposition. *Let (L, ∇) be a line bundle with connection on M. Let $\gamma : [0, 1] \to M$ be a piecewise smooth path in M. Then there exists an isomorphism of complex lines: $T_\gamma : L_{\gamma(0)} \overset{\sim}{\to} L_{\gamma(1)}$ such that for $z \in L^+_{\gamma(0)}$, $T_\gamma(z)$ is the endpoint of the unique horizontal path $\tilde{\gamma}$ in L^+ which lifts γ and starts at $x \in L^+_{\gamma(0)}$. The isomorphism T_γ is called <u>parallel transport</u> (or parallel displacement) along the path γ. The following properties hold:*

(1) *T_γ remains unchanged if γ is replaced by $\gamma \circ f$, where f is a diffeomorphism of $[0, 1]$ which fixes 0 and 1.*

(2) *For composable paths γ and γ', with composition $\gamma * \gamma'$, we have*

$$T_{\gamma * \gamma'} = T_{\gamma'} \circ T_\gamma : L_{\gamma(0)} \overset{\sim}{\to} L_{\gamma'(1)}.$$

(3) *If γ is a loop (i.e., $\gamma(1) = \gamma(0)$), and is the boundary of a smooth map $\sigma : \Delta_2 \to M$, we have*

$$T_\gamma = exp(-\int_{\Delta_2} \sigma^* K).$$

For a loop γ, the number T_γ is called the __holonomy__ *of (L, ∇) around γ.*

Proof. It is clear that the lift $\tilde{\gamma}$ of γ starting at z is unique, if it exists. To prove its existence, we may subdivide $[0, 1]$ into smaller intervals, since lifts of smaller intervals may be patched together to make a lift of γ. We may then reduce to the case L is trivial (using the compactness of $[0, 1]$), $L = M \times \mathbb{C}$, and the connection 1-form is $A = \frac{dz}{z} + B$, for a 1-form B on M. We may also assume that γ is smooth. Then the horizontal lift $\tilde{\gamma}(t) = (\gamma(t), z(t))$ is the solution of the differential equation: $\frac{dz}{dt} = -B(\gamma(t)) \cdot z(t)$, which can be solved explicitly by integration, namely,

$$z(t) = exp(-\int_0^t \langle B(\gamma(u)), d\gamma/du \rangle du).$$

Properties (1) and (2) are immediate. It suffices to show (3) for the boundary of one smooth 2-simplex, and one may subdivide the 2-simplex so as to have its image contained in an open set where L is trivialized. So again the connection will be $\frac{dz}{z} + B$, and we have

$$T_\gamma = exp(-\int_0^1 \gamma^* B) = exp(-\int_{\Delta_2} \sigma^* dB) = exp(-\int_{\Delta_2} \sigma^* K)$$

by Stokes' theorem. ∎

It is very useful to have a fancier version of parallel transport over the space PM of all piecewise smooth maps $\gamma : [0, 1] \to M$. PM itself does not have to be a manifold, but it is an increasing union of manifolds of paths which are smooth over a given subdivision of $[0, 1]$ into smaller intervals. A C^∞-line bundle over PM will be called smooth if its pullback to each such manifold is smooth.

2.4.7. Proposition. *Let $s, t : PM \to M$ be the maps $s(\gamma) = \gamma(0)$, $t(\gamma) = \gamma(1)$. Let (L, ∇) be a line bundle with connection on M. Then there exists a unique isomorphism $T : s^* L \xrightarrow{\sim} t^* L$ of pulled-back line bundles, such that for any $\gamma \in PM$, $T_\gamma : L_{s(\gamma)} \xrightarrow{\sim} L_{t(\gamma)}$ is the isomorphism described in Proposition 2.4.6.*

Proof. We must show that T_γ is a smooth function of γ (relative to a local trivialization of L). This clearly is the case. For instance, the derivative of T_γ (with the notations of the proof of 2.4.6) is the linear form on $T_\gamma(PM)$ given by $v(t) \mapsto - \int_0^1 \langle B, v(u)\rangle du$. Here the tangent vector $v(t)$ is viewed as a field of tangent vectors to M along γ. ∎

We can now describe a class of local sections of the surjective group homomorphism $q : \tilde{H} \to H$. Let U be a contractible neighborhood of 1 in H, and let $F : U \subseteq H \to PU$ be the retraction; for $g \in U$ we have a path $F(g) = F(g,t)$ which starts at 1 and ends at g. We have a natural map $\mu : PG \times M \to PM$, given by $\mu(\gamma, m) = \gamma \cdot m$. By pulling-back the line bundles s^*L and t^*L to $PU \times M$ via μ, we get two line bundles on $PU \times M$, say L_1 and L_2. From Proposition 2.4.7, we obtain an isomorphism of line bundles between L_1 and L_2. Pulling-back by $F \times Id$, we get an isomorphism between two line bundles over $U \times M$. All these line bundles are equipped with connections. We are actually only interested in the part of the connections that is in the M-direction. By this we mean that we are only considering the connections as taking values in the pull-back to $U \times M$ of the cotangent bundle T^*M. We call this a *partial connection in the M-direction*.

We have to modify the isomorphism of line bundles in order to make it compatible with the partial connections in the direction of M. We do this by following Weinstein [**Wein**]. Let us select, for $g \in U$ and $t \in [0,1]$, a hamiltonian function $f_{t,g}$ such that the hamiltonian vector field $X(f_{t,g})$ is equal to $F(g,t)^* \cdot \frac{d}{dt}F(g,t)$. Of course $f_{t,g}$ is supposed to be a smooth function of (t,g).

2.4.8. Proposition. *Let L_1 (resp. L_2) be the pull-back of L via the map $s\circ\mu\circ(F\times Id): U\times M \to M$ (resp. $t\circ\mu\circ(F\times Id)$). Let (g,x) denote a point of $U \times M$. Then $exp(-\int_0^1 2\pi\sqrt{-1}\, f_{t,g}(x)dt)\cdot T$ is an isomorphism $L_1\tilde\to L_2$ of line bundles over $U \times M$ with partial connections in the M-direction.*

Proof. The proof is essentially the same as in [**Wein**]. The question is local near $(g,x) \in U \times M$, so we may assume that L is trivialized in a neighborhood of the path $t \mapsto F(g,t) \cdot x$. Then the connection is $\nabla = d + B$ in this neighborhood. What we have to prove is that for v a vector field on M, the equality

$$v\cdot \int_0^1 \left\{2\pi\sqrt{-1}\cdot f_{t,g}(x)+\langle B, \frac{d}{dt}F(g,t)\rangle(F(g,t)\cdot x)\right\}\cdot dt = \langle B,v\rangle(g\cdot x)-\langle B,v\rangle(x)$$

holds. The action of v is given at the value t of the parameter by the value of v at the point $F(g,t) \cdot x$ of the path. So we get

$$v \cdot \int_0^1 \left\{ 2\pi\sqrt{-1} \cdot f_{t,g}(x) + \langle B, \frac{d}{dt} F(g,t) \rangle \, (F(g,t) \cdot x) \right\} dt$$

$$= \int_0^1 \left\{ 2\pi\sqrt{-1} \, i(v) \frac{dF(g,t)}{dt} + v \cdot \langle B, \frac{d}{dt} F(g,t) \rangle) \right\} (F(g,t) \cdot x) dt$$

$$= \int_0^1 \left\{ -K(v, \frac{dF(g,t)}{dt}) + v \cdot \langle B, \frac{d}{dt} F(g,t) \rangle \right\} (F(g,t) \cdot x) \, dt$$

$$= \int_0^1 \frac{d}{dt} \left(\langle B, v \rangle \, (F(g,t) \cdot x) \right) dt = \langle B, v \rangle (g \cdot x) - \langle B, v \rangle (x)$$

using the fact that v and $\frac{d}{dt} F(g,t)$ are commuting vector fields on $U \times M$. ∎

For $g \in U$, the restriction to $\{g\} \times M$ of the first line bundle is equal to L. The second line bundle has restriction to $\{g\} \times M$ equal to g^*L. Hence, for $g \in U$, we have a lifting $\tilde{g} \in \tilde{H}$ of g. The section $g \mapsto \tilde{g}$ will be smooth, specifying the smooth structure of $\tilde{H}$. To show that it is independent of the choice of the homotopy F, one notes that the quotient of two such sections is of the form $g \mapsto T_{\gamma_g}$, where γ_g is a loop which depends smoothly on $g \in U$. The $\mathbb{C}^*$-valued function is smooth according to Proposition 2.4.6.

The proof that $\tilde{H}$ is a Lie group is not difficult and is left to the reader.

2.4.9. Proposition. *The Lie group $\tilde{H}$ acts smoothly on the space $\Gamma(L)$ of sections of L by*

$$(\tilde{h} \cdot s)(x) = \tilde{h} \cdot s(h^{-1} \cdot x).$$

Hence the problem of making H act on $\Gamma(L)$ is transformed into the problem of finding a section of the central extension $\tilde{H}$ of H. We note that one can easily construct a version of the central extension in which the central subgroup is $\mathbb{T}$ instead of $\mathbb{C}^*$. One works with a hermitian connection on L and replaces $\tilde{H}$ of Theorem 2.4.5 by the Lie group of connection-preserving bundle diffeomorphisms of L_1. We refer to [G-S, §5.2] for details.

The relation with the Lie algebra central extensions of §2.3 is given by

2.4.10. Proposition. *Assume that the Lie algebra homomorphism $\mathfrak{h} \to Symp(M)$ has its image contained in $\mathcal{H}_M$. Then the central extension of Lie algebras deduced from the Lie group central extension in Theorem 2.4.5*

is the pull-back to $\mathfrak{h}$ of the central extension of Lie algebras

$$0 \to \mathbb{C} \to \mathcal{V}_L \to \mathcal{H}_M \to 0.$$

The Kostant construction of a group of quantomorphisms as a central extension of a group of symplectomorphisms allows many of the known central extensions of Lie groups to be recovered, including the case of loop groups and of $Diff^+(S^1)$.

We introduce the infinite-dimensional Grassmannian manifold of Segal and Wilson [S-W]. It is associated to a separable Hilbert space E, and an orthogonal decomposition $E = E_+ \oplus E_-$, where $E_\pm$ are separable Hilbert spaces. Let $p_\pm : E \to E_\pm$ be the orthogonal projections. Recall that for E_1, E_2 two Hilbert spaces, a linear map $T : E_1 \to E_2$ is called Hilbert-Schmidt, if for some (and thus for every) orthonormal basis (e_n) of E_1, we have $\sum_n \|T(e_n)\|^2 < \infty$. The space $H.S.(E_1, E_2)$ of Hilbert-Schmidt operators is a Hilbert space, with norm $\|T\|_2 = \sqrt{\sum_n \|T(e_n)\|^2}$. Note that we have $\|T\|_2^2 = Tr(T^*T)$, which explains why it is independent of the orthonormal basis (e_n).

2.4.11. Proposition [S-W] [P-S] (1) *Let $Gr(E)$ denote the set of closed subspaces $W \subset E$ such that $p : W \to E_-$ is a Hilbert-Schmidt operator. Then $Gr(E)$ is a Hilbert manifold, modelled on $H.S.(E_+, E_-)$.*

(2) *Let $GL_{res}(E)$ denote the group of continuous linear automorphisms of E, with block decomposition $\begin{pmatrix} A & B \\ C & D \end{pmatrix}$, such that B and C are Hilbert-Schmidt operators. Then $GL_{res}(E)$ is a Lie group which operates smoothly and transitively on $Gr(E)$.*

(3) *There is a symplectic form ω on $Gr(E)$, such that*

$$\omega_W(X, Y) = \frac{1}{2\pi\sqrt{-1}} \cdot Tr(X^*Y - Y^*X),$$

where the tangent space $T_W\ Gr(E)$ is identified with the space $H.S.(W, W^\perp)$. ω is invariant under the action of the Lie subgroup $U_{res}(E)$ of $GL_{res}(E)$ consisting of the operators in $GL_{res}(E)$ which are unitary. The cohomology class of ω is integral.

There is a natural line bundle with curvature $2\pi\sqrt{-1} \cdot \omega$, namely the *determinant line bundle* constructed in [P-S].

It is proved in [S-W] [P-S] that each component of $Gr(E)$ is simply-connected. The components of $Gr(E)$ are parametrized by the index of the operator A in the upper left corner, which is a Fredholm operator. Denote by $Gr_k(E)$ the component of $Gr(E)$ corresponding to $k \in \mathbb{Z}$. The theory of Kostant, applied to the action of the identity component $U_{res}^0(E)$ on $Gr_0(E)$, produces a central extension

$$1 \to \mathbb{T} \to \widetilde{U_{res}^0}(E) \to U_{res}^0(E) \to 1.$$

This depends on the fact that the cohomology class of ω is integral. In particular, we obtain a central extension of the identity component $L^0 U(n)$ of the smooth loop group $LU(n)$. Indeed, one can realize $LU(n)$ as a subgroup of $U_{res}(E)$ by thinking of E as $L^2(S^1, \mathbb{C}^n)$, and of E_+ as those vector-valued functions of the form $f = \sum_{n \geq 0} f_n \cdot e^{int}$, for t the angle variable on S^1. Thus $\gamma \in LU(n)$ acts on $f \in E$ by the pointwise action of $\gamma(t) \in U(n)$ on $f(t) \in \mathbb{C}^n$. In this way, $LU(n)$ is a Lie subgroup of $U_{res}(E)$, namely, the subgroup which commutes with the operator of multiplication by e^{it}. It is a remarkable fact that this central extension of $L^0 U(n)$ can be naturally enlarged to a central extension of $LU(n)$ itself. We refer to [P-S] for a detailed treatment.

Let us now look at the Lie group $Diff^+(S^1)$ of orientation-preserving diffeomorphisms of S^1. Kirillov [Ki2] showed that the homogeneous space $Diff^+(S^1)/S^1$ is a symplectic manifold on which $Diff^+(S^1)$ acts by symplectomorphisms. The value of the symplectic form ω at the base point is given by

$$\omega_1(f, h) = \int_0^1 f(t) \cdot h'''(t) dt.$$

Since $Diff^+(S^1)/S^1$ is contractible, the machinery of Kostant gives a central extension

$$1 \to \mathbb{T} \to \widetilde{Diff}^+(S^1) \to Diff^+(S^1) \to 1.$$

This Lie group $\widetilde{Diff}^+(S^1)$ is called the *Bott-Virasoro group*. The Lie algebra of $\widetilde{Diff}^+(S^1)$ is the so-called Virasoro Lie algebra, which in fact was first constructed by Gel'fand and Fuks [G-F]. Bott constructed the Lie group central extension, and gave an explicit cocycle for it [Bott1].

We now come to a result about the splitting of the central extension $\tilde{H}$ of H.

2.4.12. Theorem. *Let (L, ∇) be a line bundle with connection over M, and let H be a Lie group acting on M which preserves the isomorphism class of the pair (L, ∇). Assume there exists a point $x \in M$ which is fixed by H. Then there is a group homomorphism $s : H \to \tilde{H}$ such that $q \circ s = 1$.*

Proof. Let $s(h)$ be the unique connection-preserving diffeomorphism of L^+ which acts trivially on the fiber $p^{-1}(x) \subset M$. Then s is a smooth group homomorphism, and clearly $q \circ s = 1$. ∎

The fact that the central extension $1 \to \mathbb{C}^* \to \tilde{H} \xrightarrow{q} H \to 1$ is trivial has "numerical" consequences, as we will now explain. Assume that there are elements $(a_1, \ldots, a_g, b_1, \ldots, b_g)$ in H such that $[a_1, b_1] \cdot [a_2, b_2] \cdots [a_g, b_g] = 1$. Lift each a_i to an element $\tilde{a}_i$ of $\tilde{H}$, and each b_i to an element $\tilde{b}_i$ of $\tilde{H}$. The product of commutators $[\tilde{a}_1, \tilde{b}_1] \cdots [\tilde{a}_g, \tilde{b}_g]$ belongs to $\mathbb{C}^*$. It is easily seen to be independent of the choice of the liftings $\tilde{a}_i$, $\tilde{b}_i$. Hence it is equal to 1 if the central extension splits, producing the following consequence.

2.4.13. Corollary. *Let H be a group of diffeomorphisms of M, which preserve the isomorphism class of (L, ∇). Assume that there is a point x in M fixed by H. Let $a_1, \ldots, a_g, b_1, \ldots, b_g$ be elements of H such that $[a_1, b_1] \cdots [a_g, b_g] = 1$. Then for every $y \in M$, choose a piecewise smooth polygon P with $4g$ sides, with vertices $y, a_1^{-1} \cdot y, b_1 a_1^{-1} \cdot y, \ldots, y$, and such that two edges adjacent to a common edge are transforms of one another by b_i or $b_i^{-1} a_i b_i$. Then we have*

$$\int_P \omega \in \mathbb{Z}.$$

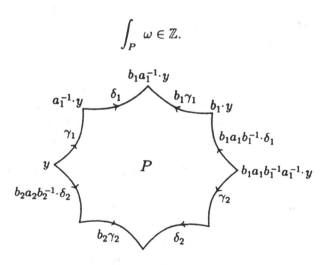

The polygon P.

Proof. We view H as a Lie group with respect to the discrete topology. According to Theorem 2.4.12 and the remarks following it, for any liftings $\tilde{a}_i$ and $\tilde{b}_i$ of a_i and b_i to $\tilde{H}$, we have $[\tilde{a}_1, \tilde{b}_1] \cdots [\tilde{a}_g, \tilde{b}_g] = 1$. We pick $z \in p^{-1}(y)$ and choose these liftings as suggested by the boundary of the polygon, i.e., we want the horizontal lift to L^+ of the boundary ∂P to be a piecewise smooth path with vertices z, $\tilde{a}_1^{-1} \cdot z$, $\tilde{b}_1 \tilde{a}_1^{-1} \cdot z$, $\ldots$, $([\tilde{a}_1, \tilde{b}_1] \cdots [\tilde{a}_g, \tilde{b}_g])^{-1} \cdot z$. This is possible because the edges of P are grouped in pairs according to the action of b_i and $b_i^{-1} a_i b_i$. The lifted path closes because $[\tilde{a}_1, \tilde{b}_1] \cdots [\tilde{a}_g, \tilde{b}_g] = 1$. The holonomy of ∂P is therefore equal to 1; on the other hand, it is also equal to $exp(-\int_P K) = exp(-\int_P 2\pi\sqrt{-1} \cdot \omega)$, hence $\int_P \omega \in \mathbb{Z}$. ∎

One can give a direct proof of Corollary 2.4.13 (as was pointed out to the author by Weinstein). Notice that it is in some sense the opposite of a fixed point theorem; if some number is not equal to 1, then some group of transformations cannot have a common fixed point. A Lefschetz type fixed point theorem usually says that if some number is non-zero, there must be a fixed point.

There is a holomorphic analog of Kostant's construction of a central extension, which is due to Mumford [Mu1].

2.4.14. Theorem. *Let M be a complex manifold such that $\mathcal{O}^*(M) = \mathbb{C}^*$. Let L be a holomorphic line bundle over M. Let H be a complex Lie group which acts holomorphically on M and preserves the isomorphism class of L. Then the group $\tilde{H}$ of holomorphic bundle automorphisms of L which lift an element of H is a complex Lie group, which is a central extension of H by $\mathbb{C}^*$. If there is a point x of M fixed by H, the central extension splits.*

There are at least two very interesting examples of this situation. First if M is an abelian variety, and L is an ample line bundle (in the sense of algebraic geometry) over M, the subgroup H of M consisting of v such that $T_v^* L \xrightarrow{\sim} L$, where $T_v : L \to L$ is translation by v, is a finite group. If L is not the pull-back of a line bundle over a quotient abelian variety M/G for some non-trivial subgroup G, then the center of $\tilde{H}$ is exactly equal to $\mathbb{C}^*$. The commutator pairing $[\,,\,] : H \times H \to \mathbb{C}^*$ can be expressed in terms of the so-called Weil pairing on points of finite order of M. The group $\tilde{H}$ is then the *Heisenberg group* associated to the finite group H with non-degenerate skew-pairing $[\,,\,]$, in the sense of A. Weil [Weil3]. The group $\tilde{H}$ acts on sections of the line bundle L, which are theta-functions. Many of the classical identities between theta-functions have proofs and interpretations in the language of representations of Heisenberg groups [Mu1] [Mu2]. This remains a powerful method in characteristic p, where the theta-functions themselves don't make sense any more.

Another example is the Grassmannian manifold $Gr(H)$ associated with a Hilbert space decomposition $H = H_- \oplus H_-$. Here we have a holomorphic line bundle, namely the determinant line bundle of Segal and Wilson [S-W], which is defined as follows. Let $\mathcal{P}$ be the complex Banach space of continuous operators $w = \begin{bmatrix} w_+ \\ w_- \end{bmatrix} : H_+ \to H$ such that w_+ is a trace class operator. For $w \in \mathcal{P}$, the image W of w is a space W of $Gr(H)$. The vectors $w_n = w(e_n)$ form a so-called *admissible basis* of W. $\mathcal{P}$ is the manifold of all possible admissible bases for all $W \in Gr(H)$.

Let $\mathcal{T}$ be the Lie group of automorphisms T of H_+ such that $T - 1$ is of trace class, i.e., $\sum_n (H(e_n), e_n) < \infty$, for $H = |T - 1|$, the positive hermitian operator which occurs in the polar decomposition $T = |T - 1| \cdot U$ (with U unitary). The topology of $\mathcal{T}$ is induced by the distance $d(T_1, T_2) = Tr(|T_2 - T_1|)$. $\mathcal{T}$ acts on $\mathcal{P}$ by right multiplication. This is a free action, with quotient space $Gr(H)$, so $\mathcal{P} \to Gr(H)$ is a principal $\mathcal{T}$-fibration. Now there is a character $det : \mathcal{T} \to \mathbb{C}^*$, since an operator in $\mathcal{T}$ has a (Fredholm) determinant. The associated holomorphic line bundle $Det = \mathcal{P} \times^{\mathcal{T}} \mathbb{C}_{det}$ is the determinant line bundle over $Gr(H)$.

The equality $\mathcal{O}(Gr(H)) = \mathbb{C}$ is proved in [**S-W**] from the fact that an increasing union of finite-dimensional Grassmann manifolds is dense in $Gr(H)$. Therefore Theorem 2.4.14 applies, and one obtains a central extension $\widetilde{GL_{res}}(H)$ of the loop group $GL_{res}(H)$ by $\mathbb{C}^*$. As $L\,GL(n, \mathbb{C})$ is a Lie subgroup of $GL_{res}(H)$, one also obtains a central extension of $L\,GL(n, \mathbb{C})$ by $\mathbb{C}^*$. We refer to [**S-W**] and [**P-S**] for details.

2.5. Generalizations of Kostant's central extension

In the preceding section, we considered a closed 2-form ω on a manifold M which satisfies the integrality condition $[\omega] \in H^2(M, \mathbb{Z})$. We chose a line bundle with connection (L, ∇) whose curvature is $2\pi\sqrt{-1} \cdot \omega$. When a Lie group H acts on M, preserving the isomorphism class of (L, ∇), we obtained (following Kostant) a central extension

$$1 \to \mathbb{C}^* \to \tilde{H} \to H \to 1.$$

In this section, we want to describe the central extension obtained when we remove the integrality condition on ω. Such a central extension was constructed by Weinstein in [**Wein**]. For the most part, we will not be concerned with the differentiable structure of $\tilde{H}$, as this requires a theory of Souriau [**So2**] which is beyond the scope of this book. So H might as well be a discrete group in what follows.

We begin with a different description of $\tilde{H}$, which uses the line bundle in a different sense, and which we will be able to generalize.

2.5.1. Proposition. *Assume H acts on a symplectic manifold, preserving* *the isomorphism class of the pair* (L, ∇). *Then the group* $\tilde{H}$ *of Theorem 2.4.5 identifies with the set of equivalence classes of pairs* (g, ϕ), *where* $g \in H$ *and* $\phi : g_*(L, \nabla) \xrightarrow{\sim} (L, \nabla)$ *is an isomorphism of line bundles with connection. The product on* $\tilde{H}$ *is given by*

$$(g_1, \phi_1) \cdot (g_2, \phi_2) = (g_1 g_2, \phi_1 \circ g_{1,*}(\phi_2)).$$

The group homomorphism $q : \tilde{H} \to H$ *is simply* $q(g, \phi) = g$.

Proof. This follows immediately from the fact that, given $g \in H$, a connection-preserving diffeomorphism $\tilde{h}$ of L^+ lifting h is the same thing as an isomorphism $\phi : g_*(L, \nabla) \xrightarrow{\sim} (L, \nabla)$ (see Lemma 2.4.3). ∎

Notice that one could take the formula of Proposition 2.5.1 as the definition of the product on $\tilde{H}$. To prove the associativity of the product on $\tilde{H}$ in this fashion, the crucial required fact is that one has a consistent isomorphism $g_{1*}g_{2*}(L, \nabla) \xrightarrow{\sim} (g_1 g_2)_*(L, \nabla)$, which is "compatible" with the associativity of the product of mappings. The inverse of (g, ϕ) is $(g^{-1}, (g^{-1})_*(\phi^{-1}))$.

The map $(g, \phi) \mapsto g$ is obviously a group homomorphism, and its kernel consists of pairs $(1, \phi)$, where ϕ is an automorphism of the pair (L, ∇); an automorphism of L is given by multiplication by an invertible function f. In order for f to yield an automorphism of (L, ∇), one must have $\nabla(f \cdot s) = f \cdot \nabla(s)$ for any local section s. By the Leibniz rule, this means $df = 0$, i.e., f is constant. Clearly $\mathbb{C}^*$ is a central subgroup of $\tilde{H}$.

For any $g \in H$, the line bundle with connection (L, ∇) and its transform $g_*(L, \nabla)$ are isomorphic; hence there exists ϕ such that $(g, \phi) \in \tilde{H}$. Hence q is surjective.

To adapt this method for the case when ω is not integral, we need some substitute for the notion of line bundle. Let $\Lambda \subset \mathbb{C}$ be a countable subgroup, which need not be discrete. Let $q : \mathbb{C} \to \mathbb{C}/\Lambda$ be the canonical projection. We will define a notion of $\mathbb{C}/\Lambda$-bundle over a manifold M.

2.5.2. Definition. *Let* $\Lambda \subseteq \mathbb{C}$ *be a subgroup. A* $\underline{\mathbb{C}/\Lambda\text{-bundle}}$ *over a smooth manifold* M *is a quadruple* (P, μ, f, T), *where*

(a) P *is a set.*

(b) $\mu : (\mathbb{C}/\Lambda) \times P \to P$ *is an action of the group* $\mathbb{C}/\Lambda$ *on* P.

(c) $f : P \to M$ *is a mapping, the fibers of which are exactly the* $\mathbb{C}/\Lambda$-*orbits.*

(d) *Let* $\mathcal{F}$ *be the sheaf of local trivializations of* f, *so that* $\mathcal{F}(U)$ *is the set of bijections* $g : U \times \mathbb{C}/\Lambda \xrightarrow{\sim} f^{-1}(U)$ *which satisfy*

(e) g is $\mathbb{C}/\Lambda$-equivariant.

(f) *The following diagram is commutative*

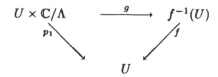

Then $\mathcal{T}$ is a subsheaf of $\mathcal{F}$ which satisfies the following two conditions:

(g) *Every point x of M has some neighborhood U such that $\mathcal{T}(U)$ is not empty.*

(h) *Let U be a contractible open set of M, and let $g \in \mathcal{T}(U)$. Then $\mathcal{T}(U)$ is the set of bijections from $U \times \mathbb{C}/\Lambda$ to $f^{-1}(U)$ of the form $x \mapsto q(\phi(x)) \cdot f(x)$, for ϕ a smooth $\mathbb{C}$-valued function on U.*

Given two $\mathbb{C}/\Lambda$-bundles $(P_1, \mu_1, f_1, \mathcal{T}_1)$ and $(P_2, \mu_2, f_2, \mathcal{T}_2)$, a bundle isomorphism from $(P_1, \mu_1, f_1, \mathcal{T}_1)$ to $(P_2, \mu_2, f_2, \mathcal{T}_2)$ is a bijection $\psi : P_1 \xrightarrow{\sim} P_2$ which satisfies:

(A) ψ is $\mathbb{C}/\Lambda$-equivariant.

(B) *We have $f_1 = f_2 \circ \psi$.*

(C) *For any open set U of M, the bijection $g : U \times \mathbb{C}/\Lambda \xrightarrow{\sim} f_1^{-1}(U)$ belongs to $\mathcal{T}_1(U)$ if and only if $\psi \circ g$ belongs to $\mathcal{T}_2(U)$.*

In this definition, the sheaf $\mathcal{T}$ consists of all *admissible* local trivializations of the bundle P. Condition (g) means that admissible local trivializations exist. Condition (h) means that two such admissible trivializations differ by a mapping $U \to \mathbb{C}/\Lambda$ which is of the form $q \circ \phi$ for a smooth $\mathbb{C}$-valued function ϕ.

One can define the contracted product $P_1 \times^{\mathbb{C}/\Lambda} P_2$ of two $\mathbb{C}/\Lambda$-bundles over M. It is the quotient of the fiber product $P_1 \times_M P_2$ by the following action of $\mathbb{C}/\Lambda$: $z \cdot (x, y) = (z \cdot x, z^{-1} \cdot y)$. The action of $\mathbb{C}/\Lambda$ on P_1 then induces an action on $P_1 \times_M P_2$. The map $P_1 \times^{\mathbb{C}/\Lambda} P_2 \to M$ is the obvious one. The sheaf of admissible local trivializations of $P_1 \times^{\mathbb{C}/\Lambda} P_2$ is characterized by the fact that it contains $x \mapsto (g_1(x), g_2(x))$, for g_1 (resp. g_2) an admissible local trivialization of P_1 (resp. P_2).

For $\Lambda = \mathbb{Z}(1)$, we can identify $\mathbb{C}/\mathbb{Z}(1)$ with $\mathbb{C}^*$ by the exponential map, and a $\mathbb{C}/\Lambda$-bundle is therefore the same thing as a $\mathbb{C}^*$-bundle.

To describe $\mathbb{C}/\Lambda$-bundles in general, we need to consider the quotient sheaf $\underline{\mathbb{C}}/\Lambda$ on M. This is the sheaf of $\mathbb{C}/\Lambda$-valued functions f which are locally of the form $q \circ \phi$, for a smooth $\mathbb{C}$-valued function ϕ. Such a function with values in $\mathbb{C}/\Lambda$ will be called smooth.

2.5.3. Proposition. *The group of isomorphism classes of $\mathbb{C}/\Lambda$-bundles over M identifies with the sheaf cohomology group $H^1(M, \underline{\mathbb{C}}_M/\Lambda)$.*

Proof. This is very similar to the proof of Theorem 2.1.3, so we will just outline the proof. Given a $\mathbb{C}/\Lambda$-bundle $(P, \mu, f, \mathcal{F})$, there is an open covering $\mathcal{U} = (U_i)_{i \in I}$ of M, and an element s_i of $\mathcal{F}(U_i)$ for each i. Each s_i is a bijection $U_i \times \mathbb{C}/\Lambda \xrightarrow{\sim} f^{-1}(U_i)$, which is compatible with the identity map on M and commutes with the $\mathbb{C}/\Lambda$-action. Over U_{ij}, we have the bijection $s_j^{-1} s_i$ from $f^{-1}(U_{ij})$ to itself, which is given by $x \mapsto g_{ij}(x) \cdot x$, for g_{ij} a smooth function with values in $\mathbb{C}/\Lambda$. Then g_{ij} is a Čech 1-cocycle of $\mathcal{U}$ with values in $\underline{\mathbb{C}}_M/\Lambda$. Its cohomology class is independent of the choice of the admissible local trivializations s_i. The whole construction is invariant under the operation of replacing $\mathcal{U}$ by a finer covering.

This defines a group homomorphism from the group of isomorphism classes of $\mathbb{C}/\Lambda$-bundles over M to $\check{H}^1(M, \underline{\mathbb{C}}_M/\Lambda) = H^1(M, \underline{\mathbb{C}}_M/\Lambda)$. To show it is injective, let P be a $\mathbb{C}/\Lambda$-bundle such that the cocycle g_{ij} is a coboundary. So we have $g_{ij} = f_i - f_j$ for some $f_i \in \Gamma(U_i, \underline{BC}_M/\Lambda)$. Then the trivializations $f_i^{-1} \cdot s_i$ glue together to a global trivialization of P. Hence P is the trivial bundle.

The surjectivity is proved by constructing a $\mathbb{C}/\Lambda$-bundle associated to a Čech 1-cocycle g_{ij}. We start with the space $W = \coprod_{i \in I} U_i \times (\mathbb{C}/\Lambda)$, and we construct the quotient space under the relation which, for $x \in U_{ij}$ and $z \in \mathbb{C}/\Lambda$, identifies $(x, z) \in U_i \times (\mathbb{C}/\Lambda)$ with $(x, z + g_{ij}(x)) \in U_j \times (\mathbb{C}/\Lambda)$. This is easily seen to be a $\mathbb{C}/\Lambda$-bundle over M, which gives the 1-cocycle $\underline{g}$. ∎

We next need to define a connection on a $\mathbb{C}/\Lambda$-bundle $(P, \mu, f, \mathcal{F})$. In case $\Lambda = \mathbb{Z}(1)$, this should give the usual notion of connection. Actually what we wish to generalize is not $\nabla(s)$ but $\frac{\nabla(s)}{s}$ (we don't have any choice). We note that for f a local section of $\underline{\mathbb{C}}_M/\Lambda$, we have the differential form df, which is a local section of $\underline{A}^1_{M, \mathbb{C}}$. Also we will need the notion of smooth section $\sigma : U \to P$ of $f : P \to M$, for $U \subseteq M$ open. This means that for an admissible local trivialization $g : V \times (\mathbb{C}/\Lambda) \xrightarrow{\sim} f^{-1}(V)$, the mapping $g^{-1} \circ \sigma$ is smooth on V. Let $\underline{P}$ denote the sheaf of smooth sections of P.

2.5.4. Definition. *Let $(P, \mu, f, \mathcal{F})$ be a $\mathbb{C}/\Lambda$-bundle over M. A <u>connection</u> D on this bundle is a homomorphism of sheaves of groups $D : \underline{P} \to \underline{A}^1_{M, \mathbb{C}}$ which satisfies*

$$D(u + f) = D(u) + df,$$

for u a local section of $\underline{P}$ and f a smooth local function with values in $\mathbb{C}/\Lambda$.

We prefer to use the notation D for the connection instead of ∇ as in §2.2, since D is really a generalization of $\frac{\nabla(s)}{s}$ rather than of $\nabla(s)$ itself. We can now proceed to establish the analogs of many of the results of §2.2 on line bundles with connection. The proofs are formal and are omitted.

2.5.5. Proposition. *The set of connections on a $\mathbb{C}/\Lambda$-bundle is an affine space under $A^1(M)_\mathbb{C}$.*

Recall that for a closed complex-valued p-form η on a manifold M, a *period* of η is an integral of the type $\int_\sigma \eta$, for $\sigma \in H_p(M, \mathbb{Z})$. The periods of η comprise a countable subgroup of $\mathbb{C}$, called the *period group* of η.

2.5.6. Proposition. *Let D be a connection on a $\mathbb{C}/\Lambda$-bundle P. There is a unique closed 2-form K such that $K = d(Du)$ for any local smooth section u of P. The periods of K belong to Λ.*

The relation of $\mathbb{C}/\Lambda$-bundles with connection to smooth Deligne cohomology is given by the following theorem, which generalizes Theorems 2.2.11 and 2.2.12.

2.5.7. Theorem. (1) *The group of isomorphism classes of pairs (P, D) of $\underline{\mathbb{C}}/\Lambda$-bundles with connection is isomorphic to the hypercohomology group*

$$H^1(\underline{\mathbb{C}}_M/\Lambda \xrightarrow{d} \underline{A}^1_{M, \mathbb{C}}).$$

If $\Lambda \subset \mathbb{R}$, it is isomorphic to the Deligne cohomology group $H^2(M, \Lambda(2)^\infty_\mathcal{D})$.

(2) *The curvature of (P, D) is the element of $H^0(M, \underline{A}^2_{M, \mathbb{C}})$ obtained by applying d to the class of (P, D) in $H^1(M, \underline{\mathbb{C}}_M/\Lambda \xrightarrow{d} \underline{A}^1_{M, \mathbb{C}})$.*

The next result is a generalization of a theorem of Weil and Kostant (Theorem 2.2.15).

2.5.8. Corollary. *Let K be a closed 2-form on the smooth manifold M, with period group contained in Λ. There exists a $\mathbb{C}/\Lambda$-bundle over M with connection (P, D), whose curvature is equal to K.*

Now we can generalize the Kostant construction of central extensions to the case where ω is not integral.

2.5.9. Theorem. [Wein] *Let K be a closed complex-valued 2-form on the connected manifold M, with the period group contained in Λ. Let (P, D) be* a $\mathbb{C}/\Lambda$-bundle over M with curvature K. Assume a group H acts on M, and preserves the isomorphism class of (P, D). Let $\tilde{H}$ be the set of pairs (g, ϕ), where $g \in H$ and $\phi : g_*(P, D) \tilde{\to} (P, D)$ is an isomorphism of $\mathbb{C}/\Lambda$-bundles with connection. Define a product on $\tilde{H}$ by

$$(g_1, \phi_1) \cdot (g_2, \phi_2) = (g_1 g_2, \phi_1 \circ g_{1,*}(\phi_2)).$$

Then $\tilde{H}$ is a group, and we have a central extension

$$1 \to \mathbb{C}/\Lambda \to \tilde{H} \to H \to 1.$$

2.5.10. Corollary [Wein] *Under the assumptions of Theorem 2.5.9, assume further that $H_1(M, \mathbb{Z}) = 0$. Let H act on M by diffeomorphisms which preserve K. Then H preserves the isomorphism class of (P, D), hence we have a central extension $\tilde{H}$ of H by $\mathbb{C}/\Lambda$.*

There is a version of Theorem 2.5.9 and its corollary adapted to a group H which preserves the isomorphism class of an $\mathbb{R}/\Lambda$-bundle with connection. Then one finds a central extension of H by $\mathbb{R}/\Lambda$.

Even if H is a Lie group, we must address the obvious problem that $\mathbb{C}/\Lambda$ (or $\mathbb{R}/\Lambda$) has no manifold structure if Λ is not discrete. However one can use the concept of *diffeology*, as developed by Souriau [So2]. Recall that a diffeology on a set X consists in giving, for each $p \geq 0$, a class of functions $f : U \to X$ from open sets U of $\mathbb{R}^p$ to X, called *p-plaques* on X, such that:

(1) A constant f is a p-plaque;

(2) If U can be covered by open sets V_i such that the restriction of f to each V_i is a p-plaque, then f is a p-plaque;

(3) Let U be an open set of $\mathbb{R}^p$, $f : U \subseteq \mathbb{R}^p \to X$ a p-plaque. If V is an open set of some $\mathbb{R}^q$, and $g : V \to U$ is a smooth mapping, then $f \circ g$ is itself a q-plaque.

A smooth manifold X has a natural diffeology, where p-plaques consist of smooth mappings from open sets of $\mathbb{R}^p$ to X. One has the notion of differentiable mapping between diffeological spaces: $h : X \to Y$ is differentiable if for any p-plaque $f : U \to X$ of X, $f \circ h$ is a p-plaque of Y. One can define the product of two diffeological spaces. Hence one can define a *diffeological group* as a group G, equipped with a diffeology, such that the product mapping $G \times G \to G$ and the inverse map are differentiable.

We first make the following simple observation, which is a special case of results of Donato and Iglesias [D-I].

2.5.11. Lemma. *Let Λ be any subgroup of $\mathbb{R}$. Then the quotient $\mathbb{R}/\Lambda$ is a diffeological group. For U an open subset of $\mathbb{R}^p$, $f : U \to \mathbb{R}/\Lambda$ is a p-plaque* *if and only if it is locally of the form $f = q \circ g$, for $q : \mathbb{R} \to \mathbb{R}/\Lambda$ the canonical projection, and $g : U \to \mathbb{R}$ a smooth function.*

Similarly $\mathbb{C}/\Lambda$ is a diffeological group.

2.5.12. Proposition. [Wein] *With the assumptions of Theorem 2.5.9, assume further that H is a Lie group which acts smoothly on M. Then $\tilde{H}$ is a diffeological group, which is a central extension of H by $\mathbb{R}/\Lambda$.*

We refer to [So2] for a discussion of central extensions of diffeological groups.

There is a construction of the Kostant central extension, due to Pressley and Segal [P-S], which does not need the 2-form K to have integral periods, but which requires M to be simply-connected (or at least $H_1(M, \mathbb{Z}) = 0$). It is not quite canonical, since it depends on the choice of a base point x. It is however very concrete, hence well-suited to computations. One defines $\tilde{H}$ to be the set of equivalence classes of triples (g, γ, λ), where $g \in H$, γ is a smooth path from x to $g \cdot x$, and $\lambda \in \mathbb{C}/\Lambda$. The equivalence relation identifies the triple (g, γ, λ) with $(g, \gamma', \lambda + v)$, if γ' is another path from x to $g \cdot x$ and $v \in \mathbb{C}/\Lambda$ is the reduction modulo Λ of the complex number $\int_\sigma K$, for $\sigma : \Delta_2 \to M$ a smooth map with boundary $\gamma - \gamma'$.

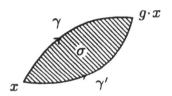

It is clear that for $\sigma' : \Delta_2 \to M$ another smooth map with the same boundary, the difference $\int_{\sigma'} K - \int_\sigma K$ belongs to Λ.

One defines a product by $(g_1, \gamma_1, \lambda_1) + (g_2, \gamma_2, \lambda_2) = (g_1 g_2, \gamma_1 * g_1 \cdot \gamma_2, \lambda_1 + \lambda_2)$, where $*$ denotes composition of paths. This gives a group structure on the set $\tilde{H}$ of equivalence classes, which is then a central extension of H by $\mathbb{C}/\Lambda$. The diffeological structure is apparent; $f : U \to \tilde{H}$ is a p-plaque if it is locally represented by a smooth map (g, γ, y) with values in $G \times PM \times \mathbb{R}$.

Chapter 3

Kähler Geometry of the Space of Knots

3.1. The space of singular knots

Let M be a smooth, paracompact, oriented manifold of dimension n. We introduce the free loop space LM of smooth maps $S^1 \to M$. We will identify S^1 with $\mathbb{R}/\mathbb{Z}$ and denote by x the parameter in $\mathbb{R}/\mathbb{Z}$. First, recall from [P-S] how the space LM of smooth loops $S^1 \to M$ is made into a smooth manifold, modelled on the topological vector space $C^\infty(S^1, \mathbb{R}^n)$. This vector space has the structure of an ILH space (inverse limit of Hilbert spaces) in the sense of §1.4. It is enough to show this for the space $C^\infty(S^1)$, whose topology is defined by the family of semi-norms $||f_n||$, where

$$||f||_n^2 = \int_0^1 \left(||f(x)||^2 + \cdots + ||\frac{d^n}{dx^n} f(x)||^2 \right) dx.$$

Thus $C^\infty(S^1)$ is the inverse limit of the Hilbert spaces $H^n(S^1)$, where $H^n(S^1)$ is the completion of $C^\infty(S^1)$ for the norm $||\ ||_n$. For these facts, we refer the reader to [T].

3.1.1. Lemma. *Condition (1-25) is satisfied by the projection map $p_j :$ $C^\infty(S^1) \to H^j(S^1)$.*

Proof. Indeed, let $B \subset H^j(S^1)$ be some open ball. Let $f \in C^\infty(S^1)$ be in the closure of $p_j^{-1}(\overline{B})$, so that $p_j(f) \in H^j(S^1)$ is the limit in $H^j(S^1)$ of a sequence of functions $f_n \in B$. Let ϕ_n be a sequence of smooth functions on S^1 which "approximate the Dirac function at the origin", i.e., for every $\epsilon > 0$, the restriction of ϕ_n to $[\epsilon, 1 - \epsilon]$ converges uniformly to 0, together with the sequence of derivatives $\frac{d^k \phi_n}{dx^k}$ of any order, and $\int_0^1 \phi_n(x)dx = 1$ for each n. We will use the convolution product $f * h(x) = \int_0^1 f(y)h(x-y)dy$ for functions on S^1; if $f \in H^n(S^1)$ and $h \in H^m(S^1)$, we have $f * h \in H^{m+n}(S^1)$ and $||f * h||_{m+n} \leq ||f||_n \cdot ||h||_m$. For any $h \in H^m(S^1)$, the sequence $\phi_n * h$ converges to h in $H^m(S^1)$. Then the function $f_n * \phi_m$ is smooth for all n, m

and we have, for each integer $k \geq j$,

$$\|f_n * \phi_m - f\|_k \leq \|f - f * \phi_m\|_k + \|(f - f_n) * \phi_m\|_k$$
$$\leq \|f - f * \phi_m\|_k + \|f - f_n\|_j \cdot \|\phi_m\|_{k-j}.$$

Let $l \in \mathbb{N}$ and $\epsilon > 0$ be given. We can find an integer $m = m(l, \epsilon)$ such that $\|f - f * \phi_m\|_k < \epsilon/2$ for $k \leq l$. Then one can find $n = n(l, \epsilon)$ such that $\|f - f_n\|_j \cdot \|\phi_m\|_{k-j} < \epsilon/2$ for $j \leq k \leq l$, and such that $f_n * \phi_m$ belongs to $p_j^{-1}(B)$. So we have

$$\|f_{n(l,\epsilon)} * \phi_{m(l,\epsilon)} - f\|_k < \epsilon \text{ for } k \leq l.$$

Thus the sequence $(f_{n(l,1/l)} * \phi_{m(l,1/l)})_l$ in $C^\infty(S^1)$, indexed by $l \in \mathbb{N}$, belongs to $p_j^{-1}(B)$ and converges to f in $C^\infty(S^1)$, proving the lemma. ∎

We will describe the manifold structure of the free loop space LM using a riemannian structure on M. Given $\gamma : S^1 \to M$ in LM, let $\gamma^* TM \to S^1$ be the pull-back to S^1 of the tangent bundle. For $\epsilon > 0$, let $T_\epsilon \subset \gamma^* TM$ be the open set consisting of tangent vectors of length $< \epsilon$. For ϵ small enough, we have the exponential mapping $exp : T_\epsilon \to M$ [K-N], which identifies T_ϵ with an open set of M. Using this mapping, we identify LT_ϵ with an open set of LM. Now the oriented vector bundle $\gamma^* TM$ over S^1 is trivial; let $\phi : \gamma^* TM \tilde{\to} S^1 \times \mathbb{R}^n$ be a trivialization. Then ϕ induces a mapping from LT_ϵ to $L \mathbb{R}^n$, which is the inclusion of an open set. This will be a typical chart for the manifold structure on LM. We summarize its properties in

3.1.2. Proposition. [P-S] [McLau] *The above charts define a smooth manifold structure on LM, which is independent of the riemannian metric on M. LM is a Lindelöf space (hence paracompact).*

In the particular case of $M = S^1$, the loop space LM contains as an open set the group $Diff^+(S^1)$ of orientation preserving diffeomorphisms of S^1; this open set consists of $f : S^1 \to S^1$ such that $f'(x) > 0$ for all $x \in [0,1]$ and $\int_0^1 f'(x) = 1$. Thus we obtain the Lie group structure on $Diff^+(S^1)$. An important remark here is that there is no way to define a Lie group structure on C^r-diffeomorphisms of S^1 for any $r < \infty$. The reason is that the operation of composition (i.e., multiplication in $Diff^+(S^1)$) would not be a smooth map. In fact, the space of vector fields of class C^r does not even constitute a Lie algebra, as the bracket of two such vector fields will in general only be of class C^{r-1}.

We are mostly interested in the case where M is a three-manifold, but most of the constructions of this section do not assume this. Let $\hat{X} \subset LM$

be the open subset consisting of *immersions* γ which have the properties
that γ induces an embedding of $S^1 \setminus A$, for A a finite subset of S^1, and that
for any two distinct points x_1 and x_2 of A, the branches of γ at $\gamma(x_1)$ and
$\gamma(x_2)$ can have tangency of at most finite order. When $n = 3$, the image of
such an immersion γ in M will be called a *singular knot*.

A singular knot

Let $X \subset \hat{X}$ be the open subset consisting of embeddings. The image $[\gamma]$
of $\gamma \in X$ in Y will simply be called a *knot*, and $\gamma : S^1 \to M$ will be called
a *parametrization* of the knot. The Lie group $Diff^+(S^1)$ of orientation-
preserving diffeomorphisms of S^1 acts smoothly on the right on LM by
$\gamma \cdot f = \gamma \circ f$. This action preserves the open sets $\hat{X}$ and X.

The next result does not require $n = 3$.

3.1.3. Proposition. *Let $\hat{Y}$ be the quotient of $\hat{X}$ by the action of $Diff^+(S^1)$.
There is a natural structure of a smooth Hausdorff paracompact manifold on
$\hat{Y}$, modelled on the Fréchet space $C^\infty(S^1, \mathbb{R}^2)$ of vector-valued functions on
S^1, such that the projection $\pi : \hat{X} \to \hat{Y}$ is a smooth principal bundle with
structure group $Diff^+(S^1)$.*

Proof.   We first observe that the stabilizer subgroup of any $\gamma \in \hat{X}$ is
trivial because any diffeomorphism of S^1 which stabilizes γ has to induce
the identity on $S^1 \setminus A$ for some finite set A; hence it has to be the identity.
We now take any $\gamma : S^1 \to M$ which belongs to $\hat{X}$. Let T be an open subset
of the normal bundle N to the immersion γ, containing the zero-section S^1
of N, such that there exists an immersion $i : T \to M$ extending γ. Let
$p : T \to S^1$ be the projection. Let U be the smooth manifold of smooth
maps $\gamma : S^1 \to T$ such that $p \circ \gamma$ is an orientation-preserving diffeomorphism
of S^1; it is clear that U is an open set of LT. On the other hand, the

immersion i induces a smooth map $\gamma \mapsto i \circ \gamma$ from U to $\hat{X}$, which is easily seen to be an open immersion.

We may identify U with an open subset of $\hat{X}$, which is stable under $Diff^+(S^1)$. A slice V for the action of this group on U is defined as follows: V is the closed subspace of U consisting of those $\gamma : S^1 \to T$ such that $p \circ \gamma$ is the identity map of S^1. We may view V as the space of smooth sections of the bundle $T \xrightarrow{p} S^1$; hence it is a smooth submanifold of U. The map $V \times Diff^+(S^1) \to U$ giving the action is a smooth bijection. The inverse map may be explicitly described as $\gamma \mapsto (\gamma \circ (p \circ \gamma)^{-1}, p \circ \gamma)$, which shows that the inverse map is also smooth; hence $V \times Diff^+(S^1) \to U$ is a diffeomorphism. In turn we now have that the quotient $U/Diff^+(S^1)$ has the structure of a smooth manifold, diffeomorphic to V, and the quotient map $U \to V$ admits a smooth section. It becomes immediately clear that $\hat{Y}$ is a manifold, with open sets V as above serving as local charts. It is not even necessary to check that transition maps are smooth, since one has the intrinsic description of smooth functions on an open set W of $\hat{Y}$ as those smooth functions on $\pi^{-1}(W) \subseteq \hat{X}$ which are $Diff^+(S^1)$-invariant. Furthermore, we have already exhibited local sections for the smooth map π; it follows that π is a locally trivial principal bundle. ∎

3.1.4. Corollary. *The quotient space $Y = X/Diff^+(S^1)$ is an open subset of $\hat{Y}$ so that $\pi : X \to Y$ is a smooth locally trivial principal $Diff^+(S^1)$-bundle. If $n = 3$, Y is called the <u>space of oriented knots in M</u>, and $\hat{Y}$ is called the <u>space of oriented singular knots in M</u>.*

For $\gamma \in \hat{X}$, we denote by $[\gamma] = \pi(\gamma)$ its image in $\hat{Y}$. To describe intrinsically the tangent space of $\hat{Y}$ at a point $C = [\gamma]$, it is convenient to introduce the *normalization* $\tilde{C} \to C$. The normalization is obtained by separating the branches of C at crossing points; more precisely, a point of $\tilde{C}$ is a pair (x, b), where $x \in C$ and b is a branch of C at x. A neighborhood of (x, b) in $\tilde{C}$ consists of the points (y, b), where y belongs to a neighborhood of x in the smooth curve b. Then $\tilde{C}$ is a smooth curve, the mapping $\tilde{C} \to M$ is an immersion, and $T_C \hat{Y}$ is the space of smooth sections of the normal bundle to $\tilde{C} \to M$.

We next establish some conventions relating to differential calculus on the manifolds $\hat{X}$ and $\hat{Y}$. We will need distributions and generalized functions on S^1, for which we refer the reader to [T]. Recall simply that a *distribution* on S^1 is a linear form on $C^\infty(S^1)$; any 1-form defines a distribution. A *generalized function* on S^1 is a linear form on the space $\underline{A}^1(S^1)_\mathbb{C}$ of smooth 1-forms. In particular, any smooth function defines a generalized function.

One can define the notion of *generalized section* (with distributional coefficients) of a vector bundle $E \to S^1$; this section may be written as a sum $\sum_i f_i \cdot v_i$, where f_i is a generalized function and v_i is a section of E.

We refer to §1.4 for the notion of a smooth function on an infinite-dimensional manifold. For a differentiable function f on $\hat{X}$, its differential df at a point γ is a section with distributional coefficients of the pull-back bundle $\gamma^* T^* M$. In the case of $\mathbb{R}^n$, we may think of df as a vector-valued distribution on S^1. The directional derivative $v \cdot f$ of f in the direction of a tangent vector v is then the evaluation $\langle df, v \rangle$. For f a smooth function on $\hat{Y}$, the differential df is a generalized section of the conormal bundle T^*_γ, which is the orthogonal bundle to the subbundle TS^1 of $\gamma^* TM$.

Local coordinate systems on $\hat{Y}$ at a point $C = [\gamma]$ may be obtained by choosing an embedding $T \hookrightarrow M$ of a tubular neighborhood T of γ, as in the proof of Proposition 3.1.3. The open set V of $\hat{Y}$ consists of all sections of the bundle $T \to S^1$. If one trivializes the normal bundle to γ, V becomes an open subset of the Fréchet space $C^\infty(S^1, \mathbb{R}^{n-1})$. Then df may be viewed as an $\mathbb{R}^{n-1}$-valued distribution on the normalization $\tilde{C}$. Given a riemannian metric on M, one may use the differential form $\frac{ds}{l(C)}$ on C, for s the arc-length parameter, to identify functions on C (resp. generalized functions) with 1-forms (resp. distributions). Note that the 1-form $\frac{ds}{l(C)}$ has integral 1. Given such local coordinates, the higher derivatives $D^j f$ may be defined; $(D^j f)_C$ is a distribution on C^j. It is often useful to consider the special class of smooth functions defined as follows.

3.1.5. Definition. *A real valued function f on $\hat{Y}$ is called super- smooth if each higher differential $D^j f$ is a finite linear combination of partial derivatives of distributions on C^j that are obtained from a smooth function on the diagonal $C \subset C^j$ depending smoothly on $C \in \hat{Y}$.*

Some interesting examples of super-smooth functions can be described as follows. Let $M = \mathbb{R}^3$, with its euclidean metric, and let $l : \hat{Y} \to \mathbb{R}$ be the function which associates to C its length $l(C)$. Then l is super-smooth; the derivative dl at C is the vector-valued 1-form $-\kappa \vec{N} \cdot ds$, where κ is the curvature, $\vec{N}$ is the unit normal vector, and s the arc-length parameter. This 1-form is identified with the function $p \mapsto -l(C) \cdot \kappa \vec{N}$. The higher differentials of l may be computed inductively.

Next consider the function on LM defined by integration of a 1-form α on M. This is the function $I_\alpha(\gamma) = \int_\gamma \alpha$. It is super-smooth, and its differential is the vector-valued 1-form $\left(i(\frac{d\gamma}{dx}) \cdot d\alpha \right) dx$, which is easily seen to be independent of the parametrization of C. If $M = \mathbb{R}^3$, $x \in [0,1]$ is a multiple of the arc-length, and we view α as a vector field, then $(dI_\alpha)_\gamma$ is the vector-valued function $\frac{d\gamma}{dx} \times \overrightarrow{curl}\,\alpha$, with the usual notations of calculus for vector product and curl of a vector field.

A more general type of function is obtained as the holonomy in a non-commutative situation. Let G be a finite-dimensional real Lie group with Lie algebra $\mathfrak{g}$. Let $\rho : G \rightarrow Aut(V)$ be a finite-dimensional representation of G, and let $d\rho : \mathfrak{g} \rightarrow End(V)$ denote the differential of ρ. Let α be a $\mathfrak{g}$-valued 1-form, and let $H : \hat{X} \rightarrow G$ denote the *holonomy* mapping $\gamma \mapsto H(\gamma)$. Precisely, $H(\gamma)$ is the holonomy of the connection $d + \alpha$ on the trivial principal G-bundle $M \times G \rightarrow M$. We will describe the element $\rho(H(\gamma))$, as equal to $M(1)$, where the matrix-valued function $M(x)$ is the solution of the equation $\frac{dM}{dx} = -\alpha \cdot M$ such that $M(0) = Id$. However $H(\gamma)$ is only invariant under reparametrizations of S^1 which fix 0; so H descends to a smooth function on the manifold $\hat{Y}_b$ of based knots. When one changes the base point, H is transformed by the inner action of some element of G (holonomy along an arc of the knot).

Therefore the function $I(\gamma) := Tr(\rho(H(\gamma)))$ is base-point independent and defines a complex-valued function on $\hat{Y}$. It is super-smooth, and its differential at γ is the 1-form on C given by

$$-Tr(M(1) \cdot M(x)^{-1} \cdot d\rho(\frac{d\gamma}{dx}\rfloor d\alpha) \cdot M(x))dx.$$

In a local coordinate system, if v and w are tangent vectors to $\hat{X}$ or $\hat{Y}$, one may introduce the directional derivative $D_v w$ of w in the direction of v. Assume M has dimension 3. The differential Dw of the vector field w has a value $(Dw)_C$ at the knot C, parametrized by γ, which is a section with distributional coefficients of a vector bundle with fiber $\mathbb{R}^3 \otimes \mathbb{R}^3$ (in the case of $\hat{X}$), or $\mathbb{R}^2 \otimes \mathbb{R}^2$ (in the case of $\hat{Y}$). Since the vector bundle over $\tilde{C}$ is trivialized, we think of Dw as a tensor-valued distribution. Of course the formula $[v, w] = D_v w - D_w v$ holds.

3.2. Topology of the space of singular knots

As in the preceding section, our main object of interest is the space of oriented knots Y in an orientable 3-manifold M. However, the topology of Y is very complicated; if $M = \mathbb{R}^3$ or $M = S^3$, the connected components of Y consist of all knots of a given isotopy type, which is generally believed to be a $K(\pi, 1)$-space (as we learned from J. Cerf and from A. Hatcher). Hatcher [**Hat**] proved that the component of Y consisting of knots isotopic to the unknot (trivial knot) in S^3 is homotopy equivalent to the space of oriented great circles. However, the topology of the space $\hat{Y}$ is readily computable from known results of differential topology. The proofs in this section often consist in references to the literature. We will use Hochschild and cyclic homology of algebras, for which the book [Lo] is a general reference. The results obtained in this section will not be used in the rest of the book.

3.2.1. Lemma. *The inclusion of $\hat{X}$ into the space $Imm(S^1, M)$ of immersions $S^1 \to M$ is a weak homotopy equivalence, i.e., it induces isomorphisms on all homotopy groups.*

Proof. The space of immersions is stratified according to the type of possible singularities. More precisely, for every $n \geq 2$ and for any function f from subsets of $\{1, 2, \ldots, n\}$ to non-negative integers, we have a closed subspace $S(n, f)$ of $Imm(S^1, M)$ consisting of those immersions $\gamma : S^1 \to M$ such that there exists a cyclically ordered subset $\{x_1, x_2, \ldots, x_n\}$ of S^1, with the property that if $1 \leq i < j \leq n$ and $f(\{i, j\}) > 0$, then $\gamma(x_i) = \gamma(x_j)$ and the two pieces of curves going through this point have a tangency of order at least $f(\{i, j\})$. We now have a stratification in the sense that the intersection of two strata is a union of such strata (and, on the open set $\hat{X}$, a locally finite union). The open set $\hat{X}$ consists of all points which do not belong to the intersection of any infinite decreasing chain of strata. There are only finitely many strata of a given codimension. Let F_k be the union of all strata of codimension $\geq k$; this is a closed subset, and for $U_k = Imm(S^1, M) \setminus F_k$, we have $\hat{X} = \cup_k U_k$. One can see from standard transversality arguments that the complementary subset of F_k has the same homotopy groups as $Imm(S^1, M)$ up to degree $k - 1$. Since we have $\pi_j(\hat{X}) = \lim_k \pi_j(U_k)$, the result follows. ∎

We observe that this well-known stratification is important in the work of Vassiliev [Va], who constructs a nice smooth simplicial resolution of the strata, and obtains deep results on the topology of the space of knots. Lemma 3.2.1 transforms the problem of understanding the topology of $\hat{X}$ into the same problem for $Imm(S^1, M)$, to which Smale's theory [Sm] [Hir] [Po] applies. Let us recall a main theorem of this theory.

3.2.2. Theorem. (*Smale,* [Sm]) *The natural map from $Imm(S^1, M)$ to the free loop space $L(SM)$ of the sphere bundle SM, which to each immersion associates its 1-jet, is a weak homotopy equivalence.*

The topology of loop spaces is well-understood for simply-connected spaces, thanks to Chen's theory of iterated integrals. This involves the *bar complex* for a differential graded algebra $A^\bullet$ over a field k. The product in

$A^\bullet$ is denoted by $\wedge$. The bar complex of $A^\bullet$ is the complex

$$B(A^\bullet) = \oplus_{l \geq 0} \underbrace{A^\bullet \otimes \cdots A^\bullet}_{(l+1) \ times}$$

with $\omega_0 \otimes \cdots \otimes \omega_l$ placed in degree $deg(\omega_0) + \cdots + deg(\omega_l) - l$. The differential is given by $d + b$, where

$$b(\omega_0 \otimes \cdots \otimes \omega_l) = \sum_{i=0}^{l-1} (-1)^i \, \omega_0 \otimes \cdots (\omega_i \wedge \omega_{i+1}) \otimes \cdots \otimes \omega_l$$

$$+ (-1)^{l + deg(\omega_l) \cdot (deg(\omega_0) + \cdots + deg(\omega_{l-1}))} \cdot (\omega_l \wedge \omega_0) \otimes \omega_1 \cdots \otimes \omega_{l-1}$$

and

$$d(\omega_0 \otimes \cdots \otimes \omega_l) = \sum_{i=0}^{l} (-1)^{i + deg(\omega_0) + \cdots deg(\omega_{i-1})} \, \omega_0 \otimes \cdots d\omega_i \otimes \cdots \otimes \omega_l.$$

The cohomology of $B(A^\bullet)$ is often called the *Hochschild homology* of the differential algebra $A^\bullet$. This requires changing the degrees to their opposites. It is often convenient to replace the bar complex with a quotient complex which has the same cohomology; to do this, one introduces $\overline{A^\bullet} := A^\bullet / k$, and, replacing, $\underbrace{A^\bullet \otimes \cdots A^\bullet}_{(l+1) \ times}$ with $A^\bullet \otimes \underbrace{\overline{A^\bullet} \otimes \cdots \overline{A^\bullet}}_{l \ times}$, the new complex is called the reduced Hochschild complex (or the reduced bar complex), and is denoted $\overline{B}(A^\bullet)$.

3.2.3.   Theorem.   (*Chen* [Ch]) *Let M be a simply-connected finite-dimensional manifold. The cohomology algebra $H^\bullet(LM)$ is isomorphic to the Hochschild homology of the de Rham algebra $A^\bullet(M)$.*

The above theorem is often used in the following form.

3.2.4. Corollary.   *Let $\mathcal{M} \subset A^\bullet(M)$ be a (Sullivan) model for M, that is to say, a differential graded subalgebra of $A^\bullet(M)$ such that we have $H^j(\mathcal{M}) \tilde{\to} H^j(A^\bullet(M))$ for all j. Then the Hochschild homology of $\mathcal{M}$ is isomorphic to the cohomology algebra $H^\bullet(LM)$.*

Proof.   This follows from Theorem 3.2.3 and the fact [F-T] that the Hochschild homology of $\mathcal{M}$ is isomorphic to that of $A^\bullet(M)$. ∎

Let us give a couple of examples. For $M = \mathbb{R}^3$, $\hat{X}$ is weakly homotopy equivalent to LS^2. In that case we may take the model $\mathcal{M}$ to be the graded

algebra $\mathbb{R}[x]/(x^2)$, where $deg(x) = 2$ and $dx = 0$. The Hochschild homology
of this algebra is well-known [Han] [Bu] [F-T] [Ka] [Lo], but one must
be sure of taking into account the grading of the algebra. One then finds
(following [Ka]) that the Hochschild homology of $\mathbb{R}[x]/(x^2)$ is the vector
space generated by the following classes in the reduced Hochschild complex:

$$x(n) = 1 \otimes \underbrace{x \otimes \cdots \otimes \cdots x}_{n \; times}$$

in degree n and

$$y(n) = \underbrace{x \otimes \cdots \otimes \cdots x}_{(n+1) \; times}$$

in degree $n + 2$. It then follows from Corollary 3.2.4 that the Poincaré series
of LS^2 is equal to $1 + t + 2t^2 + 2t^3 + 2t^4 + \cdots$.

For $M = S^3$, the unit sphere bundle of S^3 is diffeomorphic to $S^3 \times S^2$,
since S^3 is parallelizable. $\hat{X}$ is weakly homotopy equivalent to $L(S^3 \times S^2) = LS^3 \times LS^2$. The Poincaré series of LS^3 is $1 + t^2 + t^3 + t^4 + \cdots$. From the
Künneth theorem, one finds that the Poincaré series of $L(S^3 \times S^2)$ is equal
to

$$(1+t+2t^2+2t^3+2t^4+\cdots)\cdot(1+t^2+t^3+t^4+\cdots) = 1+t+3t^2+5t^3+\cdots. \quad (3-1)$$

To compute the cohomology groups of $\hat{Y} = \hat{X}/Diff^+(S^1)$, one uses
the principal fibration $\hat{X} \xrightarrow{\pi} \hat{Y}$ with structure group $Diff^+(S^1)$. One may
replace $Diff^+(S^1)$ by S^1, due to the following

3.2.5. Lemma. $Diff^+(S^1)$ *is homotopy equivalent to* S^1.

Accordingly, the $Diff^+(S^1)$-bundle $\hat{X} \xrightarrow{\pi} \hat{Y}$ may be replaced by a circle
bundle $\hat{Y}_b \to \hat{Y}$. The total space $\hat{Y}_b$ is the manifold of (oriented) *based
singular knots*, consisting of pairs (C, p), where $C \in \hat{Y}$ is a singular knot,
and $p \in \tilde{C}$ is a point of the normalization $\tilde{C}$ of C. In other words, $\hat{Y}_b \to \hat{Y}$
is the tautological bundle over the space of singular knots. One obtains

3.2.6. Lemma. $\hat{X}$ *is homotopy equivalent to the space* $\hat{Y}_b$ *of oriented base
knots, and* $\hat{Y}_b \to \hat{Y}$ *is a principal* S^1*-fibration.*

We must now use the universal principal S^1-fibration $ES^1 \to BS^1$;
concretely, this may be realized as $E \setminus \{0\} \to S(E)$, where E is a separable
Hilbert space (or $E = \mathbb{C}^\infty$) and $S(E)$ is the unit sphere in E. Recall [Bo]
that if S^1 acts on a space X, the S^1-*equivariant cohomology* $H^*_{S^1}(X)$ of X is
the cohomology of the space $ES^1 \times^{S^1} X$, which is the quotient of $ES^1 \times X$

by the free diagonal action of S^1. The space $ES^1 \times^{S^1} X$ is called the *Borel construction* associated to the S^1-action on X.

3.2.7. Proposition. *The space $\hat{Y}$ of oriented singular knots on M is weakly homotopically equivalent to the Borel construction $ES^1 \times^{S^1} L(SM)$ associated to the S^1-action on $L(SM)$. In particular, we have*

$$H^\bullet(\hat{Y}) = H^\bullet_{S^1}(L(SM)).$$

Proof. This is an easy consequence of Smale's theorem (Theorem 3.2.2). $\hat{Y}$ is the quotient of $\hat{Y}_b$ by a free action of S^1; therefore it is weakly equivalent to the Borel construction $ES^1 \times^{S^1} \hat{Y}_b$, and thus to $ES^1 \times^{S^1} Imm(S^1, M)$. Since the natural map from $Imm(S^1, M)$ to $L(SM)$ is S^1-equivariant and is a weak homotopy equivalence, it induces a weak homotopy equivalence of the Borel constructions for these spaces with S^1-actions. ∎

The equivariant cohomology of the free loop space is well understood nowadays, thanks to [Go], [F-T] , [Jo], [Bu1] and [G-J-P]. It involves the cyclic homology $HC_-(A^\bullet)$ of a differential graded algebra $A^\bullet$. For its definition, recall the reduced bar complex $\overline{B}(A^\bullet)$, with differential $b + d$, and note that there is another differential B of degree -1 such that

$$B(\omega_0 \otimes \cdots \otimes \omega_k) = \sum_{i=0}^{k} \epsilon_i \cdot (1 \otimes \omega_i \otimes \omega_{i+1} \cdots \otimes \omega_k \otimes \omega_0 \otimes \cdots \otimes \omega_{i-1}),$$

where

$$\epsilon_i = (-1)^{ki + (deg(\omega_0) + \cdots + deg(\omega_{i-1})) \cdot (deg(\omega_i) + \cdots + deg(\omega_k))}.$$

One has $B^2 = 0$, and $(b + d)B + B(b + d) = 0$, so one has a structure which Burghelea [Bu1] calls a *cyclic complex*, and Kassel [Ka] calls a *mixed complex*. At any rate, one may form the tensor product $B(A^\bullet) \otimes \mathbb{R}[u]$, where u is a variable of degree 2, and equip it with the differential $b + d + uB$. The cohomology of this complex $B(A^\bullet) \otimes \mathbb{R}[u]$ is denoted $HC_-(A^\bullet)$.

We remark here that our degree for Hochschild and cyclic homology is the opposite of the usual degree found in the literature; the reason is that we prefer to have a differential of degree 1.

3.2.8. Corollary. *The real cohomology algebra of $\hat{Y}$ is isomorphic to the equivariant cohomology algebra $H^\bullet_{S^1}(L(SM), \mathbb{R})$; hence, for simply-connected M, to the cyclic homology $HC^\bullet_-(\mathcal{M})$, where $\mathcal{M}$ is a Sullivan model for SM.*

Proof. This follows from [Go] or [Bu-Vi] or [F-T] or [G-J-P]. ∎

In the case of $\mathbb{R}^3$, for example, the sphere bundle is homotopically equivalent to S^2, and the equivariant cohomology of LS^2 is computed in [Bu1]. It is the direct sum $\mathbb{R}[u] \oplus \tilde{H}^{\bullet}_{S^1}(LS^2)$, where $\tilde{H}^{\bullet}_{S^1}(LS^2)$ denotes reduced equivariant cohomology. The Poincaré series of $\tilde{H}^{\bullet}_{S^1}(LS^2)$ is equal to $t + t^2 + t^3 + \cdots$, and h ence the Poincaré series for $H^{\bullet}_{S^1}(LS^2)$ is equal to

$$(1 + t^2 + t^4 + \cdots) + (t + t^2 + t^3 + \cdots). \qquad (3-2)$$

Alternatively, one can use the computation of the cyclic homology of $\mathbb{R}[x]/x^2$ in [Han] [Ka]. The class in degree $n + 1$ in $HC^{\bullet}_{-}(\mathbb{R}[x]/x^2)$ is obtained by applying an operator of degree -1 to the class of $y(n)$ in cyclic homology. The variable u (which is identified with the generator of $H^2(BS^1)$) acts by zero on $\tilde{H}^{\bullet}_{S^1}(LS^2)$. The cohomology of $\hat{Y}$ in the case of $\mathbb{R}^3$ is therefore given by (3-1).

In the case of $M = S^3$, we have to compute the equivariant cohomology of the free loop space of $S^3 \times S^2$, which is done using the Künneth theorem for S^1-equivariant cohomology. Only a simple case of the theorem is needed, since $u \in H^2(BS^1)$ acts by 0 on $H^{\bullet}_{S^1}(LS^3)$ and on $H^{\bullet}_{S^1}(LS^2)$. We do not give the detailed computation; the result is contained in [Bu1] in any case. The point is that we have

$$H^{\bullet}_{S^1}(LS^2 \times LS^3) = H^{\bullet}_{S^1}(LS^2) \oplus H^{\bullet}(LS^2) \otimes \tilde{H}^{\bullet}_{S^1}(LS^3).$$

The Poincaré series of $\tilde{H}^{\bullet}_{S^1}(LS^3)$ is $t^2 + t^4 + t^6 + \cdots$. Therefore we find

3.2.9. Proposition. *Let $M = S^3$. Then the Poincaré polynomial of the space of oriented knots $\hat{Y}$ is equal to*

$$(1 + t^2 + t^4 + \cdots) + (t + t^2 + t^3 + \cdots) +$$
$$(1 + t + 2t^2 + 2t^3 + \cdots) \cdot (t^2 + t^4 + t^6 + \cdots) = 1 + t + 3t^2 + 2t^3 + \cdots.$$

For degree 2 cohomology, which is of special interest to us, we have some further geometric information.

3.2.10. Proposition. (i) *For $M = S^3$, one has $H_2(\hat{X}, \mathbb{R}) = \mathbb{R}^3$ and $H_2(\hat{Y}, \mathbb{R}) = \mathbb{R}^3$. The map from the first vector space to the second has rank two, and the cokernel is generated by the homology class $[S]$ of the embedded 2-sphere S consisting of all the oriented great circles on the standard 2-sphere inside S^3.*

(ii) *For $M = \mathbb{R}^3$, one has $H_2(\hat{X}, \mathbb{R}) = \mathbb{R}^2$ and $H_2(\hat{Y}, \mathbb{R}) = \mathbb{R}$, the map from the first group to the second one is 0. Further, $H_2(\hat{Y}, \mathbb{R})$ is generated by the class $[S]$ of the same 2-sphere S as in (i).*

Proof. The dimensions of the cohomology groups have already been calculated. We will only prove (ii), as the proof of (i) is similar. To find the rank of the natural map $H_2(\hat{X}, \mathbb{R}) \to H_2(\hat{Y}, \mathbb{R})$, we look at the Gysin exact sequence (cf. §1.6) for the S^1-fibration $\hat{Y}_b \to \hat{Y}$, keeping in mind that $\hat{Y}_b$ is homotopy equivalent to $\hat{X}$:

$$\cdots H_1(\hat{Y}, \mathbb{R}) \to H_2(\hat{X}, \mathbb{R}) \to H_2(\hat{Y}, \mathbb{R}) \to H_0(\hat{Y}, \mathbb{R}) = \mathbb{R} \to 0.$$

Since $H_2(\hat{Y}, \mathbb{R})$ has dimension 1, and since the map $H_2(\hat{Y}, \mathbb{R}) \to H_0(\hat{Y}, \mathbb{R}) = \mathbb{R}$ is non-zero, it follows that the map $H_2(\hat{X}) \to H_2(\hat{Y})$ is zero. Consider the 2-sphere S described in the statement of the proposition. The restriction to S of the circle bundle $\hat{Y}_b \to \hat{Y}$ is the unit tangent bundle, which is a non-trivial line bundle over S^2. Hence the class of S has non-zero image in $H_0(\hat{Y}, \mathbb{R}) = \mathbb{R}$. Therefore $[S]$ generates $H_2(\hat{Y}, \mathbb{R})$. ∎

3.3. Tautological principal bundles

In Section 1 we associated two infinite-dimensional manifolds to a smooth, finite-dimensional manifold M: the space $\hat{X}$ of smooth immersions $\gamma : S^1 \to M$ with only finitely many crossing points, and the space $\hat{Y}$ of unparametrized, oriented singular knots, which is the quotient of $\hat{X}$ by $Diff^+(S^1)$. We proved that the projection map $\pi : \hat{X} \to \hat{Y}$ is a locally trivial principal $Diff^+(S^1)$-fibration. The choice of a riemannian metric on M allows the structure group of this fibration to be reduced to the circle group $S^1 \subset Diff^+(S^1)$. Namely, let $\hat{Z} \subset \hat{X}$ consist of all $\gamma : S^1 \to M$ such that the parameter x on S^1 is a multiple of the arc-length. Then $\hat{Z} \to \hat{Y}$ is a principal S^1-bundle, and $\hat{X}$ is equal to $\hat{Z} \times^{S^1} Diff^+(S^1)$. Note that $\hat{Z}$ identifies naturally with the manifold $\hat{Y}_b$ of based loops, the base point corresponding to the value 0 of the parameter.

We will need to review the theory of connections (in the sense of E. Cartan) on principal bundles. Let $p : P \to N$ be a principal bundle with structure group G. By convention, the Lie group G acts on P on the right, and the action of $g \in G$ on P is denoted by $x \mapsto x \cdot g$. Let $\mathfrak{g}$ be the Lie algebra of G. For $\xi \in \mathfrak{g}$, the corresponding vector field on P will be denoted by $\tilde{\xi}$. For f, a function on P, we have

$$\tilde{\xi} \cdot f(x) = \frac{d}{dt} f(x \cdot exp(t \cdot \xi))|_{t=0}.$$

The linear map $\xi \mapsto \tilde{\xi}$ is a Lie algebra homomorphism.

A *connection 1-form* (or simply, a connection form) on $p : P \to N$ will be a 1-form θ on P with values in the Lie algebra $\mathfrak{g}$ satisfying

(1) For $\xi \in \mathfrak{g}$, we have $\theta(\tilde{\xi}) = \xi$.

(2) For $x \in P$, $g \in G$, and $v \in T_x(P)$, we have

$$\theta_{x \cdot g}(g_*(v)) = Ad(g^{-1})\theta_x(v). \qquad (3-3)$$

A classical reference for the theory of connections is [K-N]. It is well-known that every principal bundle (over a paracompact manifold) admits a connection. Such a connection 1-form θ determines at each point $x \in P$ a subspace $T_x^h P$, the horizontal tangent space. We have $T_x^h P = Ker(\theta_x)$. There is a direct sum decomposition $T_x P = T_x^v P \oplus T_x^h P$, where the vertical tangent space $T_x^v P$ is the kernel of $dp_x : T_x P \to T_{p(x)} N$. The horizontal space is the fiber of a subbundle $T^h P$ of TP, the horizontal tangent bundle, which is invariant under the G-action. Conversely, such a G-invariant subbundle of TP determines a unique connection 1-form. Given a vector field v on N, there is a unique vector field v_h on P which is p-related to v and horizontal; v_h is called the *horizontal lift* of v.

We now introduce the curvature Θ of a connection θ. Rather than brutally writing down an unmotivated definition, we will (as we did in §2.2 for line bundles) try to describe the curvature as the obstruction to finding a horizontal section $s : U \to P$, i.e., a section such that $s^*\theta = 0$. Consider the case when the bundle is trivial and has the linear group as its structure group. In terms of coordinates (x, g) on $P = M \times G$, we have $\theta = g \cdot A \cdot g^{-1} + dg \cdot g^{-1}$ for some matrix-valued 1-form A on M. A section amounts to a function $g(x)$, and is horizontal if and only if $gA + dg = 0$. Taking exterior derivatives, we find $0 = d(dg) = dg \wedge A + g dA = g(A \wedge A + dA)$. Thus a necessary condition for s to be horizontal is that the 2-form $dA + A \wedge A$ be zero, or, equivalently, that the 2-form $\Theta = d\theta + \theta \wedge \theta$ be 0. This expression only makes sense for a group of matrices; nevertheless the following equivalent definition only involves the bracket on $\mathfrak{g}$ and thus makes sense for any principal bundle:

3.3.1. Definition and Proposition. *The <u>curvature</u> of the connection θ is* *the $\mathfrak{g}$-valued 2-form Θ on P defined by*

$$\Theta = d\theta + \frac{1}{2} \cdot [\theta, \theta] \qquad (3-4)$$

The 2-form Θ satisfies

$$g^*\Theta = Ad(g^{-1}) \circ \Theta. \qquad (3-5)$$

If we have a riemannian metric on M, there results a natural connection on the principal $Diff^+(S^1)$-bundle $\pi : \hat{X} \to \hat{Y}$, characterized by the property that the horizontal tangent space at γ consists of all vector fields $x \mapsto v(x)$ to M along γ which are perpendicular to the tangent vector of the curve γ. The connection 1-form θ on $\hat{X}$, with values in $Vect(S^1)$, is written as

$$\theta_\gamma(v) = (v(x) \cdot \frac{d\gamma}{dx}) \cdot \|\frac{d\gamma}{dx}\|^{-2} \cdot \frac{d}{dx},$$

where the inner product of tangent vectors is denoted by a "dot". Note that for v parallel to γ, that is, $v = (d\gamma)(f(x)\frac{d}{dx}) = \widehat{f(x) \cdot \frac{d}{dx}}$, we have $\theta_\gamma(v) = f(x) \cdot (\frac{d\gamma}{dx} \cdot \frac{d\gamma}{dx})\|\frac{d\gamma}{dx}\|^{-2}\frac{d}{dx} = f(x)\frac{d}{dx}$, i.e., property (1) is verified. Property (2) is easily checked. We will compute the curvature of this connection, i.e., the $Vect(S^1)$-valued 2-form $R = d\theta + \frac{1}{2}[\theta, \theta]$. We may restrict to horizontal vector fields v and w, in which case we have $\theta(v) = \theta(w) = 0$, so $\frac{1}{2}[\theta, \theta](v, w) = 0$. Then we have $(d\theta)_\gamma(v, w) = v \cdot \theta(w) - w \cdot \theta(v) - \theta([v, w]) = -\theta([v, w])$.

Let us describe $\theta([v, w]) = \theta(D_v w) - \theta(D_w v)$. The expression $\theta(D_v w)$ measures the horizontal component of $D_v w$. To compute this quantity, we differentiate the equality $w \cdot \frac{d\gamma}{dx} = 0$ in the direction of the tangent vector v to obtain $\theta(D_v w) + (w \cdot \frac{dv}{dx})\frac{d}{dx} = 0$. This enables us to finish the computation as follows:

3.3.2. Proposition. *The curvature Θ of the principal $Diff^+(S^1)$-bundle $\hat{X} \to \hat{Y}$ is the $Vect(S^1)$-valued 2-form on $\hat{X}$ given by*

$$\Theta(v, w) = (w \cdot \frac{dv}{dx} - v \cdot \frac{dw}{dx})\|\frac{d\gamma}{dx}\|^{-2}\frac{d}{dx}.$$

Note that if the parameter x is a multiple of the arc-length, the expression for Θ is simplified to $\Theta(v, w) = (w \cdot \frac{dv}{dx} - v \cdot \frac{dw}{dx})l^{-2} \cdot \frac{d}{dx}$, where l is the length of the knot.

To obtain a connection on the S^1-bundle $\hat{Z} \to \hat{Y}$, we use the following general method.

3.3.3. Proposition. *Let $M \to N$ be a smooth principal G-bundle, and let $Z \subset M$ be a submanifold of M which reduces the structure group to a Lie subgroup H of G. Let $\mathfrak{g}$ (resp. $\mathfrak{h}$) be the Lie algebra of G (resp. H). Assume $\mathfrak{h}$ is reductive in $\mathfrak{g}$, i.e., there exists an H-equivariant linear map $\sigma : \mathfrak{g} \to \mathfrak{h}$ such that $\sigma(\xi) = \xi$ for $\xi \in \mathfrak{h}$. Given a connection 1-form θ on M, the $\mathfrak{h}$-valued 1-form $\alpha = \sigma \circ (\theta_{/Z})$ is a connection form for the principal H-bundle $Z \to N$. Furthermore, let Θ be the curvature form of θ (a $\mathfrak{g}$-valued 2-form on M). Then the curvature of α is equal to $\sigma \circ (\Theta_{/Z})$.*

Proof. By definition of a connection, the 1-form θ satisfies:

(1) $\theta(\tilde{\xi}) = \xi$, for $\xi \in \mathfrak{g}$.

(2) $\theta(g_*(\xi)) = Ad(g^{-1}) \cdot \theta(\xi)$, for ξ a vector field on M and $g \in G$.

Thus, for $\xi \in \mathfrak{h}$, we have $\alpha(\tilde{\xi}) = \sigma(\theta(\tilde{\xi})) = \sigma(\xi) = \xi$, using the properties of θ and σ. Now α satisfies property (1). Also, for $h \in H$ and for any vector field ξ on Z, extending it to a vector field on a neighborhood of Z in M, also denoted ξ, we have

$$\alpha(h_*(\xi)) = \sigma(\theta(h_*(\xi))) = \sigma(Ad(h^{-1}) \cdot \theta(\xi))$$
$$= Ad(h^{-1}) \cdot (\sigma \circ \theta)(\xi) = Ad(h^{-1}) \cdot \alpha(\xi).$$

So α satisfies property (2) and is a connection form. The curvature of $M \to N$ is the $\mathfrak{g}$-valued 2-form $\Theta = d\theta + \frac{1}{2}[\theta, \theta]$, and the curvature of $Z \to N$ is the $\mathfrak{h}$-valued 2-form $\Omega = d\alpha + \frac{1}{2}[\alpha, \alpha]$. Let v and w be horizontal tangent vector fields to Z (on some open set), and extend them to tangent vector fields to M in a neighborhood, which are also horizontal. We have $\Omega(v, w) = -\alpha([v, w])$ for two such horizontal vectors; finally we have $\sigma(\Theta(v, w)) = -\sigma(\theta([v, w]))$, which is equal to $\Omega(v, w)$. ∎

We apply this construction to the case when $M = \hat{X}$, $N = \hat{Y}$, $G = Diff^+(S^1)$ and $H = S^1$. The Lie algebra of G is $Vect(S^1)$. We identify the Lie algebra of H with $\mathbb{R}$, using $\frac{d}{dx}$ as the generator of this one-dimensional Lie algebra. The unique H-equivariant projection $\sigma : Vect(S^1) \to \mathbb{R}$ such that $\sigma(\frac{d}{dx}) = 1$ is given by $\sigma(f \cdot \frac{d}{dx}) = \int_0^1 f(x)dx$. We have the $Diff^+(S^1)$-bundle $\hat{X} \to \hat{Y}$ and its reduction $\hat{Z} \to \hat{Y}$ to S^1. From Proposition 3.3.2, we obtain a connection on the latter bundle. The curvature Ω of this connection is given by $\Omega(v, w) = l^{-2} \cdot \int_0^1 (w \cdot \frac{dv}{dx} - v \cdot \frac{dw}{dx})dx = 2l^{-2} \cdot \int_0^1 w \cdot \frac{dv}{dx} dx$. Except for the factor l^{-2}, this is the 2-form which was introduced by Witten in his (non-rigorous) proof of the Atiyah-Singer index theorem based on Segal localization for S^1-equivariant cohomology on the free loop space LM. See [At] for a very nice exposition of Witten's method.

It is worthwhile to give an explicit description of the horizontal lift of vector fields for the connection on $\hat{Z} \to \hat{Y}$ afforded by Proposition 3.3.3. First we describe the tangent space to $\hat{Z}$ at $\gamma : S^1 \to M$. For simplicity, we assume $M = \mathbb{R}^n$ with its standard metric, in which case the parameter x is a multiple of the arc-length. Consider a smooth one-parameter family $(t, x) \mapsto \gamma(t, x)$ of knots, such that for each value of t, x is a multiple of the arc-length on the closed curve $C_t = \gamma(\{t\} \times S^1)$. This condition means that the speed $\|\frac{\partial \gamma}{\partial x}\|$ is independent of x, i.e., $\frac{\partial \gamma}{\partial x} \cdot \frac{\partial^2 \gamma}{\partial x^2} = 0$.

Take the derivative of this expression with respect to t, and evaluate at $t = 0$. Setting $v(x) = \frac{\partial \gamma}{\partial t}$, we obtain $\frac{dv}{dx} \cdot \frac{d^2\gamma}{dx^2} + \frac{d\gamma}{dx} \cdot \frac{d^2 v}{dx^2} = 0$, where γ and v are now just vector-valued functions of one variable x. This last expression is the derivative of $\frac{dv}{dx} \cdot \frac{d\gamma}{dx}$ with respect to x. Hence we find the condition $\frac{dv}{dx} \cdot \frac{d\gamma}{dx} = k$, k a constant, for $v(x)$ a tangent vector to Z at γ. Denote by v_L the longitudinal component of v, and by v_N the component of v in the direction of the unit normal vector $\vec{N}$ to C, which is parallel to $\frac{d^2\gamma}{dx^2}$. Let l be the length of C, so that $\frac{x}{l}$ is an arc-length parameter. We have $v_L = \frac{1}{l} v \cdot \frac{d\gamma}{dx}$, hence $\frac{dv_L}{dx} = \frac{1}{l}\frac{d}{dx}(v \cdot \frac{d\gamma}{dx}) = \frac{1}{l}(k + v \cdot \frac{d^2\gamma}{dx^2}) = \frac{k}{l} + l \cdot \kappa \cdot v_N$, for κ the curvature of C.

Hence, a tangent vector V to $\hat{X}$ at $\gamma \in \hat{Z}$ is tangent to $\hat{Z}$ if and only if its longitudinal and normal components satisfy

$$\frac{dv_L}{dx} = k' + l \cdot \kappa \cdot v_N, \text{ for some constant } k'. \qquad (3-6)$$

The horizontal tangent space consists of longitudinal vector fields with v_L constant. In order to find the horizontal lift v of a tangent vector w to Z, we must find the solution $v_L(x)$ of the differential equation (3-6) which has k' as a Lagrange multiplier and is periodic with integral 0. It is clear that such a solution exists and is unique.

Consequently, we have obtained the following description of the connection on $\hat{Z} \to \hat{Y}$.

3.3.4. Proposition. (i) *A tangent vector to $\hat{Z}$ at γ is a vector field $v(x)$ along γ satisfying (3-1).*

(ii) *The tangent vector field $v(x)$ as in (i) is horizontal if and only if $\int_0^1 v_L(x)dx = 0$.*

(iii) *Given a normal vector field w along γ, its horizontal lift v is the unique vector field v along C which coincides with w up to a longitudinal* *vector field, and whose longitudinal component satisfies (3-6) and has integral* 0.

It is intuitively obvious that the S^1-bundle should be non-trivial, since there is no reason that one could devise a process to continuously choose the origin of a knot. In fact, we saw in Proposition 3.2.10 that this bundle is non-trivial. One can also use the following curvature computation.

3.3.5. Proposition. *Let $M = \mathbb{R}^3$, and let $\Sigma \subset X$ be the space of oriented great circles on the standard 2-sphere inside $\mathbb{R}^3$. Then the curvature Ω of the circle bundle $\hat{Z} \to \hat{Y}$ satisfies $\int_\Sigma \Omega = 4\pi$. In particular, the first Chern class of $\hat{Z} \to \hat{Y}$ is non-torsion.*

Proof. It is clear that Ω is invariant under the group $SO(3)$ of rotations of $\mathbb{R}^3$. Its restriction to Σ is then determined by its value at one point, say the horizontal circle parametrized by $\gamma(x) = cos(2\pi x)\vec{i} + sin(2\pi x)\vec{j}$. The standard volume form ν on S^2 has integral equal to 4π. Since it is also $SO(3)$-invariant, we have $\Omega_{/\Sigma} = a \cdot \nu$, for some constant a. We have $a = \beta(u, v)$, where u and v are the tangent vector fields to the circle obtained by the infinitesimal action of $SO(3)$. Concretely, $u(x) = sin(2\pi x)\vec{k}$ and $v(x) = -cos(2\pi x)\vec{k}$. We obtain $\Omega(u, v) = \int_0^1 \{sin^2(2\pi x) + cos^2(2\pi x)\}dx = 1$, so $a = 1$ and $\int_\Sigma \Omega = 4\pi$. ∎

3.4. The complex structure

We will define a complex structure on the space $\hat{Y}$ of oriented singular knots in a three-manifold M. This complex structure will depend on a *conformal structure* on M, i.e., a class of a riemannian metric g defined up to a rescaling $x \mapsto f(x) \cdot g(x)$, for a function f on M with values in $\mathbb{R}_+^*$. Recall that for $C \in \hat{Y}$ a singular knot, we have the normalization $\tilde{C} \to C$, and a tangent vector v to $\hat{Y}$ at C is a section of the normal bundle to the immersion $\tilde{C} \to M$.

3.4.1. Definition. *(Marsden-Weinstein [M-W]) Let $C \in \hat{Y}$ be an oriented singular knot on M. Define a complex structure J on the tangent space $T_C\hat{Y}$ as follows. For a tangent vector v to $\hat{Y}$, i.e., a section of the normal bundle N over $\tilde{C}$, $J \cdot v$ is the vector obtained by rotating v by 90^0 in the normal plane to $\tilde{C}$, with respect to the orientation and the conformal structure on this plane. J can also be described as follows. Choose a metric g giving* *the conformal structure. Let $\vec{T}$ be the unit tangent vector to C; then, for every $p \in \tilde{C}$, one has $J \cdot v(p) = \vec{T} \times v(p)$ in the oriented three-dimensional euclidean space $T_p M$. J induces a smooth almost-complex structure on the manifold $\hat{Y}$.*

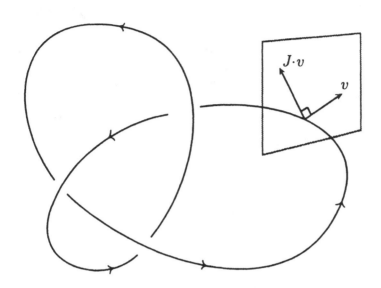

The complex structure J

It is often convenient to introduce the *Frénet frame* for M along the curve C. This is a moving oriented orthonormal frame $(\vec{T}, \vec{N}, \vec{B})$, which depends on the riemannian metric g. Say C is parametrized by $t \mapsto \gamma(t)$. $\vec{T} = \frac{\frac{d\gamma}{dt}}{\|\frac{d\gamma}{dt}\|}$ is the unit tangent vector, $\vec{N}$ is the unit vector in the direction of $\nabla_{\frac{d}{dt}} \vec{T}$, where ∇ denotes the Levi-Civita connection for the metric g. Then $\vec{B} = \vec{T} \times \vec{N}$ is the unique vector which forms an oriented orthonormal basis with the first two. Note that the construction of this frame depends on the assumption that $\frac{d\vec{T}}{dt} \neq 0$, or equivalently that the *curvature* $\kappa(t) = \|\nabla_{\frac{d}{dt}} \vec{T}\|$ does not vanish. Using this frame, we may identify the tangent space $T_C \hat{Y}$ with $C^\infty(S^1, \mathbb{C})$ as follows. To the complex valued function $f(t) + \sqrt{-1} \cdot g(t)$, we associate the section $f(t) \cdot \vec{N} + g(t) \cdot \vec{B}$ of the normal bundle to C. J then corresponds to multiplication by $\sqrt{-1}$ on $C^\infty(S^1, \mathbb{C})$.

This complex structure has been considered (for $M = \mathbb{R}^3$) in fluid mechanics, more precisely in the theory of vortex filaments [Has] [M-W] [Pe-Sp]. We will have more to say about this in §7. In the preprint [Bry3], we posed the question as to whether this complex structure is *integrable*. Recall that J

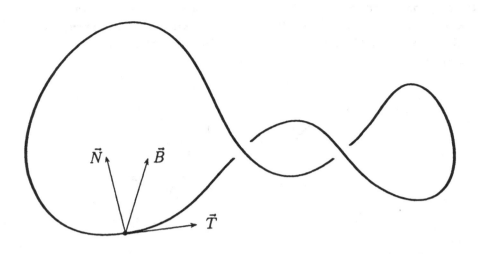

The Frénet frame.

is called integrable if, for every $C \in \hat{Y}$, there exists a diffeomorphism $\phi : U \to$ V of a neighborhood U of C to an open set V of the complex vector space $C^\infty(S^1, \mathbb{C})$, where ϕ maps the complex structure J on the first manifold to the standard complex structure of $C^\infty(S^1, \mathbb{C})$. Recently L. Lempert proved the following negative result.

3.4.2. Theorem. (*Lempert* [**Lem**, *Theorem 10.5*]) *The almost-complex structure J is not integrable on any open subset of $\hat{Y}$. In other words, no non-empty subset of $\hat{Y}$ is biholomorphically isomorphic to an open set of a complex Fréchet space.*

Lempert's proof of non-integrability involves the study of some loop spaces associated with CR structures on manifolds. There is a five-dimensional CR manifold N associated by LeBrun to the 3-manifold M, which is fibered over M with fiber S^2; Lempert embeds the manifold Y into the CR loop space of N; the image is characterized as the space of legendrian curves. The reader is referred to [**Lem**] for this beautiful geometric construction.

However, there are positive results regarding weaker notions of integrability of an almost-complex structure, which we now want to address. In finite dimensions, a famous theorem of Newlander and Nirenberg asserts that an almost-complex structure J is integrable if and only if a certain tensor, the *Nijenhuis tensor*, vanishes identically [N-N]. Let us recall how this tensor comes about.

Given an almost-complex structure J on a a manifold V, one has two distributions T_h and $\overline{T_h}$ of tangent planes inside the complexified tangent bundle $TV_{\mathbb{C}}$. The *holomorphic* tangent distribution T_h is the kernel of $J - \sqrt{-1}$; the *anti-holomorphic* tangent distribution $\overline{T_h}$ is the kernel of $J + \sqrt{-1}$. It is the complex-conjugate distribution to T_h (as the notation indicates). Note that the real part linear map $\Re : TV_{\mathbb{C}} \to TV$ induces an isomorphism of vector bundles $T_h \tilde{\to} TV$. Dually one has a decomposition

$$A_{\mathbb{C}}^1 = A^{(1,0)} \oplus A^{(0,1)}$$

of complex valued 1-forms into types. Similarly one defines the vector space $A^{(p,q)}$ of differential forms of type (p,q). For a complex manifold, the distribution $T_h = ker(J - \sqrt{-1})$ inside the complexified tangent bundle is integrable, hence satisfies the Pfaff condition $[ker(J - \sqrt{-1}), ker(J - \sqrt{-1})] \subset ker(J - \sqrt{-1})$. Note that the kernel of $J - \sqrt{-1}$ is the range of $J + \sqrt{-1}$, hence for two real vector fields ξ and η, we have the condition

$$(J - \sqrt{-1})[(J + \sqrt{-1})(\xi), (J + \sqrt{-1})(\eta)] = 0.$$

The imaginary part of this expression gives $S(\xi, \eta) = 0$, where S is the Nijenhuis tensor

$$S(\xi, \eta) = [\xi, \eta] - [J(\xi), J(\eta)] + J([J(\xi), \eta]) + J([\xi, J(\eta)]). \qquad (3-7)$$

Let us also recall why this expression is a tensor, i.e., is bilinear over smooth functions. First one has $S(\xi, \eta) = -S(\eta, \xi)$ (S is skew-symmetric). Then for f a smooth function, one has $S(f \cdot \xi, \eta) - f \cdot S(\xi, \eta) = -(\eta \cdot f)\xi + (J(\eta) \cdot f)J(\xi) - J((\eta \cdot f)J(\xi)) - J((J(\eta) \cdot f)\xi$, and this is 0 because J itself is linear over smooth functions and satisfies $J^2 = -1$.

Let us say that an almost-complex structure J is *formally integrable* if the Nijenhuis tensor S vanishes identically. T his is well-known to be equivalent to the condition $d'' \circ d'' = 0$, where $d'' : A^{(0,k)} \to A^{(0,k+1)}$ is the *Dolbeault differential* (so that if α if of type (p,q), $d''\alpha$ is the component of type $(p, q + 1)$ of $d\alpha$). This differential d'' was already introduced in §1.2.

3.4.3. Theorem. *Let M be a conformal 3-manifold. Then the almost-complex structure J on $\hat{Y}$ of Definition 3.4.1 is formally integrable.*

Proof.   It is an easy observation that $S(\xi, J \cdot \xi) = 0$ for any almost-complex structure and any tangent vector ξ; this is related to the fact that in dimension 2, any almost-complex structure is integrable. Also we recall that $S(\xi, \xi) = 0$. In our situation, for C a singular knot, one can take a tangent vector field ξ along $\tilde{C}$, which is normal to C, and such that ξ and $J \cdot \xi$ form a basis of the tangent space $T_C \hat{Y}$ as a module over the algebra $C^\infty(\tilde{C})$ of smooth functions on the normalization $\tilde{C}$. For instance, if the curvature of C nowhere vanishes, one may take $\xi = \vec{N}$. Then the vanishing of S at $C \in \hat{Y}$ is a consequence of the following lemma. ∎

3.4.4. Lemma. *The Nijenhuis tensor S at C is bilinear over $C^\infty(\tilde{C})$, that is, $S(f \cdot \xi, \eta) = f \cdot S(\xi, \eta)$, for ξ and η normal vector fields along $\tilde{C}$ and f a smooth function on $\tilde{C}$.*

Proof. Choose local coordinates on $\hat{Y}$ at C using the geodesic flows from points of C. This is accomplished by first choosing a parametrization $x \mapsto \gamma(x)$ of C; then to a small section $x \mapsto v(x)$ of the normal vector bundle to γ, we associate the immersed knot $x \mapsto exp_{\gamma(x)}(v(x))$. If we view f as a function of x, f also gives a function on nearby knots, since these are given a preferred parametrization. Hence $f \cdot \xi$ gives a vector field on a neighborhood of C in $\hat{Y}$, by taking the normal component of $x \mapsto f(x)\xi(x)$ at $C' \in \hat{Y}$. The difference between $S(f \cdot \xi, \eta)$ and $f \cdot S(\xi, \eta)$ is a sum of 4 terms, each of which involves the derivative of f along a vector field such as ξ, $J \cdot \xi$, η or $J \cdot \eta$. Evaluated at C, such a derivative becomes 0, since the vector field in question is always normal to C, and the function f only depends on the parameter x on the curve $\tilde{C}$. ∎

Even though the almost-complex structure on $\hat{Y}$ is a tensor which is a real-analytic function of C, there is no result in the literature which will show that its formal integrability implies its integrability, which is, of course, consonant with the theorem of Lempert. Note that for Banach manifolds, there is a theorem of Penot [**Pe**] which says that a formally integrable almost-complex structure J, that is real-analytic (as a section of a suitable vector bundle), is integrable. Instead of the Fréchet manifold of singular C^∞ knots, we may consider the Banach manifold of knots of class C^r, for $r \geq 1$. However, there is no almost-complex structure on this manifold, because Definition 3.4.1 imposes "loss of differentiability" in that it features the derivative $\frac{d\gamma}{dt}$, which is only of class C^{r-1}. In [**Bry3**], the author expressed some hope to extend Penot's method by using the inverse function theorem for good Fréchet spaces. This hope has obviously been dashed by Theorem 3.4.2 of Lempert.

We will explain below results of Drinfeld, LeBrun and Lempert, which give a weaker form of integrability of the complex structure J for the space of *real-analytic* knots. Let us first introduce this weaker concept of integrability.

3.4.5. Definition. *An almost-complex structure J on a manifold V is called <u>weakly integrable</u> if for any $x \in V$, there exists a weakly dense subset E of $T^*_{h,x} V$ such that each $\xi \in E$ is the differential at x of some holomorphic function f defined in a neighborhood of x.*

The article [**Lem**] gives a large class of examples of almost-complex structures which are formally integrable but not integrable. This is why a weaker notion of integrability appears appropriate. If one is interested in having an ample supply of local holomorphic functions rather than an explicit local model for the manifold, this concept may be of use.

By a direct geometric argument reminiscent of twistor theory, Drinfeld and LeBrun have proved the weak integrability of the almost-complex structure on the manifold $\hat{Y}_{an}$ of real-analytic knots on a real-analytic conformal 3-fold M. Then Lempert [**Lem**] went on to prove that J is weakly integrable in case the manifold M and the metric are real-analytic.

3.4.6. Theorem. (*Lempert*) [**Lem**] *Let M be a conformal real-analytic 3-fold. The almost-complex structure J on $\hat{Y}$ is <u>weakly integrable</u>.*

We refer to [**Lem**] for the proof, and will instead give an informal discussion of the geometric ideas of Drinfeld and LeBrun involving some twistor theory. We start by noting various easy generalizations of the construction of J and of Theorem 3.4.2. First of all, one may work with spaces of embeddings of $\mathbb{R}$ into M, hence with a space of open immersed curves in M. Next one may consider a lorentzian metric instead of a riemannian one. By this we mean a smooth non-degenerate symmetric bilinear form on the tangent spaces of M with signature $(2, 1)$. One then restricts to *time-like* curves, which are such that their tangent vector has negative squared length. For such time-like curves, the induced metric on the normal plane is positive-definite, and one may define J just as in Definition 3.4.1. Theorem 3.4.2 applies to this situation with exactly the same proof.

The manifold $\hat{Y}_{an}$ of real-analytic knots is defined in the same way as $\hat{Y}$, as the quotient of the manifold $\hat{X}_{an}$ of real-analytic immersions $\gamma : S^1 \to M$ by the action of the Lie group of orientation preserving real-analytic diffeomorphisms of S^1. Choose a complexification $M \hookrightarrow M_{\mathbb{C}}$ of the real-analytic manifold M. Note that the topology of the "model" vector space of real-analytic functions $S^1 \to \mathbb{R}^2$ is the direct limit topology of the Fréchet topology of the space E_ϵ consisting of those continuous functions in the closed annulus $A_\epsilon = \{z \in \mathbb{C} : 1 - \epsilon \le |z| \le 1 + \epsilon\}$ which are analytic in the interior of A_ϵ.

We may choose the complexification small enough so that a given metric g in the class of the conformal structure on M extends to a non-degenerate complex-valued symmetric bilinear form on the tangent space to $M_{\mathbb{C}}$. One chooses next a complexification $Y_{\mathbb{C}}$ of a neighborhood of a knot C in $\hat{Y}_{an}$; the points of $Y_{\mathbb{C}}$ are holomorphic mappings $\gamma : A \to M_{\mathbb{C}}$ from an annulus A of $\mathbb{C}$ containing S^1 to the complexification $M_{\mathbb{C}}$ of M (the annulus is not fixed).

We choose the complexification small enough so that the holomorphic tangent vector to an annulus $\gamma : A \to M_{\mathbb{C}}$ in $Y_{\mathbb{C}}$ is non-isotropic at every point of the annulus. We will produce, by a direct geometric argument, integral submanifolds for the holomorphic and anti-holomorphic distributions, passing through an arbitrary $\gamma : A \to M_{\mathbb{C}}$. For every $z \in A$, since the vector $\frac{d\gamma}{dz}$ is non-isotropic, the normal plane N_z to γ at $\gamma(z)$ is a two-dimensional complex vector space, which inherits a non-degenerate symmetric bilinear form. Hence there are two isotropic normal lines at $\gamma(z)$.

The crucial observation is that one of these lines, say L_+, is in $Ker(J - \sqrt{-1})$, and the other one, say L_-, is in $Ker(J + \sqrt{-1})$. This is very easy to check: if u is a vector of length 1 in the normal plane, u and $J \cdot u$ form an orthonormal basis of this plane. The $\sqrt{-1}$ and $-\sqrt{-1}$-eigenspaces of J are spanned by $u - \sqrt{-1} \cdot J \cdot u$ and $u + \sqrt{-1} \cdot J \cdot u$ respectively, and these vectors are isotropic. Now from each $\gamma(z)$ draw the so-called *null-geodesic* l_+ in the direction of L_+; this is a geodesic which starts at $\gamma(z)$ with tangent vector parallel to L_+. The parametrized geodesic curve $u(t)$ is defined (at least for small complex numbers t) as the solution of the differential equation

$$\nabla_{\frac{du}{dt}} \frac{du}{dt} = 0$$

with the given initial conditions. Let Σ_+ be the complex manifold of dimension 2 comprising the union of all these null-geodesics. Let Λ_+ be the germ at γ of submanifold of $Y_{\mathbb{C}}$ which consists of all holomorphic mappings $\gamma : A \to \Sigma_+$ (for A some annulus around $|z| = 1$), modulo holomorphic reparametrizations.

We claim that Λ_+ is an integral manifold for the holomorphic distribution of tangent planes; this is clear, since for γ' close to γ, hence with non-isotropic tangent vector, the normal vector to γ' is the tangent vector to the null-geodesic l_+, hence the tangent space to Λ_+ consists of all fields of normal vectors which are parallel to L_+, and is thus exactly $ker(J - \sqrt{-1})$. Using the null-geodesics l_-, one similarly constructs a two-dimensional complex manifold Σ_- inside $M_{\mathbb{C}}$, and the manifold Λ_- of annuli inside Σ_- is an integral manifold for the anti-holomorphic distribution. The holomorphic and anti-holomorphic distributions of tangent planes are therefore integrable.

Locally, at $\gamma \in \hat{Y}_{an,\mathbb{C}}$, we have found integral manifolds for the holomorphic and anti-holomorphic distributions. We have a holomorphic projection map $p : Y_{an,\mathbb{C}} \to \Lambda_+$, which is constant along the leaves of the anti-holomorphic foliation. The value of p on the annulus parametrized by $\gamma' : A \to M_{\mathbb{C}}$ is obtained as follows. For each $z \in A$, follow the null-geodesic l_- through $\gamma'(z)$ to its point $h(z)$ of intersection with the surface Σ_+. Then set $p(\gamma') = h$.

The existence of this point $h(z)$, for γ' close to γ, follows from Rouché's theorem. From the holomorphic implicit function theorem, one sees that $z \mapsto h(z)$ is indeed holomorphic, and so h indeed describes a point of Λ_+. It is clear that p is holomorphic and is constant on the anti-holomorphic leaves. One now verifies that the restriction of p to $Y_{an} \subset Y_{an,\mathbb{C}}$ is a smooth mapping which is compatible with the complex structures of both manifolds. The differential of this mapping at γ is an isomorphism. Hence one can find a germ of holomorphic function at γ with any given differential at this point, implying that J is weakly integrable.

Let us give examples of complex submanifolds of the space of knots which have finite codimension or finite dimension. For $p \in M$, the set D_p of singular knots which pass through p is a complex analytic subvariety of codimension 1, i.e., a divisor; for a singular knot C which contains p as a smooth point, the tangent space to D_p at C consists of all sections v of the normal bundle which vanish at p; this is a complex subspace of $T_C \hat{Y}$ of codimension 1. If p is an ordinary double point of C, the divisor D_p has a normal crossing singularity with two components. For distinct points $p_1, \ldots, p_n$ of M, the intersection $D_{p_1} \cap \cdots \cap D_{p_n}$ is a complex-analytic subvariety of pure codimension n. For $M = S^3$ and p the point at infinity, the complementary subset to the divisor D_p is the space of knots in $\mathbb{R}^3$.

There are a few examples of finite-dimensional complex submanifolds of Y. Assume M is a riemannian 3-manifold, and that there is a smooth action of S^1 on M which preserves the metric and is fixed point free. Then M/S^1 is a smooth manifold and the mapping $q : M \to M/S^1$ is a smooth S^1-fibration. There is a riemannian metric on M/S^1. In fact, the tangent space at a point $q(x)$ of M/S^1 identifies with the orthogonal to the tangent space of $S^1 \cdot x$ inside $T_x M$. One can view M/S^1 as a submanifold of Y, parametrizing the S^1-orbits.

3.4.7. Proposition. *(Lempert) The space M/S^1 of S^1-orbits is a one-dimensional complex submanifold of Y.*

Proof. As M/S^1 is an oriented surface with a riemannian structure, it acquires a complex structure in the usual manner. Then the inclusion $M/S^1 \hookrightarrow Y$ is compatible with the complex structures, both of which consist of a $\frac{\pi}{2}$-rotation in the 2-dimensional normal bundle to the S^1-orbit. ∎

A nice example is obtained for $S^3 = \{(z_1, z_2) \in \mathbb{C}^2 : |z_1|^2 + |z_2|^2 = 1\}$. The element x of the circle $S^1 = \mathbb{R}/\mathbb{Z}$ acts by multiplication by $e^{2\pi\sqrt{-1}\cdot x}$ on both complex coordinates. The quotient S^3/S^1 is the projective line $\mathbb{CP}_1$, which identifies with S^2. The metric on S^2 is the standard one. The fibration $S^3 \to S^2$ is the well-known *Hopf fibration*. One thus gets a complex Riemann sphere S inside Y. Note that $SU(2)$ acts on S^3, preserving the metric and commuting with the S^1-action. Hence one recovers an $SU(2)$-action on S which factors through $SU(2)/\{\pm 1\}\tilde{\to}SO(3)$.

It would be very interesting to construct a universal family of embedded 2-spheres containing S. One would expect this to be a very large family of 2-spheres. This is justified by the following formal computation of the tangent space to the "universal family" (assuming it exists). This tangent space identifies with the space of holomorphic sections over S of the normal bundle for the embedding $S \hookrightarrow Y$.

We will use the notations of algebraic geometers for holomorphic line bundles on $S^2 = \mathbb{CP}_1$; thus $\mathcal{O}(n)$ will be the line bundle corresponding to the integer n. $\mathcal{O}(n)$ has non-zero holomorphic sections when $n \geq 0$. The tangent bundle is $\mathcal{O}(2)$.

3.4.8. Proposition.   *The $SO(3)$-equivariant holomorphic normal bundle N_S to $S \hookrightarrow Y$ contains the infinite direct sum $\oplus_{n\in\mathbb{Z},n\neq 1}\mathcal{O}(2n)$ as a dense holomorphic subbundle. Hence the space $\Gamma(S, N_S)$, as a representation of $SO(3)$, contains $\oplus_{n\geq 0,n\neq 1}V_{2n}$ as a dense subspace, where V_k is the representation $Sym^k(\mathbb{C}^2)$ of $SU(2)$.*

Proof.   Let $C \in S$ be the circle of radius 1 in the complex line $z_2 = 0$. The stabilizer of C is the diagonal subgroup $\{\begin{pmatrix} z & 0 \\ 0 & z^{-1} \end{pmatrix}, |z| = 1\}$ of $SU(2)$, which identifies with $\mathbb{T}\tilde{\to}S^1$. Note that the S^1-action on the circle C is given by rotations, or, in other words, by translation in the parameter $x \in [0, 1]$. As an $SO(3)$-equivariant holomorphic bundle on S, N_S is determined by its fiber at the base point C, together with the action of the stabilizer $\mathbb{T}$. The same holds for the restriction to S of the tangent bundle TY. We first determine $TY_{/S}$ as an equivariant vector bundle over S. Let $v(x)$ be the normal vector to C. Then v and $J \cdot v$ form a basis of the normal bundle to C. The space of normal vector fields has a topological basis $(u_n)_{n\in\mathbb{Z}}$ (over $\mathbb{C}$) consisting of eigenvectors for the action of S^1, namely, $u_n(x) = cos(2\pi nx)v + sin(2\pi nx)J \cdot v$. Note that S^1 acts only on the parameter x, since the vector fields v and $J \cdot v$ are S^1-invariant. We verify easily that

$x \in S^1 = \mathbb{R}/\mathbb{Z}$ acts on u_n by the character $e^{2\pi\sqrt{-1}\cdot nx}$. Hence the fiber of $(TY)_{/S}$ at C decomposes as a (completed) direct sum of the characters $x \mapsto e^{2\pi\sqrt{-1}\cdot nx}$, for $n \in \mathbb{Z}$. To compare with the standard line bundles $\mathcal{O}(k)$, note that $\mathcal{O}(k)$ is always $SU(2)$-equivariant, and is $SO(3)$-equivariant exactly when k is even. The tangent bundle to S is the line bundle $\mathcal{O}(2)$, which corresponds to the character $x \mapsto e^{-2\pi\sqrt{-1}\cdot x}$. Hence the normal bundle N_S contains $\oplus_{n\in\mathbb{Z},n\neq 1}\mathcal{O}(2n)$ as a dense subbundle. The space of global holomorphic sections of $\mathcal{O}(2n)$ is equal to 0 if $n < 0$, and to V_{2n} if $n \geq 0$. This finishes the proof. ∎

An example of a complex submanifold of dimension 2 is given by the space of geodesics (i.e., great circles) in S^3. This complex surface, isomorphic to the tangent bundle of S^2, has been studied by Hitchin [Hi1] [Hi2] and by LeBrun [LB], in connection with the Bogomolny equation and the Einstein equation. The definition of the complex structure given by Hitchin coincides with ours.

More generally, Hitchin defines a complex structure on the space of geodesics of a riemannian 3-manifold with constant curvature, which is a four-dimensional manifold. As shown by Hitchin, the significance of the "constant curvature" condition is that it is equivalent to the condition that the tangent space to the space of geodesics be invariant under J.

We conclude this section with a simple observation

3.4.9. Proposition. *The self-map ι of $\hat{Y}$, which maps the singular oriented singular knot C to the knot $-C$ (same knot, with opposite orientation), is an anti-holomorphic involution.*

3.5. The symplectic structure

Let M be a smooth 3-manifold equipped with a 3-form ν. We will construct a non-degenerate closed 2-form β on the space $\hat{Y}$ of singular knots. First we describe a 2-form β on LM obtained from ν by transgression (or fiber integration). Let $ev : S^1 \times LM \to M$ be the evaluation map $ev(x,\gamma) = \gamma(x)$. We set

$$\beta = \int_{S^1} ev^*(\nu) \tag{3-7}$$

where $\int_{S^1}$ denotes integration along the fibers of the projection $p_2 : S^1 \times$

$LM \to LM$. In other words, β is given by the formula

$$\beta_\gamma(u,v) = \int_0^1 \nu_{\gamma(x)}(\frac{d\gamma}{dx}, u(x), v(x))dx \qquad (3-8)$$

with respect to an arbitrary parametrization $x \mapsto \gamma(x)$ of the loop.
    We observe the following.

3.5.1. Proposition. *For a differential form θ on M of degree k, let $\tau(\theta)$ be the degree $(k-1)$-differential form on LM defined by*

$$\tau(\theta) = \int_{S^1} ev^*(\theta). \qquad (3-9)$$

Then τ gives a morphism of complexes $A^\bullet(M) \to A^{\bullet-1}(LM)$.

Proof. This was already discussed in §1.6, in connection with the Leray spectral sequence. ∎

3.5.2. Proposition. (i) *For any 3-form ν on M, β is a closed 2-form on LM. If ν is exact, then so is β.*
    (ii) *If the class of ν in $H^3(M, \mathbb{R})$ is integral, the class of β in $H^2(LM, \mathbb{R})$ is also integral.*
    (iii) *For any 3-form ν on M, the 2-form β on LM is $Diff^+(S^1)$-basic.*

Proof. As every 3-form on a 3-fold is closed, (i) follows from Proposition 3.5.1. To prove (ii), one notes that there is a similar map $H^3(M, \mathbb{Z}) \to H^2(LM, \mathbb{Z})$. To see this, one uses the pull-back map $ev^* : H^3(M, \mathbb{Z}) \to H^3(S^1 \times LM, \mathbb{Z})$, and the Künneth decomposition

$$H^3(S^1 \times LM, \mathbb{Z}) = H^3(LM, \mathbb{Z}) \oplus [H^2(LM, \mathbb{Z}) \otimes H^1(S^1, \mathbb{Z})]$$
$$\xrightarrow{\sim} H^3(LM, \mathbb{Z}) \oplus H^2(LM, \mathbb{Z})$$

which gives the integration map $H^3(S^1 \times LM, \mathbb{Z}) \to H^2(LM, \mathbb{Z})$. The map on singular cohomology is compatible with the map on de Rham cohomology, showing that the integrality of the cohomology class $[\nu]$ implies the integrality of $[\beta]$. To prove (iii), since β is closed, one need only observe that $i(f \cdot \frac{d\gamma}{dx}) \cdot \beta = 0$. This is clear from (3-8), since the two tangent vectors $f\frac{d\gamma}{dx}$ and $\frac{d\gamma}{dx}$ are parallel at every point of $\gamma(x)$. ∎

Since β is $Diff^+(S^1)$-basic, we must pass to the quotient of $\hat{X}$ by $Diff^+(S^1)$ in order to arrive at a non-degenerate 2-form. We obtain

3.5.3. Corollary. *Assume that the 3-form ν on M is nowhere vanishing. Then the restriction of β to the open set $\hat{X}$ of smooth immersions with finitely many singularities descends to a 2-form on the space of singular knots $\hat{Y}$, which we will also denote by β. The 2-form β on $\hat{Y}$ is* weakly symplectic, *i.e., it induces an injection of $T_C \hat{Y}$ into $T_C^* \hat{Y}$ with dense image.*

Proof. We may assume that the volume form is obtained from a riemannian metric on M. We observe that if we parametrize the singular knot C by a multiple x of arc-length, and we choose a normal vector field u of length 1, then the 2-form β may be written

$$\beta(f_1(x)u + g_1(x)J \cdot u, f_2(x)u + g_2(x)J \cdot u) = l(C) \cdot \int_0^1 (f_1 g_2 - g_1 f_2)dx,$$

which has the stated properties. ∎

If M is endowed with a riemannian metric, a tangent vector to $\hat{Y}$ at γ is given by a vector field $x \mapsto u(x)$, where $u(x)$ is orthogonal to $\frac{d\gamma}{dx}$. If ν is the associated volume form on M, we now have the formula

$$\beta(u, v) = \int_\gamma A(u, v)ds = \int_\gamma (J \cdot u, v)ds \qquad (3 - 10)$$

where s is the arc-length parameter, A denotes the (oriented) surface area of the plane parallelogram with sides u and V, and J is the complex structure defined in §4.

This symplectic form has been used by Marsden and Weinstein [M-W] in their study of vortex filaments; they deduce it from an embedding of the space of curves ("filaments") into the dual of the Lie algebra $\mathfrak{g}$ of divergence-free vector fields on M. Further discussion of this embedding will take place in §7.

Because β is symplectic only in a weak sense, one cannot define the hamiltonian vector field X_f for an arbitrary smooth function f on $\hat{Y}$. However, if f is super-smooth (cf. Definition 3.1.4), and if one chooses locally a riemannian metric for which ν is the volume form, one may define X_f as $-l(\gamma)^{-1} \cdot J(df)$, where df is viewed as a field of tangent vectors. Indeed, we have

$$\beta(\xi, l(\gamma)^{-1}J(df)) = \int_0^1 A(\xi, l(\gamma)^{-1}J(df))\, dx = \int_0^1 (\xi \cdot df)dx = \xi(f),$$

for ξ a vector field along γ, s the arc-length parameter, and $x = \frac{s}{l(\gamma)}$. Then one can define the Poisson bracket $\{f, g\}$ of f with any smooth function g. If g is also super-smooth, then $\{f, g\}$ is super-smooth. We have the following formula

$$\{f, g\}_C = -l(\gamma)^{-2} \int_C \nu(\frac{d\gamma}{dx}, df, dg) dx. \qquad (3-11)$$

The compatibility of the symplectic form with the complex structure studied in §4 is expressed by the following:

3.5.4. Theorem. *Let M be a riemannian manifold, let J be the (formally integrable) almost-complex structure on $\hat{Y}$ discussed in §4, and let β be the symplectic form on $\hat{Y}$ given by Corollary 3.5.3, for the associated volume form ν on M. Then β is of type $(1,1)$ and the anti-holomorphic involution ι of $\hat{Y}$ transforms β into $-\beta$.*

Proof. This simply means that $\beta(J \cdot v, J \cdot w) = \beta(v, w)$, which is obvious. ∎

3.5.5. Theorem. *Assume M is compact and $\int_M \nu \in \mathbb{Z}$. Then the cohomology class of the 2-form β on $\hat{Y}$ is integral.*

Proof. This is analogous to the proof of Proposition 3.5.2. Consider the evaluation mapping $ev : S^1 \times \hat{X} \to M$ obtained by restriction of ev to the open subset $\hat{X}$ of LM. There is a natural smooth action of $Diff^+(S^1)$ on $S^1 \times \hat{Y}$, given by $f \cdot (x, \gamma) = (f^{-1}(x), \gamma \circ f)$ (diagonal action). We have $ev(f \cdot (x, \gamma)) = ev(x, \gamma)$ for any $f \in Diff^+(S^1)$. Therefore ev factors through a smooth mapping $e : (S^1 \times \hat{X})/Diff^+(S^1) \to M$. We note that the quotient space $(S^1 \times \hat{X})/Diff^+(S^1)$ is a smooth manifold, and that there is a S^1-bundle $(S^1 \times \hat{X})/Diff^+(S^1) \overset{q}{\to} \hat{Y}$. We then can describe the 2-form β as $\beta = \int_q e^* \nu$. As these cohomological operations are defined in integer-valued cohomology, the theorem follows. ∎

According to Chapter 2, this implies that there exists a line bundle with connection over $\hat{Y}$, whose curvature is exactly equal to $-2\pi\sqrt{-1} \cdot \beta$. It is always much better to explicitly construct a line bundle rather than to use an abstract existence theorem. Such an explicit construction (using the abstract machinery of this book) will be given in Chapter 6.

The symplectic manifold $\hat{Y}$ has some remarkable lagrangian submanifolds. Recall that if β is a symplectic form on a vector space E, a subspace

F of E is called *isotropic* if $F \subseteq F^{\perp}$, *coisotropic* if $F^{\perp} \subseteq F$, lagrangian if it is maximal isotropic. A submanifold Λ of a symplectic manifold (N, ω) is said to be isotropic (resp. lagrangian) if its tangent space at each point is isotropic (resp. lagrangian). Λ is said to be *involutive* if its tangent space at each point is coisotropic. Λ is involutive if and only if the ideal of $C^{\infty}(N)$, consisting of functions which vanish on Λ, is stable under Poisson brackets. If Λ is involutive, we have the subbundle $T_x \Lambda^{\perp}$ of the tangent bundle $T\Lambda$, which is an integrable distribution of tangent spaces (at least for N finite-dimensional). This is the *null foliation* of Λ.

3.5.6. Proposition. *Let $\Sigma \subset M$ be a smooth surface. Then $\Lambda_{\Sigma} := \{C \in \hat{Y}, C \subset \Sigma\}$ is a lagrangian submanifold of $\hat{Y}$.*

Proof. Let $\gamma : S^1 \to \Sigma$ be a parametrization of $C \in \Lambda_{\Sigma}$. Let $v(x)$ be the normal vector to C inside Σ; then $T_C \Lambda_{\Sigma}$ is the space of normal vector fields to C which are parallel to v at each point. It is equal to its orthogonal subspace, hence is lagrangian. ∎

This should be significant in connection with geometric quantization for $\hat{Y}$.

Codimension one foliations on M lead to involutive submanifolds of $\hat{Y}$.

3.5.7. Proposition. *Let $\mathcal{F}$ be a (possibly non-integrable) distribution of tangent planes of dimension 2 on M. Let V be the subspace of $\hat{Y}$ consisting of knots C which are tangent to $\mathcal{F}$ at every point. Let $C \in V$. Then, in each of the two cases (a) and (b) described below, V is a submanifold at C, the tangent space to V is coisotropic, and the space $T_C V / T_C V^{\perp}$ is two-dimensional.*

(a) $\mathcal{F}$ *is integrable in a neighborhood of C.*

(b) $M = \mathbb{R}^3$ *with the standard volume form, and $\mathcal{F}$ is defined by the contact form $\alpha = x_1 dx_2 + dx_3$, where the x_i are the usual coordinates on $\mathbb{R}^3$.*

Proof. In the first case, replacing, if necessary, C and a tubular neighborhood of it by a double cover, we may assume that $\mathcal{F}$ is orientable in a neighborhood of C, hence is defined by a smooth function f with nowhere vanishing differential. Let C be parametrized by $x \to \gamma(x)$, for x a multiple of the arc-length, and let $v(x)$ be the unit normal vector to C inside the leaf $f^{-1}(a)$ which contains it. Then $T_C V$ consists of normal vector fields of the type $x \to h(x)v(x) + C \cdot \frac{\vec{grad} f}{\|grad f\|^2}$, for some constant C. This is coisotropic, and the orthogonal subspace consists of all normal vector fields of the type

$x \to h(x)v(x)$, where the function h satisfies $\int_0^1 \frac{h}{\|grad f\|} dx = 0$. The quotient $T_C V / T_C V^\perp$ is two-dimensional.

The leaves of the codimension two null-foliation on the involutive manifold V are (locally) the level sets of a pair of smooth functions; one is f itself, the other one any function g with differential equal to v. If C bounds a surface inside the leaf which contains it, a choice of g is: $g(C') = area\ (S)$, for S a surface inside the leaf with boundary C'.

In case (b), the tangent space to V consists of the normal vector fields $u(x)$ along C such that

$$(D_u \alpha \cdot \frac{d\gamma}{dx}) + (\alpha \cdot \frac{du}{dx}) = 0. \qquad (3-12)$$

Any u parallel to v satisfies this equation, since $\alpha \cdot \frac{d\gamma}{dx} = 0$, hence, taking derivatives with respect to x, $\alpha \cdot \frac{dv}{dx} = 0$. So we look for a solution of (3-12) of the form $v(x) = h(x)\alpha_{\gamma(x)}$; in terms of h, (3-12) becomes: $h(x)(D_\alpha \alpha \cdot \frac{d\gamma}{dx}) + (\alpha \cdot \alpha) \cdot \frac{dh}{dx} = 0$. The necessary and sufficient condition for the existence of a periodic solution is that the integral $\int_0^1 \frac{D_\alpha \alpha \cdot \frac{d\gamma}{dx}}{\alpha \cdot \alpha} dx$ vanish. Since $\alpha = x_1 dx_2 + dx_3$, we have $D_\alpha \alpha = 0$, since α is independent of x_2 and x_3. So $h = const.$ is a solution of (3-12). This proves the proposition in this case. ∎

Let us now specialize to the case $M = \mathbb{R}^3$ and study some remarkable properties of the hamiltonian vector field X_l for the length functional l. It is easy to see that the differential dl is given by $dl_\gamma = -l(\gamma)^{-1} \cdot \frac{d^2\gamma}{dx^2}$, for x a multiple of arc-length. Then we have

$$X_l = -l(\gamma)^{-1} \cdot J(dl) = l(\gamma)^{-2} \cdot J(\frac{d^2\gamma}{dx^2}) = \kappa \cdot J(\vec{N}) = \kappa \cdot \vec{B} \qquad (3-13)$$

where $\vec{B}$ is the unit *binormal vector*. The flow of C along the vector field X_l is given by the following evolution equation, where t is time and x is the parameter on the curve (which, at all times, is a multiple of arc-length).

$$\frac{\partial \gamma}{\partial t} = \kappa(t, x) \frac{\partial \gamma}{\partial x} \times \frac{\partial^2 \gamma}{\partial x^2}. \qquad (3-14)$$

Now assume for simplicity that C has length 1 (which can be achieved by a dilation). Taking the partial derivative of (3-14) with respect to x, we obtain

$$\frac{\partial \vec{T}}{\partial t} = \vec{T} \times \frac{\partial^2 \vec{T}}{\partial x^2}. \qquad (3-15)$$

This evolution equation for $\vec{T} = \frac{1}{l(C)} \frac{\partial \gamma}{\partial x}$ is known as the *continuous Heisenberg chain*. Equation (3-15) was derived by Hasimoto [Has] and Lakshmanan [Lak]. As shown by Hasimoto [Has], this can, in turn, be transformed into evolution equations for the curvature and the torsion of the curve C_t parametrized by x. We use a method of I. Anderson to write down these equations. Denote the partial derivative of a function with respect to x by f', and the partial derivative with respect to t by $f^{\bullet}$. Use the Frénet equations: $\frac{d\vec{T}}{dx} = \kappa \vec{N}$, $\frac{d\vec{N}}{dx} = -\kappa \vec{T} + \tau \vec{B}$, and $\frac{d\vec{B}}{dx} = -\tau \vec{N}$. (3-14) gives $\frac{\partial \vec{T}}{\partial t} = \kappa' \cdot \vec{B} - \kappa \tau \cdot \vec{N}$. Taking the derivative with respect to x, we obtain

$$\frac{\partial(\kappa \vec{N})}{\partial t} = \frac{\partial \vec{T}}{\partial t \partial x} = \kappa'' \vec{B} - 2\kappa' \tau \vec{N} - \kappa \tau' \vec{N} + \kappa^2 \tau \vec{T} - \kappa \tau^2 \vec{B}.$$

The component in the direction of $\vec{N}$ gives

$$\kappa^{\bullet} = \frac{\partial \kappa}{\partial t} = -2\kappa' \tau - \kappa \tau'. \tag{3-16}$$

The component in the plane spanned by $\vec{T}$ and $\vec{B}$ gives: $\frac{\partial \vec{N}}{\partial t} = \kappa \tau \vec{T} + \kappa''/\kappa \vec{B} - \tau^2 \vec{B}$. Taking the derivative with respect to x once again, we find

$$\frac{\partial(-\kappa \vec{T} + \tau \vec{B})}{\partial t} = (\kappa \tau)' \vec{T} + \tau \kappa^2 \vec{N} + \frac{\kappa'''}{\kappa} \vec{B} - \frac{\kappa' \kappa''}{\kappa^2} \vec{B} + \kappa''/\kappa \frac{\partial \vec{B}}{\partial x} - 2\tau \tau' \vec{B} - \tau^2 \frac{\partial \vec{B}}{\partial x}.$$

Looking at the $\vec{B}$-components of both sides, we obtain

$$-\kappa \kappa' + \frac{\partial \tau}{\partial t} = \frac{\kappa'''}{\kappa} - \frac{\kappa' \kappa''}{\kappa^2} - 2\tau \tau' = \frac{\partial}{\partial x}(\kappa''/\kappa) - \frac{\partial}{\partial x}(\tau^2),$$

which may be written

$$\tau^{\bullet} = \frac{\partial \tau}{\partial t} = \frac{\partial}{\partial x}(\kappa''/\kappa + 1/2\kappa^2 - \tau^2). \tag{3-17}$$

It follows immediately that $\int_0^1 \kappa^2(x)dx$ and $\int_0^1 \kappa \tau^2(x)dx$ are conserved quantities. In fact, Hasimoto proved that the flow is equivalent to the *nonlinear Schrödinger equation* (NLS) for the complex-valued function

$$\psi(x, t) = \kappa(x, t) \cdot exp(i \cdot \int_0^x \tau(t, u)du). \tag{3-18}$$

We give Hasimoto's proof, emphasizing the relation with ideas from gauge theory.

We first recall that if a vector bundle E on a manifold V decomposes as a direct sum $E = F \oplus G$ of two vector bundles, then any connection ∇ on E induces a connection ∇_F on F, such that

$$\nabla_F(s) = p_F(\nabla(s)),$$

for a germ s of section of F, where p_F denotes the projection of an E-valued differential form to a F-valued differential form. We apply this to the vector bundle $\gamma^* T\mathbb{R}^3_\mathbb{C}$ on S^1, which is the pull-back of the complexified tangent bundle of $\mathbb{R}^3$. It decomposes as $T_\mathbb{C} \oplus N_\mathbb{C}$, where T is the tangent bundle of S^1 and N is the normal bundle to γ. This is not yet the decomposition we want; we further decompose $N_\mathbb{C}$ as $N_h \oplus \overline{N_h}$, where N_h is the holomorphic part of $N_\mathbb{C}$, in the sense of the complex structure on the normal bundle N. In other words, N_h is the $\sqrt{-1}$-eigenspace of J, and is spanned by $\vec{N} - \sqrt{-1} \cdot \vec{B}$. The decomposition of $\gamma^* T\mathbb{R}^3_\mathbb{C}$ we want is $\gamma^* TM_\mathbb{C} = F \oplus G$, with $F = T_\mathbb{C} \oplus \overline{N_h}$ and $G = N_h$. It is a decomposition which is orthogonal with respect to the hermitian form on $\gamma^* T\mathbb{R}^3_\mathbb{C}$.

We use the trivial connection on the tangent bundle of $\mathbb{R}^3$ and on its pull-back. The derivative $\nabla_{F, \frac{d}{dx}}$ for the connection ∇_F on F will simply be denoted by $s \mapsto s'$, so as to keep the notations reader-friendly. The vector fields $\vec{T}$ and $\vec{N} + \sqrt{-1} \cdot \vec{B}$ form a basis of the vector bundle F. We have

$$(\vec{N} + \sqrt{-1} \cdot \vec{B})' = -\sqrt{-1} \cdot \tau (\vec{N} + \sqrt{-1} \cdot B) - \kappa \vec{T}$$

and

$$\vec{T}' = \frac{1}{2}\kappa (\vec{N} + \sqrt{-1} \cdot \vec{B}).$$

In other words, the matrix of the connection in the basis $(\vec{N} + \sqrt{-1}\vec{B}, \vec{T})$ is

$$\begin{pmatrix} -\sqrt{-1} \cdot \tau & \frac{\kappa}{2} \\ -\kappa & 0 \end{pmatrix}.$$

This is not a pleasant matrix form for a connection. One wants to improve it by a change of basis for the vector bundle (which gives a *gauge transformation* for the connection matrix). Of course one is most happy with a connection that is represented by a diagonal matrix, but in our case it is easier to put a connection matrix into the off-diagonal form $\begin{pmatrix} 0 & b \\ c & 0 \end{pmatrix}$. Such an off-diagonal form allows for a rather easy solution of the matrix-valued differential equation $\frac{dM}{dx} + M \cdot A = 0$; one obtains ordinary second order differential equations for the entries of M.

In our case, we have to introduce the new basis $(\vec{u}, \vec{T})$ of F, where $\vec{u} = \lambda \cdot (\vec{N} + \sqrt{-1} \cdot \vec{B})$. The function $\lambda(x)$ must be chosen so that $\frac{d\lambda}{dx} - \sqrt{-1}\tau\lambda = 0$, hence $\lambda(x) = exp(\int_0^x \sqrt{-1} \cdot \tau(s)ds)$. In this new basis, setting $\psi(x) = \kappa \cdot exp(\int_0^x \sqrt{-1} \cdot \tau(s)ds)$, we have

$$\begin{cases} \vec{u}' = -\psi \cdot \vec{T} \\ \vec{T}' = 1/2\overline{\psi}\vec{u} \end{cases}.$$

Thus the function ψ of Hasimoto appears in the connection matrix.

We now introduce the time evolution of the curve γ according to the vector field $\kappa\vec{B}$. We will study directly the evolution of the function $\psi(x, t)$.

3.5.8. Theorem. (*Hasimoto* [**Has**]) *There exists a function $A(t)$ such that the function $\Psi(x, t) = exp(\sqrt{-1} \cdot \int_0^t A(y)dy) \cdot \psi(x, t)$ satisfies the non-linear Schrödinger equation*

$$-\sqrt{-1}\Psi^{\bullet} = \Psi'' + 1/2(|\Psi|^2 + C) \cdot \Psi \qquad (NLS)$$

for some constant C.

Proof. We use the formulas for $\kappa^{\bullet}$ and $\tau^{\bullet}$. We have

$$\frac{\psi^{\bullet}}{\psi} = \frac{\kappa^{\bullet}}{\kappa} + \sqrt{-1} \cdot \int_0^x \tau^{\bullet}(s)ds$$

$$= -2\frac{\kappa'}{\kappa}\tau - \tau' + \sqrt{-1}\int_0^x \frac{\partial}{\partial s}(\frac{\kappa''}{\kappa} + 1/2\kappa^2 - \tau^2)ds$$

$$= -2\frac{\kappa'}{\kappa}\tau - \tau' + \sqrt{-1} \cdot (\frac{\kappa''}{\kappa} + 1/2\kappa^2 - \tau^2 + A(t))$$

where $A(t)$ is the value of $-\frac{\kappa''}{\kappa} + 1/2\kappa^2 - \tau^2$ at the origin of the curve (i.e., for $x = 0$). On the other hand, we compute that

$$\frac{\psi''}{\psi} = \frac{\kappa''}{\kappa} - \tau^2 + \sqrt{-1} \cdot \tau' + 2\sqrt{-1} \cdot \tau\frac{\kappa'}{\kappa}$$

and $|\psi|^2 = \kappa^2$.

From this we obtain

$$-\sqrt{-1}\psi^{\bullet} = \psi'' + 1/2|\psi|^2 \cdot \psi - A(t)\psi,$$

which is the NLS equation, except for the fact that $A(t)$ depends on t. The dependence on t is erased by multiplication of ψ by the t-dependent phase factor $exp(\sqrt{-1} \cdot \int_0^t A(y)dy)$. ∎

The importance of this theorem is due to the fact that Zakharov and Shabat [Z-S] used the inverse scattering method to show that the NLS equation is completely integrable. They also show that there is an infinite number of effectively computable integral invariants for our flow, which may be written in the form: $\int_0^1 P(\kappa, \frac{d\kappa}{dx}, \dots, \tau, \frac{d\tau}{dx}, \dots) dx$, where P is a polynomial in κ, τ and their derivatives. The geometry of the corresponding commuting hamiltonian systems has been studied in [L-P]. Langer and Perline also study the evolution of filaments which correspond to solitons for the NLS equation.

There are other interesting functionals on knots which have been investigated recently, namely, various notions of "energy of a knot". We refer the reader to [F-H] and [O'H].

3.6. The riemannian structure

In this section, we consider an oriented three-manifold M equipped with a riemannian structure g. We will introduce a riemannian structure on the space $\hat{Y}$ of (singular) oriented knots, and we will study the Levi-Civita connection and the Riemann curvature tensor, as well as give an example of a geodesic in $\hat{Y}$. Recall that we already have the following structures on $\hat{Y}$:

(1) The almost complex structure J of §4, which is formally integrable.

(2) The symplectic form β on $\hat{Y}$, which is given in terms of the volume form ν on M by

$$\beta_\gamma(v, w) = \int_0^1 \nu(\frac{d\gamma}{dx}, v(x), w(x)) dx.$$

This 2-form is of type $(1, 1)$ with respect to the complex structure J (see §5).

We would like to fit these structures into a sort of kählerian structure on $\hat{Y}$. We take this to mean that there is a positive-definite hermitian form h on the tangent bundle $T\hat{Y}$ such that $\Im(h) = -\beta$. Thus h is complex-linear in the first variable and antilinear in the second variable. The hermitian form h is determined by its real part $\langle \ , \ \rangle$, which must be positive-definite and satisfy

$$\langle v, w \rangle = \beta(v, J \cdot w).$$

We therefore define a riemannian structure $\langle \ , \ \rangle$ on $\hat{Y}$ by

$$\langle v, w \rangle = \int_0^1 g(v(x), w(x)) \cdot \|\frac{d\gamma}{dx}\| dx = \int_\gamma g(v, w) \cdot ds, \qquad (3-19)$$

for two tangent vectors v and w to $\hat{Y}$ at γ, given by fields of tangent vectors along γ which are perpendicular to γ at every point. The same formula defines a riemannian structure $\langle \, , \, \rangle$ on LM.

3.6.1. Proposition. (1) *The symmetric bilinear form* $\langle \, , \, \rangle$ *on* $\hat{X}$ *(resp.* $\hat{Y}$*) is positive-definite and defines an injection of* $T_\gamma \hat{X}$ *(resp.* $T_\gamma \hat{Y}$*) into its continuous dual, with dense image.*

(2) *We have the formula*

$$\langle v, w \rangle = \beta(v, J \cdot w), \qquad (3-20)$$

relating the riemannian, symplectic and almost complex structures on $\hat{Y}$.

(3) *The projection map of riemannian manifolds* $\pi : \hat{X} \to \hat{Y}$ *is a* <u>*riemannian submersion*</u>. *This means (see* [O'N]*) that for every* $\gamma \in \hat{X}$, *the differential* $d\pi_\gamma$ *induces an isometry between the horizontal tangent space (i.e., the orthogonal space* N_γ *to the kernel of* $d\pi_\gamma$*) and the tangent space to* $\hat{Y}$.

We note that this riemannian metric is familiar from the theory of embeddings of riemannian manifolds, and of minimal submanifolds (see [**Law**]).

In this section, we will explore the Levi-Civita connection and the Riemann curvature tensor of $\hat{Y}$ in the flat case $M = \mathbb{R}^3$. We will exploit the fact that π is a riemannian submersion. According to [O'N], this gives strong relations between the riemannian geometry of both manifolds; on the other hand, $\hat{X}$ is an open subset of the vector space $L\mathbb{R}^3 = C^\infty(S^1, \mathbb{R}^3)$, and certain computations are easier on $\hat{X}$. We will use the dot product notation $v \cdot w$ for the inner product of tangent vectors on $\mathbb{R}^3$. Recall from §3 that a tangent vector $v(x)$ to $\hat{X}$ at γ is vertical if and only if $v(x)$ is parallel to $\frac{d\gamma}{dx}$ for every x, and is horizontal if and only if $v(x)$ is perpendicular to $\frac{d\gamma}{dx}$ for every x.

The Levi-Civita connection on a riemannian manifold is the unique connection on the tangent bundle which satisfies:

(a) ∇ is compatible with the riemannian metric.

(b) The *torsion tensor* is zero, that is: $\nabla_\xi \eta - \nabla_\eta \xi = [\xi, \eta]$ for any vector fields ξ and η.

We will use the well-known formula for the *Christoffel symbol* $\langle \nabla_\xi \eta, v \rangle$, where ξ, η and v are vector fields on a riemannian manifold

$$2\langle \nabla_\xi \eta, v \rangle = \xi \langle \eta, v \rangle + \eta \langle v, \xi \rangle - v \langle \xi, \eta \rangle - \langle \xi, [\eta, v] \rangle + \langle \eta, [v, \xi] \rangle + \langle v, [\xi, \eta] \rangle.$$
$$(3-21)$$

In his fundamental study of riemannian submersions [O'N], O'Neill introduces several tensors. First we fix notations by calling $\tilde{\nabla}$ the Levi-Civita connection on $\hat{X}$ and ∇ the Levi-Civita connection on $\hat{Y}$. Similarly, $\tilde{R}$ (resp. R) will be the Riemann curvature tensor of $\hat{X}$ (resp. $\hat{Y}$). For a vector field ξ on $\hat{X}$, we denote by ξ_H its horizontal component (which belongs to $ker\ (d\pi)$), and by ξ_V its vertical component (which is perpendicular to $ker\ (d\pi)$). The first tensor T of O'Neill is a tensor on $\hat{X}$, which to a pair (ξ, η) of vector fields on $\hat{X}$, associates a vector field $T_\xi(\eta)$, where

$$T_\xi(\eta) = (\tilde{\nabla}_{\xi_V}(\eta_V))_H + (\tilde{\nabla}_{\xi_V}(\eta_H))_V. \qquad (3-22)$$

3.6.2. Lemma. [O'N] *The tensor T satisfies* (1) T_ξ *is zero if ξ is horizontal.*

(2) *For given ξ, the endomorphism of $T_\gamma \hat{X}$: $T_\xi : \eta \mapsto T_\xi \eta$ is skew-symmetric; it maps vertical tangent vectors to horizontal vectors and conversely.*

(3) *For vertical vector fields ξ and η, one has $T_\xi \eta = T_\eta \xi$.*

Proof. (1) is immediate. To prove (2), it is enough to show that $\langle T_\xi \eta, v \rangle = -\langle T_\xi v, \eta \rangle$ for ξ vertical, η vertical and v horizontal. But we have

$$\langle T_\xi \eta, v \rangle = \langle \tilde{\nabla}_\xi \eta, v \rangle = -\langle \eta, \tilde{\nabla}_\xi v \rangle = -\langle \eta, T_\xi v \rangle.$$

To prove (3), one notes that for ξ and η vertical vector fields, $\tilde{\nabla}_\xi \eta - \tilde{\nabla}_\eta \xi = [\xi, \eta]$ is vertical, so its horizontal component $T_\xi \eta - T_\eta \xi$ is zero. ∎

We will now compute the tensor T. First we compute $\tilde{\nabla}_\xi \eta$ when ξ and η are both vertical vector fields. So we may consider ξ of the form $\xi_\gamma = d\gamma(f(x) \cdot \frac{d}{dx})$ and similarly $\eta_\gamma = d\gamma(g(x) \cdot \frac{d}{dx})$, for f and g smooth functions on S^1. We use formula (3-21) for $\langle \tilde{\nabla}_\xi \eta, v \rangle$.

In case v is horizontal and $Diff^+(S^1)$-invariant (hence is the horizontal lift of a vector field on $\hat{Y}$), the vector fields $[\eta, v]$ and $[\xi, v]$ vanish, and consequently all terms in (3-21) vanish except for the third term. We have $\langle \xi, \eta \rangle = \int_0^1 f(x)g(x)\|\frac{d\gamma}{dx}\|^3 dx$, hence $v \cdot \langle \xi, \eta \rangle = 3 \int_0^1 f(x)g(x)(\frac{d\gamma}{dx} \cdot \frac{dv}{dx})\|\frac{d\gamma}{dx}\| dx$. Since v is vertical, we have $v \cdot \frac{d\gamma}{dx} = 0$, hence $\frac{d\gamma}{dx} \cdot \frac{dv}{dx} = -v \cdot \frac{d^2\gamma}{dx^2}$. Thus $v \cdot \langle \xi, \eta \rangle = -3 \int_0^1 fg\ (v(x) \cdot \frac{d^2\gamma}{dx^2})\|\frac{d\gamma}{dx}\| dx$.

Consequently we find

3.6.3. Lemma.

$$T_\xi \eta(x) = \frac{3}{2} \cdot f(x)g(x)(\frac{d^2\gamma}{dx^2})^\perp,$$

where ξ (resp. η) is the vector field on $\hat{X}$ induced by $f(x)\frac{d}{dx}$ (resp. $g(x)\frac{d}{dx}$), and where the superscript $\perp$ denotes the component in the normal plane of a field of vectors along γ.

The second tensor field used by O'Neill is called A; for ξ and η vector fields on $\hat{X}$, we have the vector field

$$A_\xi\eta = (\tilde{\nabla}_{\xi_H}(\eta_H))_V + (\tilde{\nabla}_{\xi_H}(\eta_V))_H. \qquad (3-23)$$

3.6.4. Lemma. [O'N] (1) *$A_\xi\eta$ is equal to 0 unless ξ is horizontal.*

(2) *A_ξ is a skew-symmetric endomorphism of the tangent bundle, i.e.,*

$$\langle A_\xi\eta, v\rangle = -\langle A_\xi v, \eta\rangle.$$

(3) *If ξ and η are both horizontal, $A_\xi\eta$ is equal to $\frac{1}{2}\cdot[\xi,\eta]_V$. The vertical component of $[\xi,\eta]$ is equal to $-d\gamma(\Theta(\xi,\eta))$, where Θ is the curvature of the principal $Diff^+(S^1)$-bundle $\hat{X} \to \hat{Y}$.*

Proof. (1) is obvious. To prove (3), we may assume that the vector fields ξ and η on $\hat{X}$ are basic. Let v be the vertical vector field induced by an element of $Diff^+(S^1)$. Then we have, from (3-21),

$$2\langle A_\xi\eta, v\rangle = 2\langle \tilde{\nabla}_\xi\eta, v\rangle$$
$$= -\langle\xi,[\eta,v]\rangle + \langle\eta,[v,\xi]\rangle + \langle v,[\xi,\eta]\rangle = \langle v,[\xi,\eta]\rangle$$

since $[v,\xi] = [v,\eta] = 0$. This is equal to $-d\gamma(\Theta(\xi,\eta))$ as we showed in §3. For (2), we may as well assume that ξ and η are basic and v is vertical. Then we have

$$2\langle A_\xi v, \eta\rangle = 2\langle \tilde{\nabla}_\xi v, \eta\rangle = \langle v, [\eta,\xi]\rangle = -2\langle A_\xi\eta, v\rangle$$

using $A_\xi\eta = 1/2\cdot[\xi,\eta]_V$. ∎

The curvature Θ of the $Diff^+(S^1)$-bundle $\hat{X} \to \hat{Y}$ was computed in Proposition 3.3.2, so we find:

3.6.5. Lemma. *We have $A_v w = 1/2(v\cdot\frac{dw}{dx} - w\cdot\frac{dv}{dx})\|\frac{d\gamma}{dx}\|^{-2}\frac{d\gamma}{dx}$, for v and w horizontal vector fields.*

We next need to compute $A_v\xi$, for v horizontal and ξ vertical, say $\xi = f(x)\frac{d\gamma}{dx}$. We will use the fact that A_v is a skew-symmetric operator, which gives $\langle A_v\xi, w\rangle = -\langle A_v w, \xi\rangle$. We have

$$\langle A_v\xi, w\rangle = -1/2 \int_0^1 (v \cdot \frac{dw}{dx} - w \cdot \frac{dv}{dx})\|\frac{d\gamma}{dx}\| f(x) dx.$$

Thus we obtain:

3.6.6. Lemma. *Assume that $\|\frac{d\gamma}{dx}\|$ is constant (i.e., the parameter x is a multiple of the arc-length). Then, for a horizontal vector field v and for ξ the vertical vector field given by $\xi = f(x)\frac{d}{dx}$, we have*

$$A_v\xi = f(x) \cdot (\frac{dv}{dx})^\perp + \frac{1}{2}\frac{df}{dx}v.$$

We also need to compute $(\tilde{\nabla}_\xi v)_H$, for v horizontal and $\xi = f(x)\frac{d\gamma}{dx}$ vertical. If v is $Diff^+(S^1)$-invariant, then $[\xi, v] = 0$ and $\tilde{\nabla}_\xi v = \tilde{\nabla}_v\xi$, which is given by Lemma 3.6.6. This gives the following

3.6.7. Lemma. *Let v be a basic vector field, and let ξ be a vertical vector field. Let $\gamma \in \hat{Z}$. Then $(\tilde{\nabla}_\xi v)_H$ at the loop γ is equal to $A_v\xi = f(x) \cdot (\frac{dv}{dx})^\perp + \frac{1}{2}\frac{df}{dx}v$, where $\xi_\gamma(x) = f(x)\frac{d\gamma}{dx}$.*

Next we compute the Christoffel symbols $\langle \tilde{\nabla}_u v, w\rangle$, for constant coefficient vector fields on $\hat{X}$ induced by elements of $C^\infty(S^1, \mathbb{R}^3)$. Again we use (3-21), where now all brackets of the relevant vector fields are 0. We then get

$$2\langle\tilde{\nabla}_u v, w\rangle = \int_0^1 (v \cdot w)(\frac{du}{dx} \cdot \frac{d\gamma}{dx}) \cdot \|\frac{d\gamma}{dx}\|^{-1} dx$$

$$+ \int_0^1 (u \cdot w)(\frac{dv}{dx} \cdot \frac{d\gamma}{dx}) \cdot \|\frac{d\gamma}{dx}\|^{-1} dx - \int_0^1 (u \cdot v)(\frac{dw}{dx} \cdot \frac{d\gamma}{dx}) \cdot \|\frac{d\gamma}{dx}\|^{-1} dx$$

Integration by parts enables us to determine $\tilde{\nabla}_u v$.

3.6.8. Proposition. *Assume that $\|\frac{d\gamma}{dx}\|$ is constant. Let u and v be smooth functions $S^1 \to \mathbb{R}^3$, viewed as vector fields on $\hat{X}$. Then we have, at a point of $\hat{Z}$,*

$$2\tilde{\nabla}_u v = \|\frac{d\gamma}{dx}\|^{-2}\left((\frac{du}{dx}\cdot\frac{d\gamma}{dx})v + (\frac{dv}{dx}\cdot\frac{d\gamma}{dx})u + (\frac{du}{dx}\cdot v)\frac{d\gamma}{dx} + (\frac{dv}{dx}\cdot u)\frac{d\gamma}{dx} + (u\cdot v)\frac{d^2\gamma}{dx^2}\right)$$

We now want to compute the covariant derivative $\nabla_{u_H} v_H$, for u and v as in Proposition 3.6.8, and u_H, v_H their horizontal components. We note that u_H and v_H are in fact basic vector fields on $\hat{X}$. It is clear that $\nabla_{u_H} v_H$ is the horizontal part of $\tilde{\nabla}_{u_H} v_H$. We will again restrict ourselves to $\gamma \in \hat{Z}$, i.e., $\|\frac{d\gamma}{dx}\|$ constant. The vertical components of u and v are $u_V = u_L \|\frac{d\gamma}{dx}\|^{-1} \cdot \frac{d\gamma}{dx}$, for $u_L = \|\frac{d\gamma}{dx}\|^{-1}(u \cdot \frac{d\gamma}{dx})$ the longitudinal component of u, and $v_V = v_L \|\frac{d\gamma}{dx}\|^{-1} \cdot \frac{d\gamma}{dx}$.

We have $u_H = u - u_V$, $v_H = v - v_V$, hence we obtain $\tilde{\nabla}_{u_H} v_H = \tilde{\nabla}_u v - \tilde{\nabla}_{u_H} v_V - \tilde{\nabla}_{v_H} v_V - \tilde{\nabla}_{u_V} v_V$. We then take the horizontal parts, using the equality $\frac{du_L}{dx} = \|\frac{d\gamma}{dx}\|^{-1}(\frac{du}{dx} \cdot \frac{d\gamma}{dx} + u \cdot \frac{d^2\gamma}{dx^2})$ which we established in §3.

3.6.9. Proposition. *For u and v constant vector fields on $\hat{X}$, we have the equality of vector fields on $\hat{Y}$:*

$$\nabla_{u_H} v_H = \frac{1}{2} l^{-2}(u \cdot v)\frac{d^2\gamma}{dx^2} - l^{-1} u_L (\frac{dv}{dx})^\perp - l^{-1} v_L (\frac{du}{dx})^\perp$$
$$+ \frac{1}{2} l^{-2} u_L v_L \|\frac{d\gamma}{dx}\|^{-2} \frac{d^2\gamma}{dx^2} - \frac{1}{2}\kappa u_N v^\perp - \frac{1}{2}\kappa v_N u^\perp,$$

where $u_N = u \cdot \vec{N}$ is the component of u along the normal vector $\vec{N}$.

In principle, Proposition 3.6.9 allows us to compute the Levi-Civita connection on $\hat{Y}$ and to write down the differential equation satisfied by a geodesic $t \mapsto \gamma_t$. Note that the existence of a geodesic $\gamma(x,t)$ with given initial data γ_0 and $(\frac{d\gamma}{dt})_{t=0}$ is guaranteed by Proposition 3.6.9 and Cauchy's theorem for ordinary differential equations. It would be interesting to study these geodesics geometrically, and in particular to understand the shape of the surface they span in 3-space. We are only able to describe a very special geodesic. It is natural to think of a family $t \mapsto \gamma_t$ of knots as a map $(x,t) \mapsto \gamma(x,t)$. The parametrization independent equation characterizing a geodesic is then, as usual in riemannian geometry,

$$\nabla_v v = \lambda v, \quad \text{for} \quad v = (\frac{\partial\gamma}{\partial t})^\perp. \tag{3-24}$$

Let us consider a simple example, where $\gamma(x,0) = \cos(2\pi x)\vec{j} + \sin(2\pi x)\vec{k}$ (a vertical circle), and $\frac{\partial\gamma}{\partial t} = \vec{i}$ at $t = 0$. Since the initial data are rotationally symmetric about the x-axis, so will be the solution to (3-24). Hence, for each t, γ_t will be a circle parametrized by a multiple of arc-length.

Since we have the freedom to change the time parameter as we please, we may as well take the x coordinate of the circle as parameter; so γ_t is the circle of center $(t,0,0)$, say, of radius $f(t)$. Then if $\vec{u}(x,t)$ is the outward pointing normal vector to this circle, we have $\vec{v} = \vec{i} + \frac{df}{dt}\vec{u}$. Note that $\vec{i}$ and

$\vec{u}$ are constant vector fields. Hence we may apply Proposition 3.6.9 to find

$$\nabla_{\vec{i}}\,\vec{i} = -\frac{1}{2f(t)}\vec{u},$$

$$\nabla_{\vec{u}}\,\vec{i} = \nabla_{\vec{i}}\vec{u} = \frac{1}{2f(t)}\vec{i},$$

$$\nabla_{\vec{u}}\,\vec{u} = \frac{1}{2f(t)}\vec{u}$$

It follows that $\nabla_v v = \frac{df/dt}{f(t)}\vec{i} + (\frac{d^2 f}{dt^2} + \frac{(\frac{df}{dt})^2}{2f(t)} - \frac{1}{2f(t)})\vec{u}$. This must be proportional to $v = \vec{i} + (df/dt)\vec{u}$; hence we obtain the ordinary differential equation (where f' is the derivative of f, f'' the second derivative, and so on):

$$f'' + \frac{(f')^2}{2f} - \frac{1}{2f} = \frac{(f')^2}{f},$$

or $f f'' - f'^2/2 = 1/2$. Differentiating both sides of this equation gives $f''' = 0$, hence f is a polynomial of degree ≤ 2. Since $f(0) = 1$ and $f'(0) = 0$, we have $f(t) = at^2 + 1$ for some a. The only value of a for which the equation $f f'' - f'^2/2 = 1/2$ holds is $a = 1/4$. So we have proven

3.6.10. Proposition. *Let γ_t be the circle of radius $t^2/4 + 1$, centered at the point $(t, 0, 0)$. Then the curve $t \mapsto \gamma_t$ is a geodesic of $\hat{Y}$.*

We will next compute the Riemann curvature tensor R of $\hat{Y}$. We use the following expression for $\langle R_{u,v}w, \xi \rangle$, where we denote by the same letter a vector field on $\hat{Y}$ and the corresponding basic vector field on $\hat{X}$ (see [**K-N**] or [**He**] for the Riemann curvature tensor).

$$\langle R_{u,v}w, \xi \rangle = u \cdot \langle \tilde{\nabla}_v w, \xi \rangle - v \cdot \langle \tilde{\nabla}_u w, \xi \rangle$$
$$- \langle (\tilde{\nabla}_v w)^{\perp}, (\tilde{\nabla}_u \xi)^{\perp} \rangle + \langle (\tilde{\nabla}_u w)^{\perp}, (\tilde{\nabla}_v \xi)^{\perp} \rangle$$

In this formula for the curvature tensor at a given point γ, we may now replace the basic vector fields on $\hat{X}$ by constant vector fields which take the same value at γ. We now replace u, v, w and ξ by constant vector fields which are perpendicular to $\frac{d\gamma}{dx}$ at every point of γ. We also restrict to a parameter of γ of the form $x = \frac{1}{l} \cdot s$, for l the length of γ and s the arc-length. With these new conventions, we obtain

$$2u \cdot \langle \tilde{\nabla}_v w, \xi \rangle = \int_0^1 (w \cdot \xi)(v' \cdot u') \|\gamma'\|^{-1} dx + \int_0^1 (v \cdot \xi)(w' \cdot u') \|\gamma'\|^{-1} dx$$

$$- \int_0^1 (w \cdot v)(\xi' \cdot u') \|\gamma'\|^{-1} dx - \int_0^1 (w \cdot \xi)(v' \cdot \gamma')(\gamma' \cdot u') \|\gamma'\|^{-3} dx$$

$$- \int_0^1 (v \cdot \xi)(w' \cdot \gamma')(\gamma' \cdot u') \|\gamma'\|^{-3} dx$$

$$+ \int_0^1 (v \cdot w)(\xi' \cdot \gamma')(\gamma' \cdot u') \cdot \|\gamma'\|^{-3} dx.$$

Note that since $v \cdot \gamma' = 0$, we have $v' \cdot \gamma' = -v \cdot \frac{d^2\gamma}{dx^2} = -\kappa l^2 \cdot v_N$, for κ the curvature, l the length of the knot, and $v_N = v \cdot \vec{N}$ the component of v along the normal vector $\vec{N}$. Hence we obtain:

3.6.11. Lemma. *For a knot γ of length 1 parametrized by arc-length, and for u, v, w, ξ constant vector fields which are perpendicular to γ, we have*

$$2u \cdot \langle \tilde{\nabla}_v w, \xi \rangle = \int_0^1 (w \cdot \xi)(v' \cdot u') dx + \int_0^1 (v \cdot \xi)(w' \cdot u') dx$$

$$- \int_0^1 (v \cdot w)(\xi' \cdot u') dx - \int_0^1 (w \cdot \xi)\kappa^2 v_N u_N dx$$

$$- \int_0^1 (v \cdot \xi)\kappa^2 w_N u_N dx + \int_0^1 (v \cdot w)\kappa^2 \xi_N u_N dx.$$

Putting all this information together, we derive the curvature tensor of $\hat{Y}$.

3.6.12. Proposition. *Let u, v, w, ξ be smooth sections of the normal bundle to γ, viewed as tangent vectors to $\hat{Y}$ at γ. Then we have the following formula for the coefficient $\langle R_{u,v} w, \xi \rangle$ of the curvature tensor R:*

$$\langle R_{u,v} w, \xi \rangle = \int_0^1 F(x) dx,$$

where $F(x)$ is the following function:

$$F(x) = 1/2(v \cdot \xi)(w' \cdot u') - 1/2(v \cdot w)(\xi' \cdot u') - 1/2(u \cdot \xi)(w' \cdot v') +$$
$$1/2(u \cdot w)(\xi' \cdot v') - 1/4(v \cdot w)(u \cdot \xi)\kappa^2 + 1/4(u \cdot w)(v \cdot \xi)\kappa^2$$
$$- 5/4(v \cdot \xi)w_N u_N \kappa^2 + 5/4(v \cdot w)\xi_N u_N \kappa^2$$
$$+ 5/4(u \cdot \xi)w_N v_N \kappa^2 - 5/4(u \cdot w)\xi_N v_N \kappa^2.$$

3.6.13. Remark.   It is easy to use dilations of $\mathbb{R}^3$ to find the value of the curvature tensor at any knot of arbitrary length. Indeed, the dilation of ratio λ multiplies lengths in $\hat{Y}$ by λ, so it preserves the Levi-Civita connection on $\hat{Y}$ and the Riemann curvature tensor.

It would be of interest to investigate further the properties of the curvature tensor. For example, can one define a Ricci tensor, and what is its value? One has a rather nice formula (from Proposition 3.6.12) for the holomorphic sectional curvature. We note that the one-parameter family of Kähler structures on the homogeneous space $Diff^+(S^1)/S^1$, studied by Kirillov and Yurev [K-Y] have beautiful curvature properties. One might hope that the space of knots does too, but the size of the computations seems formidable.

3.7. The group of unimodular diffeomorphisms

Let M be a smooth 3-fold with a volume form ν. Recall the space of oriented singular knots $\hat{Y}$ on which we have the non-degenerate 2-form β of §5. Let G be the group of unimodular diffeomorphisms of M. According to [O], G is a Lie group with Lie algebra $\mathfrak{g}$, the Lie algebra of *divergence-free* vector fields on M. Recall that the divergence of a vector field ξ is the function $div(\xi)$ such that $d(i(\xi)\nu) = div(\xi) \cdot \nu$. The condition $div(\xi) = 0$ is equivalent to $\mathcal{L}(\xi) \cdot \nu = 0$, i.e., ξ preserves the volume form. The Lie group G operates smoothly on $\hat{Y}$, and preserves the symplectic form β. This is why it is more natural to consider this group in our context, rather than the group of all diffeomorphisms. In the case of $\mathbb{R}^3$, the Lie group G occurs in the formulation of the Euler equations for an incompressible fluid; this fact was first explicitly noted by Arnold [Ar].

In the work of Marsden and Weinstein on fluid mechanics [M-W], a variant of the space of knots appears, namely, the space of vortex filaments in 3-space. Such a vortex filament is really a bunch of open curves in $\mathbb{R}^3$. From the point of view of physics, the theory of vortex filaments addresses itself to the ideal situation where the fluid motion is concentrated entirely on this bunch of curves (which are called vortices). Marsden and Weinstein

[M-W] showed that the methods of hamiltonian mechanics, in particular moment maps, apply very well to this situation. They showed the existence of a moment map for the action of G on the space of vortex filaments.

Let us recall first the notion of moment mapping.

3.7.1. Definition. *Let (Y, β) be a symplectic manifold, and let G be a Lie group which acts symplectically on Y. For an element ξ of the Lie algebra $\mathfrak{g}$, we denote also by ξ the resulting vector field on Y. Then a* moment map *for this action is a mapping $\mu : Y \to \mathfrak{g}^*$ which satisfies*

(a) *For any $\xi \in \mathfrak{g}$, introducing the function $\mu_\xi = \langle \xi, \mu \rangle$ on Y, the derivative $d\mu_\xi$ is equal to $I(\xi)$, where $I : TY \to T^*Y$ is the linear map induced by β.*

(b) *For any ξ and η in $\mathfrak{g}$, we have $\{\mu_\xi, \mu_\eta\} = \mu_{[\xi,\eta]}$.*

When $\mathfrak{g}$ is finite-dimensional, finding a moment map for the G-action amounts to finding a Lie algebra homomorphism $\mathfrak{g} \to C^\infty(Y)$, denoted by $\xi \mapsto \mu_\xi$, such that the hamiltonian vector field X_{μ_ξ} is equal to the vector field corresponding to ξ. If $\mathfrak{g}$ is infinite-dimensional, there is a further requirement that the mapping $Y \to \mathfrak{g}^*$, with "coordinates" the μ_ξ, is a smooth mapping.

We now present the result of Marsden and Weinstein, and generalize it to simply-connected 3-folds. Denote by $\mathcal{S}_j(M)$ the space of compactly-supported currents of degree $3 - j$, i.e., the continuous dual of the space of differential forms of degree j on M. There is a differential $d : \mathcal{S}_j(M) \to \mathcal{S}_{j-1}(M)$. The topological dual of $\mathfrak{g}$ is the quotient $\mathcal{S}_2(M)/d\mathcal{S}_3(M)$, since the orthogonal of the space of divergence-free vector fields consists of exact currents of degree 1.

Because β is symplectic only in a weak sense, one may not define the hamiltonian vector field X_f for an arbitrary smooth function f on $\hat{Y}$. However, if f is super-smooth — cf. §3.1. —, and if one chooses locally a riemannian metric for which ν is the volume form, then one may define X_f as $-l(\gamma)^{-1}J(df)$, where df is viewed as a field of tangent vectors. Then one can define the Poisson bracket $\{f, g\}$ with any smooth function g as in §3.5. If g is also super-smooth, $\{f, g\}$ is also super-smooth.

3.7.2. Theorem. [M-W] *Assume that $H^1(M, \mathbb{R}) = 0$. Let $\mu : \hat{Y} \to \mathfrak{g}^*$ be the mapping defined by*

$$\langle \mu(C), \xi \rangle = \int_C \alpha,$$

for any 1-form α on M such that $d\alpha$ is the 2-form corresponding to ξ by duality. μ is a smooth mapping which satisfies the following properties:

(1) *μ is injective and equivariant under the Lie group G of volume and orientation-preserving diffeomorphisms of M, where G acts on $\mathfrak{g}^*$ by the coadjoint action;*

(2) *μ is a moment map for the action of G on Y.*

Proof. Injectivity of μ follows from the fact that, if C_1 and C_2 are different knots, there exists a 1-form α which vanishes identically on C_1 and is tangent to C_2, with $\int_{C_2} \alpha = 1$, hence C_1 and C_2 induce different linear forms on the space of 1-forms on M. As the construction of μ only depends on the volume form ν, μ is clearly G-equivariant. This proves (1).

We now prove properties (a) and (b) of a moment map for μ. We note that a divergence-free vector field ξ corresponds to a closed 2-form ψ, in the sense that $\psi = i(\xi) \cdot \nu$, and we have $\mu_\xi(C) = \int_C \alpha$, for any 1-form α with $d\alpha = \psi$. We obtain:

$$\langle d\mu_\xi, v \rangle = \int_C i(v)(d\alpha) = \int_C i(v)\psi.$$

On the other hand, we have

$$\langle I(\xi), v \rangle = \beta(\xi, v) = \int_0^1 \nu(\frac{d\gamma}{dx}, \xi, v)dx = -\int_0^1 i(\xi)(\nu)(\frac{d\gamma}{dx}, v)dx$$

$$= -\int_0^1 \psi(\frac{d\gamma}{dx}, v) = \int_C i(v)\psi.$$

This proves (a).

To prove (b), let ξ and η be divergence free vector fields. Let $F = \mu_\xi$, $G = \mu_\eta$. Then

$$\{F, G\} = \beta(\eta, \xi) = \int_0^1 \nu(\frac{d\gamma}{dx}, \eta, \xi)dx$$

$$= \int_0^1 \langle (\eta \times \xi), \frac{d\gamma}{dx} \rangle dx.$$

On the other hand, the bracket $[\xi, \eta]$ of these vector fields is equal to $D_\xi \eta - D_\eta \xi$. We have the following formula

$$d(\xi \times \eta) = -div(\xi)\eta + div(\eta)\xi + D_\eta \xi - D_\xi \eta,$$

which in the case of the standard volume form amounts to a well-known formula for the curl of the crossed product of two vector fields. In the present case, as $div(\xi) = div(\eta) = 0$, we obtain $[\xi, \eta] = -d(\xi \times \eta)$, so that the function $\mu_{[\xi,\eta]}$ is $C \mapsto - \int_C (\xi \times \eta) \cdot d\gamma = -\beta(\xi, \eta) = \{\mu_\xi, \mu_\eta\}$, as claimed. This finishes the proof. ∎

From this theorem is it clear that we have gained a lot by replacing all diffeomorphisms by unimodular diffeomorphisms. We want to show that we have not lost anything. Recall that two knots C and C' are called *isotopic* if there is a one-parameter family g_t of diffeomorphisms, with $g_0 = Id$ and $g_1 \cdot C = C'$. In other words, C and C' are in the same orbit of the connected component of the group of all diffeomorphisms of M. A basic elementary fact is that a neighborhood of a given knot C consists of knots isotopic to C; so all orbits are open. This is the reason why the classification of knots up to isotopy is a combinatorial problem. We will show that the same thing holds for the group of unimodular diffeomorphisms.

3.7.3. Lemma. *Let D_r be the closed disc of radius r in $\mathbb{R}^2$, equipped with the standard 2-form. Let D_r^0 be the open disc. For every $z \in D_r^0$, there exists a symplectic diffeomorphism ϕ_z of D_r such that*

(1) $\phi_z(0) = z$.

(2) ϕ_z *induces the identity in a neighborhood of the boundary ∂D_r.*

Furthermore, one may choose ϕ_z to depend smoothly on $z \in D_r^0$.

Proof. One can easily reduce to the unit disc D. We use complex numbers. For $z \in D^0$, $z \neq 0$, let u be the unit vector in the direction of z, and $v = \sqrt{-1} \cdot u$. Suppose we can find for each $z \in D^0$ a real-valued function f_z on D such that

(a) for $z \neq 0$, we have $u \cdot f_z(t \cdot z) = 0$ and $v \cdot f_z(t \cdot z) = |z|$ for $0 \leq t \leq 1$.

(b) $f_z(w) \equiv 0$ if $|w|^2 > \frac{3 + |z|^2}{4}$.

(c) f_z depends smoothly on $z \in D^0$.

Then the hamiltonian vector field X_{f_z} is equal to u on the segment $[0, z]$, and it is zero in a neighborhood of the boundary. It follows that $\phi_z = exp(X_{f_z})$ is a symplectic diffeomorphism of D, which is the identity in a neighborhood of ∂D, and such that $f_z(0) = z$. Clearly ϕ_z depends smoothly on $z \in D^0$.

So we have to find functions f_z satisfying (a), (b) and (c). Start with a smooth function h from $\mathbb{R}$ to $[0, 1]$ such that $h(x) = 1$ for $x < 1/4$ and $h(x) = 0$ for $x > 3/4$. Then put

$$f_z(w) = h(\frac{|w|^2 - |z|^2}{1 - |z|^2}) \cdot \Im(\bar{z}w)$$

Properties (a), (b) and (c) are easily verified. ∎

The following theorem was first proved by Marsden and Weinstein in the case of $\mathbb{R}^3$.

3.7.4. Theorem. (1) *The orbits of the connected component G^0 in the space Y of oriented knots coincide with the connected components of Y.*

(2) *Let $\mathcal{C}$ be a component of Y. Then $\mathcal{C}$ is diffeomorphic to the homogeneous space G^0/G_C^0, for any $C \in \mathcal{C}$.*

Proof. To prove (1), one need only show that the G^0-orbit of any oriented knot C is open in Y. Let T be a closed tubular neighborhood of C in M, which is diffeomorphic to $S^1 \times D$, for D the closed unit disc in $\mathbb{R}^2$. It is enough to show that if C' is a knot contained in the interior T^0, which is the graph of a smooth function $f : S^1 \to D_{1/2}$ (where $D_{1/2}$ is the disc of radius $1/2$), then there exists a unimodular diffeomorphism g of $S^1 \times D$, which is the identity near the boundary ∂T, and such that $g(C) = C'$. First of all, we may use a theorem of Moser [Mo] which says that there exists a diffeomorphism h of $S^1 \times D$, which satisfies

(1) h is the identity near the boundary of $S^1 \times D$

(b) the restriction of h to $S^1 \times D_{2/3}$ transforms ν into the standard volume form.

The theorem in [Mo] actually deals only with closed manifolds, but the proof there applies also to manifolds with boundary. Using h, we may assume that ν coincides with standard volume form over $S^1 \times D_{2/3}$. We can define

$$g(z, y) = (z, \phi_{f(z)}(y)),$$

where $\phi_{f(z)}$ is the diffeomorphism of $D_{2/3}$ given by Lemma 3.7.3. Note that this is a diffeomorphism g of $S^1 \times D_{2/3}$, which is the identity near the boundary. g is unimodular, because for each $z \in S^1$, the jacobian of $\phi_{f(z)}$ is equal to 1. So it extends to a unimodular diffeomorphism g of M, which is the identity outside of $D_{2/3}$. The diffeomorphism g maps $(z, 0)$ to $(z, f(z))$, hence g transforms C into C'. This proves (1).

To prove (2) for the G^0-orbit $\mathcal{C}$ of some knot C, it is enough to show that the mapping $G \to \mathcal{C}$ given by the G^0-action is a principal H-fibration, where $H = G_C^0$. For this purpose, it will suffice to show the existence of a local section for the mapping $q : G \to \mathcal{C}$. Such a local section was implicit in the proof of (1), namely we put $s(C') = g$, where g is the unimodular diffeomorphism used in that proof. In fact, g belongs to G^0, since it can be connected to the identity inside G. This finishes the proof. ∎

There is a smooth bijective mapping from C to the coadjoint orbit $\mu(C)$ (with the topology induced from $\mathfrak{g}^*$), but we do not know whether this is a diffeomorphism. There are strange phenomena happening on the boundary of these coadjoint orbits, as one can show that any connected component of the space of singular knots with one crossing point is in bijection with some coadjoint orbit, which has codimension 1 in the closure of a coadjoint orbit corresponding to an isotopy class of smooth knots. This is counter-intuitive, as in finite dimensions all coadjoint orbits have even dimension (being symplectic manifolds). This phenomenon was found and explained by Weinstein.

We note the following result on the topology of G.

3.7.5. Proposition. *Let M be a closed smooth 3-fold. Then the inclusion of the group of unimodular diffeomorphisms of M into the group of all orientation-preserving diffeomorphisms is a homotopy equivalence.*

In fact, assuming ν has volume 1, one can prove using [Mo] that there is a fibration

$$
\begin{array}{c}
G \hookrightarrow Diff^+(M) \\
\downarrow \\
V_1(M)
\end{array}
$$

where $V_1(M)$ is the manifold of volume forms on M with volume equal to 1.

Finally we point out the suggestion by Rasetti and Regge [**Ra-Re**] and by Penna and Spera [**Pe-Sp**] that the center of the enveloping algebra of $\mathfrak{g}$ would be sort of universal source for knot invariants. This idea would certainly be worth a detailed mathematical investigation. Especially intriguing is the relation of the enveloping algebra of $\mathfrak{g}$ to the iterated integral calculus of Chen.

U. Schäper has obtained independently numerous results on the riemannian geometry of the space of curves [**Sc1**], [**Sc2**].

Chapter 4

Degree 3 Cohomology:
The Dixmier-Douady Theory

Much of this book is devoted to a geometric description of the degree 3 cohomology $H^3(M, \mathbb{Z})$ of a manifold M. Recall in the case of degree 2 cohomology, a theorem of Weil and Kostant (Corollary 2.1.4) which asserts that $H^2(M, \mathbb{Z})$ is the group of isomorphism classes of line bundles over M. Moreover, given a line bundle L, the corresponding class $c_1(L)$ in $H^2(M, \mathbb{Z})$ is represented by $\dfrac{1}{2\pi\sqrt{-1}} \cdot K$, where K is the curvature of a connection on L. We wish to find a similar description for $H^3(M, \mathbb{Z})$, which is a more difficult task. One theory, due to Dixmier and Douady [D-D], involves so-called continuous fields of elementary C^*-algebras. We will describe this theory in the present chapter and develop, in particular, the notion of curvature, which will be a closed 3-form associated to such a field of C^*-algebras.

4.1. Infinite-dimensional algebra bundles

Given a paracompact space M, we wish to describe bundles over M which represent classes in $H^3(M, \mathbb{Z})$. Actually it is more convenient to work with the isomorphic group $H^3(M, \mathbb{Z}(1))$, where $\mathbb{Z}(1) = 2\pi\sqrt{-1} \cdot \mathbb{Z}$. First, let us use the exponential exact sequence of sheaves $0 \to \mathbb{Z}(1) \to \underline{\mathbb{C}}_M \xrightarrow{exp} \underline{\mathbb{C}}_M^* \to 0$ (cf. Chapter 1) to identify $H^3(M, \mathbb{Z}(1))$ with $H^2(M, \underline{\mathbb{C}}_M^*)$. In this section, $\underline{\mathbb{C}}_M^*$ is used to denote the sheaf of continuous $\mathbb{C}^*$-valued functions on open sets of M, as opposed to smooth functions in the rest of the book. We will need degree 1 Čech cohomology with values in a sheaf of nonabelian groups, which was introduced by Grothendieck, Dedecker and Frenkel [Gr1] [Gr2] [Ded] [Fr]. Degree 1 cohomology is only a pointed set in general, not a group. For an exact sequence of sheaves of nonabelian groups, one obtains a short exact sequence of cohomology for cohomology in degrees 0 and 1.

The study of principal bundles naturally leads to such a "nonabelian cohomology", which makes this the starting point. Let M be a manifold and let G be a Lie group; let us consider the set of isomorphism classes of principal G-bundles $q : P \to M$. Given an open covering $(U_i)_{i \in I}$ and a continuous section s_i of q over U_i, we have a continuous function $g_{ij} : U_{ij} \to G$ such that $s_j = s_i \cdot g_{ij}$ (recall that G acts on the right on P). These transition functions satisfy the equality $g_{ik} = g_{ij}g_{jk}$ over U_{ijk}. If we replace the section s_i over U_i by $s_i' = s_i \cdot h_i$, then g_{ij} is replaced by $h_i^{-1}g_{ij}h_j$.

We are guided now to the appropriate notions of 1-cocycles and coboundaries. For a sheaf A of (nonabelian) groups on a space M, and $(U_i)_{i \in I}$ an open covering, we define a Čech 1-*cocycle* with values in A to consist of a family $a_{ij} \in \Gamma(U_{ij}, A)$ such that the cocycle relation

$$a_{ik} = a_{ij}a_{jk} \qquad\qquad (4-1)$$

holds. Next, two Čech 1-cocycles a_{ij} and a_{ij}' are called *cohomologous* if there exists a section h_i of A over U_i such that

$$a_{ij}' = h_i^{-1}a_{ij}h_j, \qquad\qquad (4-2)$$

thus defining an equivalence relation on the set of Čech 1-cocycles.

4.1.1. Definition. (1) *For a sheaf A of groups on the space M, and an open covering $\mathcal{U} = (U_i)_{i \in I}$ of M, the first cohomology set $\check{H}^1(\mathcal{U}, A)$ is defined as the quotient of the set of 1-cocycles with values in A by the equivalence relation: "$\underline{a}$ cohomologous to $\underline{a}'$".*

(2) *The first cohomology set $H^1(M, A)$ is defined as the direct limit $\varinjlim_{\mathcal{U}} \check{H}^1(\mathcal{U}, A)$, where the limit is taken over the set of all open coverings of M, ordered by the relation of refinement.*

Note that the set $H^1(M, A)$ has a distinguished element, namely, the class of the trivial 1-cocycle 1. So it is a *pointed set*. The definition of first cohomology is designed to make the following result automatic.

4.1.2. Proposition. *Let G be a Lie group, and let $\underline{G}$ be the sheaf of germs of continuous G-valued functions. Then $H^1(M, \underline{G})$ identifies naturally, as a pointed set, with the set of isomorphism classes of principal G-bundles over M.*

We will denote by $[P]$ the cohomology class of a G-bundle $P \to M$.

There is an analog for sheaves of nonabelian groups of the long exact sequence for sheaf cohomology. Let $1 \to A \to B \xrightarrow{p} C \to 1$ be an exact sequence of sheaves of groups. There is a mapping of sets $\delta_0 : H^0(M, C) \to H^1(M, A)$ defined as follows. Let c be a section of C over M, and let (U_i) be an open covering of M such that there exists $b_i \in \Gamma(U_i, B)$ with $p(b_i) = c_{/U_i}$. Now form the expression $a_{ij} = b_i^{-1}b_j$. Actually, this is a section of A over

U_{ij}, since $p(a_{ij}) = 1$. Also we have

$$a_{ij}a_{jk} = b_i^{-1}b_j b_j^{-1}b_k = a_{ik}$$

over U_{ijk}. So a_{ij} is a Čech 1-cocycle with values in A; its cohomology class is independent of the choices of the open covering and of b_i.

4.1.3. Theorem. [Gr1] [Gr2] *If* $1 \to A \to B \xrightarrow{p} C \to 1$ *is an exact sequence of sheaves of groups on a space M, there is an exact sequence of pointed sets*

$$1 \to H^0(M,A) \to H^0(M,B) \to H^0(M,C) \qquad (4-3)$$
$$\xrightarrow{\delta_0} H^1(M,A) \to H^1(M,B) \to H^1(M,C)$$

We recall that a sequence $(X,x) \xrightarrow{f} (Y,y) \xrightarrow{g} (Z,z)$ of pointed sets and base-point preserving maps is called exact if $f(X)$ is equal to $g^{-1}(z)$, as subsets of Y. Following [Gr2], one can then obtain a stronger statement concerning exactness at $H^1(M,A)$. To explain this, we note that there is a right action of the group $H^0(M,C)$ on the set $H^1(M,A)$. Let $c \in \Gamma(M,C)$ as above, and let a_{ij} be a 1-cocycle with values in A. Choosing $b_i \in \Gamma(U_i,B)$ such that $p(b_i) = c$, we set $a'_{ij} = b_i^{-1}a_{ij}b_j$, which is another 1-cocycle with values in A. This defines the action of the cohomology class of c on the cohomology class of a, which will be denoted by $a * c$. The statement, therefore, is that two classes a and a' in $H^1(M,A)$ have the same image in $H^1(M,B)$ if and only if there exists $c \in H^0(M,C)$ such that $a' = c * a$.

An important case of an exact sequence $1 \to A \to B \to C \to 1$ of sheaves of groups is obtained from a sheaf of $\mathbb{C}$-algebras R over M. Then one has the exact sequence

$$1 \to Z(R)^* \to R^* \to R^*/Z(R)^* \to 1, \qquad (4-4)$$

where $Z(R)$ is the center of R, and R^* denotes the sheaf of groups of invertible elements in R. A special feature of the exact sequence (4-4) is that the first sheaf of groups is central in the middle one. Such an exact sequence is called *central*. In that case, one can continue the exact sequence of Theorem 4.1.3 one step to the right.

4.1.4. Theorem. *In the situation of Theorem 4.1.3, assume furthermore that A is central in B. Then one has the exact sequence of pointed sets*

$$1 \to H^0(M,A) \to H^0(M,B) \to H^0(M,C) \qquad (4-5)$$
$$\xrightarrow{\delta_0} H^1(M,A) \to H^1(M,B) \to H^1(M,C) \xrightarrow{\delta_1} H^2(M,A)$$

Here $\delta_1 : H^1(M, C) \to H^2(M, A)$ is defined as follows. Let c_{ij} be a Čech 1-cocycle of the open covering (U_i) with values in C, such that $c_{ij} = p(b_{ij})$, for $b_{ij} \in \Gamma(U_{ij}, B)$. Then $a_{ijk} := b_{ik}^{-1} b_{ij} b_{jk}$ gives a Čech 2-cocycle $\underline{a}$ with values in A. δ_1 then maps the cohomology class of $\underline{c}$ to that of $\underline{a}$.

Note that since A is a sheaf of abelian groups, $H^2(M, A)$ is defined.

We wish to apply this theorem to the exact sequence (4-4) in the case of a suitable sheaf of algebras R. As we are interested in the cohomology group $H^2(M, \underline{C}_M^*)$, we want the center of R to be the sheaf $\underline{C}_M$. Let us take the sheaf of algebras R to be the sheaf $\underline{L}_M$ of continuous functions with values in some Banach algebra L, with center equal to $\mathbb{C}$. The best choice will turn out to be the algebra $L = \mathcal{L}(E)$ of bounded operators in a separable Hilbert space E; the norm is the sup-norm of an operator. It is well-known (and easy to see) that the center of L is equal to $\mathbb{C}$. The quotient Lie group $L^*/\mathbb{C}^*$ identifies with the Lie group $G = Aut(\mathcal{K})$ of continuous algebra automorphisms of the Banach algebra $\mathcal{K}$ of *compact* operators of E. In other words, one has an exact sequence of Lie groups

$$1 \to \mathbb{C}^* \to L^* = Aut(E) \to G = Aut(\mathcal{K}) \to \infty. \qquad (4-6)$$

Since the mapping $Aut(E) \to G$ is a principal $\mathbb{C}^*$-fibration, one has an exact sequence of sheaves of continuous functions with values in (4-6), to wit:

$$1 \to \underline{C}_M^* \to \underline{L}_M^* \to \underline{G}_M \to 1. \qquad (4-7)$$

The following result is the basic reason behind the choice of this Banach algebra L.

4.1.5. Proposition. [D-D, Lemme 3] *The Lie group L^* of invertible continuous operators in E is contractible.*

4.1.6. Corollary. [D-D, Lemme 4] *The sheaf $\underline{L}_M^*$ of continuous L^*-valued functions is <u>soft</u>.*

Proof. Let Y be a closed subset of M, and let f be a section of $(\underline{L}_M^*)_{/Y}$. According to Lemma 1.4.4, f extends to a section g of $\underline{L}_M^*$ over a neighborhood U of Y. So g is a continuous function from U to L^*. As M is paracompact, there exists a neighborhood V of Y such that $\overline{V} \subset U$. As L^* is contractible, there exists a continuous map $r : L^* \times [0, 1] \to L^*$ such that $r(x, 1) = x$ and $r(x, 0) = 1 \in L^*$. Let ϕ be a continuous function from M to $[0, 1]$, with support in U, such that $\phi(x) = 1$ for $x \in V$. Then $h(x) = r(g(x), \phi(x))$

extends to a continuous function from M to L^*, such that $h(x) = g(x)$ for $x \in V$. ∎

The fact that $\underline{L}^*_M$ is soft will make the exact sequence (4-5) collapse. For this we need a couple of general results.

4.1.7. Proposition. *Let B be a soft sheaf of groups on M. Then $H^1(M, B) = 1$.*

Proof. We follow [D-D, Lemme 22]. Let $\mathcal{U} = (U_i)_{i \in I}$ be an open covering of M, and let b_{ij} be a Čech 1-cocycle of this covering with values in B. We may assume that $\mathcal{U}$ is locally finite. Choose an open covering $(Y_i)_{i \in I}$ such that $\overline{Y_i} \subset U_i$. Let S be the partially ordered set of pairs (J, v), where $J \subseteq I$ and $v = (v_i)_{i \in J}$, and the v_i belong to $\Gamma(\overline{Y_i}, B)$, such that $b_{ij} = v_i v_j^{-1}$ over $\overline{Y_i} \cap \overline{Y_j}$. The set S is partially ordered as follows: $(J, v) < (J', v')$ means that $J \subset J'$ and $v_i = v'_i$ for $i \in J$. Then any ordered chain of elements of S admits a maximal element. By Zorn's lemma, there exists a maximal element (J, v) of S. We show that $J = I$; this will prove that the Čech 1-cocycle b_{ij} for the covering (V_i) is a coboundary. We will derive a contradiction from the assumption that $J \neq I$. Say $i \in I \setminus J$. Set $Y = \overline{Y_i} \cap (\cup_{j \in J} \overline{Y_j})$. As the covering (U_i) is locally finite, Y is a closed subset of M. For each $j \in J$, we have the section $w_j = b_{ij} v_j$ of A over $\overline{Y_i} \cap \overline{Y_j}$. If $j, k \in J$, then over $\overline{Y_i} \cap \overline{Y_j} \cap \overline{Y_k}$, we have

$$w_k = b_{ik} v_k = b_{ij} b_{jk} v_k = b_{ij} v_j = w_j.$$

Hence there exists a section w of B over Y which has restriction to $\overline{Y_i} \cap \overline{Y_j}$ equal to w_i. Since B is a soft sheaf, there exists a section v_i of A over $\overline{Y_i}$ which coincides with w on Y. By construction $(J \cup \{i\}, (v_j))$ is strictly larger than (J, v), which contradicts the maximality of (J, v). ∎

4.1.8. Proposition. [D-D, Lemme 22] *Let $1 \to A \to B \to C \to 1$ be a central exact sequence of sheaves of groups over M. If B is a soft sheaf, then $\delta_1 : H^1(M, C) \to H^2(M, A)$ is a bijection.*

Proof. The injectivity of δ_1 is proved by the same method as Proposition 4.1.7, so we will skip its proof. Let (a_{ijk}) be a Čech 2-cocycle of a locally finite open covering (U_i) with values in A. Choose an open covering (Y_i) such that $\overline{Y_i} \subset U_i$. Let S be the ordered set of pairs (J, b), where J is a subset of I and $b_{ij} \in \Gamma(U_{ij}, C)$ such that $b_{ij} b_{jk} = b_{ik} a_{ijk}$ holds over $\overline{Y_i} \cap \overline{Y_j} \cap \overline{Y_k}$. Let (J, b)

be a maximal element of S. If there exists $i \in I \setminus J$, we introduce another ordered set T, consisting of pairs $(K, \hat{b})$, where $K \subseteq J$, and $\hat{b} = (\hat{b}_{ik})_{k \in K}$, with $\hat{b}_{ik} \in \Gamma(\overline{Y_i} \cap \overline{Y_k}, A)$, such that $\hat{b}_{ij} b_{jk} = \hat{b}_{ik} a_{ijk}$ holds on $\overline{Y_i} \cap \overline{Y_j} \cap \overline{Y_k}$. Denote again by $(K, \hat{b})$ a maximal element of T. We will show that $K = J$. Otherwise let $j \in J \setminus K$. For any given $k \in K$, there is a unique section β_k of B over $\overline{Y_i} \cap \overline{Y_j} \cap \overline{Y_k}$ such that

$$\beta_k b_{jk} = \hat{b}_{ik} a_{ijk}. \tag{4-8}$$

Then over $\overline{Y_i} \cap \overline{Y_j} \cap \overline{Y_k} \cap \overline{Y_l}$ one has

$$\beta_l = \hat{b}_{il} a_{ijl} b_{jl}^{-1} = \hat{b}_{ik} b_{kl} a_{ikl}^{-1} a_{ijl} b_{jl}^{-1}$$
$$= \hat{b}_{ik} b_{jk}^{-1} a_{jkl} a_{ijl} a_{ikl}^{-1}.$$

Using the cocycle relation for a, this is equal to

$$\hat{b}_{ik} b_{jk}^{-1} a_{ijk} = \beta_k.$$

Hence the β_k give a section β of B over $\overline{Y_i} \cap \overline{Y_j} \cap (\cup_{k \in K} \overline{Y_k})$. Since B is a soft sheaf, one can extend β to a section of B over $\overline{Y_i} \cap \overline{Y_j}$, which contradicts the fact that $(K, \hat{b})$ is maximal. The contradiction shows that $K = J$, implying that there exists a family $(\hat{b}_{ij})_{j \in J}$. Set $\hat{b}_{ji} = \hat{b}_{ij}^{-1}$ for $j \in J$. Combining the b_{jk} for $j, k \in J$, the $\hat{b}_{ij}$ $(j \in J)$ and the $\hat{b}_{ji}$ $(j \in J)$, one obtains an element $(J \cup \{i\}; b, \hat{b})$ of S strictly bigger than (J, B). This contradiction shows that $J = I$. Then the Čech 1-cocycle $c_{ij} = p(b_{ij})$ of (Y_i) with coefficients in C satisfies $\delta_1(\underline{c}) = \underline{a}$. ∎

Applying 4.1.8 to the exact sequence of sheaves of groups (4-7), we obtain:

4.1.9. Theorem. (*Dixmier-Douady* [D-D]) *For the Lie group G of algebra automorphisms of the Banach algebra $\mathcal{K}$ of compact operators on a separable Hilbert space E, we have natural bijections*

$$H^1(M, \underline{G}_M) \xrightarrow{\sim} H^2(M, \underline{\mathbb{C}}_M^*) \xrightarrow{\sim} H^3(M, \mathbb{Z}(1)).$$

This gives the geometric description of $H^3(M, \mathbb{Z}(1))$ we were looking for. A class in $H^3(M, \mathbb{Z}(1))$ corresponds to an isomorphism class of principal G-bundles over M. A natural question arises: what is the group structure on $H^1(M, \underline{G})$ arising from the one on $H^2(M, \underline{\mathbb{C}}_M^*) \xrightarrow{\sim} H^3(M, \mathbb{Z}(1))$? To describe it, we fix an isomorphism $\phi : E \otimes E \xrightarrow{\sim} E$ of Hilbert spaces. Here $E \otimes E$ denotes the Hilbert space tensor product, so that for $(e_n)_{n\geq 0}$ a hilbertian basis of E, the $e_m \otimes e_n$ form a hilbertian basis of $E \otimes E$. There is a corresponding algebra homomorphism $\psi : L \times L \to L$ such that $\psi(a,b) = a \otimes b$. This induces a Lie group homomorphism $L^* \times L^* \to L^*$, and since $\mathbb{C}^* \times \mathbb{C}^* \subset L^* \times L^*$ maps to $\mathbb{C}^* \subset L^*$, we have an induced Lie group homomorphism $\psi : G \times G \to G$. In turn, a map is induced on cohomology

$$H^1(M, \underline{G}_M \times \underline{G}_M) = H^1(M, \underline{G}_M) \times H^1(M, \underline{G}_M) \to H^1(M, \underline{G}_M),$$

which will be denoted by $\otimes$.

4.1.10. Proposition. *The operation $\otimes$ on $H^1(M, \underline{G}_M)$ gives $H^1(M, \underline{G}_M)$ the structure of an abelian group. The bijection $\delta_1 : H^1(M, \underline{G}_M) \to H^2(M, \underline{\mathbb{C}}_M^*)$ is a group isomorphism.*

Proof. Let c_{ij} and c'_{ij} be 1-cocycles of an open covering (U_i) with values in $\underline{G}_M$. We may assume that $c_{ij} = p(b_{ij})$, for a continuous function $b_{ij} : U_{ij} \to L^*$ (resp. $c'_{ij} = p(b'_{ij})$). Then the 2-cocycle $a_{ijk} = b_{ik}^{-1} b_{ij} b_{jk}$ with values in $\underline{\mathbb{C}}_M^*$ represents the cohomology class $\delta_1(\underline{c})$ (resp. $a'_{ijk} = b_{ik}'^{-1} b'_{ij} b'_{jk}$ represents the class $\delta_1(\underline{c}')$). The tensor product of the classes in $H^1(M, \underline{G}_M)$ is represented by the image in $\underline{G}_M$ of the Čech 1-cochain

$$\gamma_{ij} = b_{ij} \otimes b'_{ij}.$$

The corresponding $\underline{\mathbb{C}}_M^*$-valued 2-cocycle is

$$\begin{aligned}
\alpha_{ijk} &= (b_{ik} \otimes b'_{ik})^{-1}(b_{ij} \otimes b'_{ij})(b_{jk} \otimes b'_{jk}) \\
&= (b_{ik}^{-1} b_{ij} b_{jk}) \otimes (b_{ik}'^{-1} b'_{ij} b'_{jk}) \\
&= a_{ijk} \otimes a'_{ijk} = a_{ijk} a'_{ijk}. \quad\blacksquare
\end{aligned}$$

It is useful to develop several geometric interpretations of $H^1(M, \underline{G}_M)$. First of all, since G is the automorphism group of $\mathcal{K}$, we have

4.1.11. Lemma. $H^1(M, \underline{G}_M)$ *is in a natural bijection with the set of isomorphism classes of locally trivial algebra bundles $\mathcal{A} \to M$ with fiber $\mathcal{K}$.*

We note that since there is an algebra map $\mathcal{K} \otimes \mathcal{K} \to \mathcal{K}$, there is a notion of tensor product for such algebra bundles with fiber $\mathcal{K}$. This gives the same group structure on $H^1(M, \underline{G})$.

Another interpretation of principal G-bundles involves *projective space bundles*. By this we mean a locally trivial fiber bundle $p : P \to M$ with fiber the projective space $\mathbf{P}(E) = (E \setminus \{0\})/\mathbb{C}^*$ of lines in E. When E is a separable Hilbert space, we call $\mathbf{P}(E)$ a separable projective space. As was explained in Chapter 2, $\mathbf{P}(E)$ is a smooth manifold, and the usual notions of projective geometry make sense on $\mathbf{P}(E)$. A line in $\mathbf{P}(E)$ is defined to be $\mathbf{P}(F)$, for a two-dimensional vector subspace F of E. A diffeomorphism $f : \mathbf{P}(E) \to \mathbf{P}(E)$ is said to be a *projective automorphism* if f transforms lines in $\mathbf{P}(E)$ to lines, and induces a biholomorphic isomorphism from each line to its image. Recall that $\mathbf{P}(E)$ is covered by open subsets which identify with a Hilbert space; a subset of $\mathbf{P}(E)$ is said to be *bounded* if its intersection with any such open set is bounded in the corresponding Hilbert space.

4.1.12. Lemma. *The group of projective automorphisms of* $\mathbf{P}(E)$, *equipped with the topology of uniform convergence on bounded subsets of* $\mathbf{P}(E)$, *is homeomorphic to* G.

A separable projective space bundle (with typical fiber $\mathbf{P}(E)$) over M is a locally trivial fiber bundle for which the transition functions are required to be projective automorphisms of $\mathbf{P}(E)$. We then have

4.1.13. Proposition. *There is a natural bijection between the set of isomorphism classes of separable projective space bundles over* M *and the set* $H^1(M, \underline{G}_M)$ *of isomorphism classes of principal G-bundles over* M.

In terms of projective space bundles, the product has a geometric interpretation as the *algebraic join* of two projective spaces P_1 and P_2. This join $P_1 * P_2$ is defined as the union of projective lines $l_{a,b}$ through a point a of P_1 and a point b of P_2. If $P_1 = \mathbf{P}(E_1)$ and $P_2 = \mathbf{P}(E_2)$ for vector spaces E_1, E_2, $P_1 * P_2$ identifies with $\mathbf{P}(E_1 \oplus E_2)$, and this gives a notion of lines in $P_1 * P_2$. One similarly defines the join of two projective space bundles; this gives a projective space bundle with fiber $\mathbf{P}(E \oplus E) \xrightarrow{\sim} \mathbf{P}(E)$.

We note the following interesting feature of a projective space bundle $P \to M$: if one can find a vector bundle $\mathcal{E} \to M$ such that the associated projective space bundle is isomorphic to $P \to M$, $\mathcal{E}$ is a trivial vector bundle (since the cohomology set $H^1(M, \underline{L}_M^*)$ is trivial), hence P is a trivial projective space bundle. Therefore the obstruction (in $H^2(M, \underline{\mathbb{C}}_M^*)$) to trivializing a projective space bundle is just the obstruction to finding a corresponding vector bundle with fiber E.

The Dixmier-Douady theory we have summarized is very natural and quite powerful. It does, as we have seen, make essential use of infinite-dimensional algebra bundles and perhaps for this reason has not become very familiar to topologists. It is difficult in general to explicitly construct a principal G-bundle corresponding to a given class in $H^2(M, \underline{\mathbb{C}}_M^*)$.

We should briefly explain what happens if one tries to use finite-dimensional matrix algebras $M_n(\mathbb{C})$ instead of L. Of course the linear group $GL_n(\mathbb{C})$ is not contractible. But one can use the central extension of Lie groups

$$1 \to \mathbb{C}^* \to GL(n, \mathbb{C}) \to PGL(n, \mathbb{C}) \to 1$$

and the corresponding coboundary map $\delta_1 : H^1(M, \underline{PGL(n, \mathbb{C})}_M) \to H^2(M, \underline{\mathbb{C}}_M^*)$ to construct classes in $H^2(M, \underline{\mathbb{C}}_M^*)$. However any such class is killed by n. This is because of the following commutative diagram of central extensions, in which $\mu_n \subset \mathbb{C}^*$ denotes the group of n-th roots of unity.

$$
\begin{array}{ccccccccc}
1 & \to & \mu_n & \to & SL(n, \mathbb{C}) & \to & PGL(n, \mathbb{C}) & \to & 1 \\
 & & \downarrow & & \downarrow & & \downarrow & & \\
1 & \to & \mathbb{C}^* & \to & GL(n, \mathbb{C}) & \to & PGL(n, \mathbb{C}) & \to & 1
\end{array}
$$

This diagram shows that the image of $\delta_1 : H^1(M, \underline{PGL(n, \mathbb{C})}_M) \to H^2(M, \underline{\mathbb{C}}^*)$ is contained in the image of the natural homomorphism $H^2(M, \mu_n) \to H^2(M, \underline{\mathbb{C}}_M^*)$, hence is killed by n.

If one worked with finite-dimensional matrix algebras, one would therefore capture the torsion subgroup of $H^2(M, \underline{\mathbb{C}}_M^*) = H^3(M, \mathbb{Z}(1))$. This torsion subgroup is called the *topological Brauer group*. The Brauer group is very important in algebraic geometry [Gr4].

It is time to actually construct some non-trivial examples of G-bundles over some manifolds in order to show that the whole theory is not hopelessly abstract. We will be able to construct such a bundle for a class in $H^3(M, \mathbb{Z}(1))$ which is of the form $\alpha \cup \beta$, where $\alpha \in H^1(M, \mathbb{Z})$ and $\beta \in H^2(M, \mathbb{Z}(1))$. First we use Theorem 2.1.8 to find a hermitian line bundle over M with first Chern class equal to β. Let $Q \to M$ be the corresponding principal $\mathbb{T}$-bundle. We think of α as giving a character $\chi : \pi_1(M) \to \mathbb{Z}$; let $\tilde{M} \to M$ be the corresponding covering with group $\mathbb{Z}$. Then $Q \times_M \tilde{M}$ is a principal bundle over M, with structure group $\mathbb{T} \times \mathbb{Z}$. We now observe that $\mathbb{Z}$ is the Pontryagin dual $\mathbb{T}^{\char`\^} = Hom(\mathbb{T}, \mathbb{T})$ of $\mathbb{T}$. We are therefore led to introduce a generalized Heisenberg group H (in the sense of A. Weil [Weil3]) which is a central extension of $\mathbb{Z} \times \mathbb{T}$ by $\mathbb{T}$.

Weil's construction applies to any locally compact abelian group (countable at infinity), but we will only need the case of $\mathbb{T}$ and its dual $\mathbb{Z}$. The Hilbert space $L^2(\mathbb{T})$ has a basis $(z^n)_{n \in \mathbb{Z}}$; the group $\mathbb{T}$ operates on $L^2(\mathbb{T})$ by translation: for $w \in \mathbb{T}$ and $f \in L^2(\mathbb{T})$, we set $(\sigma(w) \cdot f)(z) = f(wz)$. The group $\mathbb{Z}$ operates on $L^2(\mathbb{T})$ by multiplication; for $n \in \mathbb{Z}$ and $f \in L^2(\mathbb{T})$, we set $(\sigma(n) \cdot f)(z) = z^n \cdot f(z)$. Then σ defines a projective representation of the product group $\mathbb{T} \times \mathbb{Z}$, since we have the Heisenberg type commutation relation

$$[\sigma(w), \sigma(n)] = w^n \cdot Id. \qquad (4-9)$$

To get an actual group representation, one has to introduce the Heisenberg group H; as a set, $H = \mathbb{T} \times \mathbb{T} \times \mathbb{Z}$. We define the product by

$$(w_1, z_1, n_1) \cdot (w_2, z_2, n_2) = (w_1 w_2 z_1^{n_2}, z_1 z_2, n_1 + n_2). \qquad (4-10)$$

Then we define a representation ρ of H on $L^2(\mathbb{T})$ by

$$\rho(w, z, n) \cdot f(\zeta) = w \cdot \zeta^n \cdot f(z \cdot \zeta). \qquad (4-11)$$

This gives a Lie group homomorphism $\rho : H \to Aut(E) = L^*$, where we take $E = L^2(\mathbb{T})$ to be our separable Hilbert space. Passing to the quotient by $\mathbb{C}^*$, we have a Lie group homomorphism $\rho : \mathbb{T} \times \mathbb{Z} \to G = L^*/\mathbb{C}^*$. Hence by applying ρ to the principal bundle $Q \times_M \tilde{M} \to M$ with group $\mathbb{T} \times \mathbb{Z}$, we obtain a new principal bundle $S \to M$ with structure group G.

4.1.14. Theorem. *Let $[Q] \in H^1(M, \underline{\mathbb{C}}^*_M)$ be the class of the line bundle associated with $Q \to M$, and let $[S] \in H^1(M, \underline{G}_M)$ be the class of the G-bundle $S \to M$. Then the class $\delta_1([S]) \in H^2(M, \underline{\mathbb{C}}^*_M)$ is equal to $\alpha \cup [Q]$.*

Proof. We may pick an open covering (U_i) of M such that α is represented by a 1-cocycle n_{ij} with values in $\mathbb{Z}$, and $[Q]$ is represented by the transition 1-cocycle g_{ij} with values in $\underline{\mathbb{C}}^*_M$. According to formula (1-17), the cup-product $[Q] \cup \alpha$ is represented by the 2-cocycle $g_{ij}^{n_{jk}}$. On the other hand, the bundle S is represented by the image in $\underline{G}_M$ of the 1-cocycle (g_{ij}, n_{ij}) with values in $\underline{\mathbb{T} \times \mathbb{Z}}_M$. A corresponding Čech 1-cochain with values in $L^* = Aut(L^2(\mathbb{T}))$ is $u_{ij} = \rho(1, g_{ij}, n_{ij})$. We have to compute the Čech coboundary $a_{ijk} = u_{ik}^{-1} u_{ij} u_{jk}$ with values in $\underline{\mathbb{C}}^*_M$. Using the cocycle relations for the g_{ij} and the n_{ij}, we have

$$a_{ijk} = \rho(1, g_{ij} g_{jk}, n_{ij} + n_{jk})^{-1} \rho(1, g_{ij}, n_{ij}) \rho(1, g_{jk}, n_{jk}) = g_{ij}^{-n_{jk}}.$$

Therefore the class $\delta_1([S])$, which is represented by the 2-cocycle $\underline{a}$, is equal to $-[Q] \cup \alpha = \alpha \cup [Q]$. ∎

4.1.15. Corollary. *The class in $H^3(M, \mathbb{Z}(1))$ corresponding to the G-bundle $S \to M$ is equal to $-\alpha \cup \delta'(\beta)$, the cup-product of $\alpha \in H^1(M, \mathbb{Z})$ and $\delta'(\beta) \in$*

$H^2(M, \mathbb{Z}(1))$, where $\delta' : H^1(M, \underline{\mathbb{C}}^*) \to H^2(M, \mathbb{Z}(1))$ is the coboundary for the exponential exact sequence of sheaves.

Proof. We note that the fact that cup-product on Čech cochains gives a morphism of complexes implies that $\delta'(\alpha \cup \beta) = -\alpha \cup (\delta'(\beta))$. The result follows from Theorem 4.1.14. ∎

This leads, in particular, to a G-bundle corresponding to the generator of $H^3(S^2 \times S^1, \mathbb{Z}(1))$, which will be studied in §3. It seems more difficult to construct a G-bundle corresponding to the generator of $H^3(M, \mathbb{Z}(1))$ for M a compact oriented 3-manifold.

We observe that the theory of Dixmier-Douady applies to a larger class of fields of C^*-algebras than the locally trivial algebra bundles considered here. In fact, these authors study continuous fields of C^*-algebras, which need not be locally trivial. This generality is very important in the theory of C^*-algebras with a continuous trace, which appear as the algebra of L^2-sections of a continuous field (not always locally trivial) over their center.

4.2. Connections and curvature

In §4.1 we showed that classes in $H^3(M, \mathbb{Z}(1))$ are realized geometrically by principal G-bundles over M, where G is the projective linear group of a separable Hilbert space E. In this section, we identify E with $C^\infty(\mathbb{T})$. We wish to develop notions of connection and curvature for G-bundles, such that the curvature will be a complex-valued 3-form Ω. The cohomology class of $[\Omega]$ should be the image in $H^3(M, \mathbb{C})$ of the class of the G-bundle in $H^3(M, \mathbb{Z}(1))$ (see Corollary 4.2.8). However it is not possible to develop a reasonable theory of smooth G-bundles. The reason is simply that the Lie algebra $\mathfrak{g} = L/\mathbb{C}$ is much too small; thus most familiar one-parameter subgroups are not even differentiable anywhere (their "derivative" exists only as an unbounded operator).

Instead of the algebra L of continuous endomorphisms of E, we must consider the algebra L^∞ of continuous endomorphisms of the Fréchet space $E^\infty = C^\infty(S^1)$. Now we run into another problem; we may introduce the group $(L^\infty)^*$ of invertible elements of L^∞, but this is not open in L^∞, so it is not clear whether there is a Lie group structure on L^∞. Therefore we will have to consider the following abstract situation: we give ourselves a central extension of Lie groups

$$1 \to \mathbb{C}^* \to \tilde{B} \to B \to 1 \qquad (4-12)$$

and a smooth linear action of $\tilde{B}$ on E^∞, such that $z \in \mathbb{C}^*$ acts by the scalar $z \cdot Id$. We get a diagram of central extensions of Lie algebras

$$
\begin{array}{ccccccccc}
0 & \to & \mathbb{C} & \to & \tilde{\mathfrak{b}} & \to & \mathfrak{b} & \to & 0 \\
 & & \downarrow{\scriptstyle Id} & & \downarrow & & \downarrow & & \\
0 & \to & \mathbb{C} & \to & L^\infty & \to & L^\infty/\mathbb{C} & \to & 0
\end{array}
\tag{4-13}
$$

Examples of such B and $\tilde{B}$ include:

(1) B is the loop group $B = LU(n)$, $\tilde{B}$ is its central extension discussed in §2.2.

(2) $\tilde{B}$ is the Heisenberg group discussed in §1.

To be precise, in both cases we have a central extension of a Lie group B by $\mathbb{T}$, and $\tilde{B}$ should be defined as the associated central extension of B by $\mathbb{C}^*$.

We will assume given a principal B-bundle $p : Q \to M$ and a connection on this bundle. Recall from §3.3 that the connection form is a 1-form θ on Q with values in the Lie algebra $\mathfrak{b}$ satisfying

(1) For $x \in Q$, $g \in B$, and $v \in T_x(Q)$, we have $\theta_{x \cdot g}(g_*(v)) = Ad(g^{-1})\theta_x(v)$.

(2) For $\xi \in \mathfrak{b}$, we have $\theta(\tilde{\xi}) = \xi$.

Here for $\xi \in \mathfrak{b}$, we denote by $\tilde{\xi}$ the corresponding vector field on Q. Now we want to measure by differential-geometric means the obstruction to lifting the structure group of $p : Q \to M$ to the Lie group $\tilde{B}$, which is a central extension of B by $\mathbb{C}^*$. Such a lifting would be a principal $\tilde{B}$-bundle $\tilde{p} : \tilde{Q} \to M$, together with an isomorphism $f : \tilde{Q}/\mathbb{C}^* \tilde{\to} Q$ of B-bundles over M. Let us consider such a lifting over some open set U of M and analyze possible connections over it. f amounts to a B-equivariant mapping $\tilde{Q} \to Q$, which we also denote by f; of course f must satisfy $p \circ f = \tilde{p}$. A connection on $\tilde{Q}$ is a $\tilde{\mathfrak{b}}$-valued 1-form $\tilde{\theta}$ on $\tilde{Q}$, which satisfies conditions analogous to (1) and (2) above. It is reasonable to require that $\tilde{\theta}$ be *compatible* with θ, in the sense that $q \circ \tilde{\theta} = f^*\theta$, as 1-forms with values in $\mathfrak{b} = \tilde{\mathfrak{b}}/\mathbb{R}(1)$.

4.2.1. Lemma. *The set of connections on $\tilde{Q} \to U$ which are compatible with θ forms an affine space under the vector space $A^1(M)_\mathbb{C}$ of complex-valued 1-forms on M.*

Now if we have a connection θ for $Q \to M$ and a compatible connection $\tilde{\theta}$ for $\tilde{Q} \to M$, then the curvature $\tilde{\Theta}$ of $\tilde{\theta}$ is compatible with the the curvature Θ of θ, in the sense that $q \circ \tilde{\Theta} = f^*\Theta$, as $\mathfrak{b}$-valued 2-forms on Q. Furthermore, we note the following behavior of the curvature of $\tilde{\theta}$ as we translate $\tilde{\theta}$ by a complex-valued 1-form on M.

4.2.2. Lemma. *Let $\tilde{\theta}$ be a connection on $\tilde{Q}$ compatible with θ, and let α be a complex-valued 1-form on M. Then the curvature of $\tilde{\theta} + \alpha$ is equal to $\tilde{\Theta} + d\alpha$, where $\tilde{\Theta}$ is the curvature of $\tilde{\theta}$.*

We would prefer the curvature of the connection $\tilde{\theta}$ to be defined as a scalar-valued 2-form, rather than a 2-form with values in the Lie algebra $\tilde{\mathfrak{b}}$. We therefore wish to introduce the notion of the "scalar component" of a 2-form on $\tilde{Q}$ with values in $\tilde{\mathfrak{b}}$. For this purpose, we need to consider some vector bundles on M associated to representations of B. Recall that for a representation W of B, one has the associated vector bundle $\mathcal{V}(W) = Q \times^G W$, the quotient of $Q \times V$ by the diagonal action of G. One has an exact sequence of representations of G:

$$0 \to \mathbb{C} \to \tilde{\mathfrak{b}} \to \mathfrak{b} \to 0$$

where B acts on $\tilde{\mathfrak{b}}$ and on $\mathfrak{b}$ by the adjoint action (note that the adjoint action of $\tilde{B}$ on $\tilde{\mathfrak{b}}$ factors through an action of B on $\tilde{\mathfrak{b}}$). So we get a corresponding exact sequence of the associated vector bundles over M. We may view the curvature Θ as a 2-form on M with values in the vector bundle $\mathcal{V}(\mathfrak{b})$: this is just a compact way of encoding the transformation properties of Θ under the B-action. Similarly $\tilde{\Theta}$ is a 2-form on M with values in the vector bundle $\mathcal{V}(\tilde{\mathfrak{b}})$.

What we need is a splitting of the exact sequence of vector bundles, i.e., a mapping $l : \mathcal{V}(\tilde{\mathfrak{b}}) \to \tilde{\mathbb{C}} = M \times \mathbb{C}$ of vector bundles over M, which is the identity over the trivial subbundle $M \times \mathbb{C}$. We can view l as a function $l : \mathcal{V}(\tilde{\mathfrak{b}}) \to \mathbb{C}$, which is linear on the fibers, and which satisfies $l(x, 1) = 1$ for any $x \in M$.

4.2.3. Definition. *Let $p : Q \to M$ be a principal B-bundle, with connection θ. Let $l : \mathcal{V}(\tilde{\mathfrak{b}}) \to \mathbb{C}$ be a splitting of the exact sequence of vector bundles*

$$0 \to \mathbb{C} \times M \to \mathcal{V}(\tilde{\mathfrak{b}}) \to \mathcal{V}(\mathfrak{b}) \to 0. \qquad (4-14)$$

For U an open set of M, for a $\tilde{B}$-bundle $\tilde{Q} \to U$ lifting $Q_{/U}$, and for $\tilde{\theta}$ a connection on $\tilde{Q}$ compatible with θ, we define the <u>scalar curvature</u> of $\tilde{\theta}$ to be the complex-valued 2-form $K(\tilde{\theta}) = l \circ \tilde{\Theta}$, where $\tilde{\Theta}$ is the $\tilde{\mathfrak{b}}$-valued curvature of $\tilde{\theta}$, viewed as a 2-form with values in the vector bundle $\mathcal{V}(\tilde{\mathfrak{b}})$.

4.2.4. Proposition. *For a given $\tilde{B}$-bundle $\tilde{Q} \to U$ lifting $Q_{/U}$, and for a complex-valued 1-form α on M, we have*

$$K(\tilde{\theta} + \alpha) = K(\tilde{\theta}) + d\alpha. \qquad (4-15)$$

Proof. We have $K(\tilde{\theta}) = l(\tilde{\Theta}) = l(d\tilde{\theta} + \frac{1}{2}[\tilde{\theta}, \tilde{\theta}])$. Similarly

$$K(\tilde{\theta} + \alpha) = l(d\tilde{\theta} + d\alpha + \frac{1}{2}[\tilde{\theta} + \alpha, \tilde{\theta} + \alpha]) = l(d\tilde{\theta} + \frac{1}{2}[\tilde{\theta}, \tilde{\theta}]) + d\alpha$$

$$= K(\tilde{\theta}) + d\alpha \qquad \blacksquare$$

We now have succeeded in defining a scalar 2-form $K(\tilde{\theta})$ associated to a connection on a $\tilde{B}$-bundle $\tilde{Q} \to U$ lifting $Q_{/U}$. It has the property (4-15) expected of the curvature, but note that it need not be a closed 2-form. Instead, we have the following result.

4.2.5. Proposition. *Let the B-bundle $p : Q \to M$ with the connection θ be given, as well as the fiberwise linear function $l : \mathcal{V}(\tilde{\mathfrak{b}}) \to \mathbb{C}$ such that $l(x, 1) = 1$ for $x \in M$. There exists a unique closed $\mathbb{C}$-valued 3-form Ω on M such that for every $\tilde{B}$-bundle $\tilde{Q} \to U$ over an open set U of M lifting Q, and any connection $\tilde{\theta}$ on $\tilde{Q}$ compatible with θ, we have*

$$\Omega_{/U} = d(K(\tilde{\theta})), \qquad (4-16)$$

where $K(\tilde{\theta})$ is the scalar curvature of $\tilde{\theta}$.

Proof. It is enough to show that over an open set U of M, $d(K(\tilde{\theta}))$ is independent of the $\tilde{B}$-bundle $\tilde{Q} \to M$ lifting Q, and of the connection $\tilde{\theta}$ compatible with θ. The independence of $\tilde{\theta}$ follows immediately from (4-15), since $d(K(\tilde{\theta} + \alpha)) = d(K(\tilde{\theta}) + d\alpha) = d(K(\tilde{\theta}))$. The independence of $\tilde{Q}$ is a local question. Locally all such liftings to $\tilde{B}$ of the structure group of Q are isomorphic to one another, and the result is obvious. $\blacksquare$

The 3-form Ω defined in Proposition 4.2.5 will be called the 3-*curvature* of the B-bundle $Q \to M$, equipped with the connection θ and the fiberwise linear function $l : \mathcal{V}(\tilde{\mathfrak{b}}) \to \mathbb{C}$.

We may also consider central extensions by $\mathbb{T}$ instead of $\mathbb{C}^*$. In 4.2.1, 4.2.2 and 4.2.4, replacing $A^1(M)_{\mathbb{C}}$ by the space $\sqrt{-1} \cdot A^1(M)$ of purely imaginary 1-forms, the 3-curvature is then purely imaginary.

We now will relate the cohomology class of this curvature 3-form Ω to the class in $H^3(M, \mathbb{Z}(1))$ of the B-bundle $Q \to M$. First, we will reformulate the construction of the class in $H^2(M, \mathbb{C}_M^*)$ in terms of the (local) liftings $\tilde{Q} \to M$ of the structure group to $\tilde{B}$. Note that over some open set U of M these liftings are the objects of a category C_U. An object of C_U is a principal $\tilde{B}$-bundle $\tilde{p} : \tilde{Q} \to U$, together with an isomorphism of B-bundles $f : \tilde{Q}/\mathbb{C}^* \xrightarrow{\sim} Q_{/U}$. Given two liftings $(\tilde{Q}_1, f_1)$ and $(\tilde{Q}_2, f_2)$, a morphism from

$(\tilde{Q}_1, f_1)$ to $(\tilde{Q}_2, f_2)$ is a $\tilde{B}$-bundle isomorphism $g : \tilde{Q}_1 \tilde{\to} \tilde{Q}_2$ such that the following diagram commutes

$$
\begin{array}{ccc}
\tilde{Q}_1 & \xrightarrow{\quad g \quad} & \tilde{Q}_2 \\
& \searrow_{f_1} \quad \swarrow_{f_2} & \\
& Q &
\end{array}
\qquad (4-17)
$$

This category is a *groupoid*, i.e., every morphism in C_U is invertible. An important point is that the group of automorphisms of an object $(\tilde{Q}, f)$ is equal to $\Gamma(U, \underline{\mathbb{C}}_M^*)$. This means that an automorphism of $(\tilde{Q}, f)$ is given by the action of a smooth $\mathbb{C}^*$-valued function, which is clear from the definitions. The category C_U is much easier to work with than bundles with non-abelian structure group, although it is more abstract.

Given an inclusion $V \subseteq U$ of open sets, there is a restriction functor $C_U \to C_V$. The restriction functor is not quite associative, but it is associative up to some canonical isomorphisms. The following gives a purely "abelian" description of the class in $H^2(M, \underline{\mathbb{C}}_M^*)$.

4.2.6. Proposition. *Let $p : Q \to M$ be a principal B-bundle, and $[Q]$ its class in $H^1(M, \underline{B})$. The corresponding class $\delta_1([Q]) \in H^2(M, \underline{\mathbb{C}}^*_M)$ admits the following description. Choose an open covering $(U_i)_{i \in I}$ of M such that there exists a lifting $\tilde{p}_i : \tilde{Q}_i \to U_i$ of the structure group to $\tilde{B}$, with given isomorphism $f_i : \tilde{Q}_i/\mathbb{C}^* \tilde{\to} Q_{/U_i}$. Assume there exists an isomorphism $u_{ij} : (\tilde{Q}_j, f_j) \tilde{\to} (\tilde{Q}_i, f_i)$ in the category $C_{U_{ij}}$. For any $i, j, k \in I$, we have a section h_{ijk} of $\underline{\mathbb{C}}^*_M$ over U_{ijk} such that*

$$
h_{ijk} = u_{ik}^{-1} u_{ij} u_{jk}, \qquad (4-18)
$$

an equality in the automorphism group of the object $(\tilde{Q}_k, f_k)$ of $C_{U_{ijk}}$. Then $\underline{h}$ is a $\underline{\mathbb{C}}^*_M$-valued Čech 2-cocycle which represents $\delta_1([Q])$.

Proof. It is first of all clear that changing the isomorphisms u_{ij} into $u_{ij} \cdot \alpha_{ij}$, for $\alpha_{ij} \in \Gamma(U_{ij}, \underline{\mathbb{C}}^*_M)$, will modify $\underline{h}$ by the coboundary of α_{ij}. One therefore only needs to prove the statement for a particular choice of the u_{ij}. First we note that the construction of the cohomology class $[Q] \in H^1(M, \underline{B})$ may be described in terms of trivializations $t_i : U_i \times B \tilde{\to} Q_{/U_i}$ of the principal bundle over the open sets U_i. Then $g_{ij} = t_i^{-1} t_j$ is an automorphism of the trivial principal bundle over U_{ij}, which amounts to a B-valued function g_{ij} (acting on $M \times U_{ij}$ by left translation on the first factor). The g_{ij} is a 1-cocycle in the class $[Q]$. To compute the Čech coboundary of g_{ij}, we need to choose a

$\bar{B}$-valued-function b_{ij} which lifts g_{ij}. Then, according to Theorem 4.1.4, the 2-cocycle in the class $\delta_1([Q])$ is $a_{ijk} = b_{ik}^{-1} b_{ij} b_{jk}$. Now (after shrinking the open covering if necessary), we may assume that t_i lifts to an isomorphism $\tilde{t}_i : U_i \times \bar{B} \xrightarrow{\sim} \tilde{Q}_i$ of $\bar{B}$-bundles, such that the diagram

$$
\begin{array}{ccc}
U_i \times \bar{B} & \xrightarrow{\tilde{t}_i} & \tilde{Q}_i \\
{\scriptstyle can}\downarrow & & \downarrow{\scriptstyle f_i} \\
U_i \times B & \xrightarrow{t_i} & Q_{/U_i}
\end{array}
$$

commutes, where *can* is the canonical projection. Now we have an isomorphism $u_{ij} : \tilde{Q}_j \xrightarrow{\sim} \tilde{Q}_i$ of bundles over U_{ij}, which makes the diagram

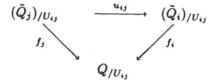

commute, namely ,

$$ u_{ij} = \tilde{t}_i b_{ij} \tilde{t}_j^{-1}. $$

The corresponding Čech 2-cochain described in the statement of the proposition is

$$
\begin{aligned}
h_{ijk} &= u_{ik}^{-1} u_{ij} u_{ik} \\
&= \tilde{t}_k b_{ij}^{-1} \tilde{t}_i^{-1} \tilde{t}_i b_{ij} \tilde{t}_j^{-1} \tilde{t}_j b_{jk} \tilde{t}_k^{-1} \\
&= \tilde{t}_k a_{ijk} \tilde{t}_k^{-1} = a_{ijk}
\end{aligned}
$$

as $a_{ijk} \in \underline{\mathbb{C}}_M^*$ commutes with everything. This finishes the proof. ∎

We are now in a completely abelian world with respect to our description of the cohomology class of $H^2(M, \underline{\mathbb{C}}_M^*)$. But we have paid a small price, namely, we are working with objects of the categories C_U for open sets U of M. We can now start relating our constructions with smooth Deligne cohomology. We will start from the Čech 2-cocycle $\underline{h}$ of Proposition 4.2.6 and build a Čech cocycle $\underline{c}$ of (U_i), with values in the complex of sheaves

$$ \underline{\mathbb{C}}_M^* \xrightarrow{d \, log} \underline{A}_{\mathbb{C}}^1 \to \underline{A}_{\mathbb{C}}^2 \qquad (4-19) $$

in which $\underline{\mathbb{C}}_M^*$ is placed in degree 0. We first pick a connection $\tilde{\theta}_i$ on $\tilde{Q}_i$, compatible with θ. Then we look at the isomorphism $u_{ij} : (\tilde{Q}_j, f_j) \xrightarrow{\sim} (\tilde{Q}_i, f_i)$

in the category $C_{U_{ij}}$. This is an isomorphism of principal $\tilde{B}$-bundles, compatible with $f_i : \tilde{Q}_i/\mathbb{C}^* \overset{\sim}{\to} Q_{U_i}$ and with f_j. Then, on $\tilde{Q}_{i/U_{ij}}$, we have two connections compatible with θ, namely $\tilde{\theta}_i$ and $(u_{ij})_*(\tilde{\theta}_j)$, the transform of $\tilde{\theta}_j$ under the isomorphism u_{ij}. According to Lemma 4.2.1, there is a 1-form α_{ij} on U_{ij} such that

$$\tilde{\theta}_i = (u_{ij*})(\tilde{\theta}_j) + \alpha_{ij}. \qquad (4-20)$$

For $i, j, k \in I$, we have on U_{ijk}:

$$\tilde{\theta}_i = (u_{ij})_*(\tilde{\theta}_j) + \alpha_{ij} = (u_{ij})_*(u_{jk})_*(\tilde{\theta}_k) + \alpha_{jk} + \alpha_{ij} \qquad (4-21)$$

On the other hand, we have

$$\tilde{\theta}_i = (u_{ik})_*\tilde{\theta}_k + \alpha_{ik}.$$

Combining (4-20) and (4-21), and using the definition of h_{ijk}, we get

$$\alpha_{ik} - \alpha_{ij} - \alpha_{jk} = (h_{ijk})_*(\tilde{\theta}_k) - \tilde{\theta}_k.$$

However, it is easy to see that

$$g_*\tilde{\theta} = \tilde{\theta} - g^{-1}dg \qquad (4-22)$$

for any connection $\tilde{\theta}$ and any $\mathbb{C}^*$-valued function g. So we obtain

$$\alpha_{jk} - \alpha_{ij} + \alpha_{jk} = d\,log(h_{ijk}), \qquad (4-23)$$

which means that $(\underline{h}, -\underline{\alpha})$ together make a Čech 2-cocycle with values in the complex of sheaves $\underline{\mathbb{C}}_M^* \overset{d\ log}{\longrightarrow} \underline{A}_{M,\mathbb{C}}^1$.

We now look at the scalar curvatures. Let K_i be the scalar curvature of the connection $\tilde{\theta}_i$ on $\tilde{Q}_i$. Then from (4-20) and (4-15) we obtain $K_i = K_j + d\alpha_{ij}$, i.e $\delta(\underline{K}) = -d\underline{\alpha}$.

Hence we have proved

4.2.7. Proposition. $(\underline{h}, -\underline{\alpha}, \underline{K})$ *is a Čech 2-cocycle of the covering with coefficients in the complex of sheaves (4-19). Equivalently it defines a cohomology class in* $H^3(M, \mathbb{Z}(2)_D^\infty) \otimes \mathbb{Z}(-1)$, *where* $H^3(M, \mathbb{Z}(2)_D^\infty)$ *is a smooth Deligne cohomology group. The smooth Deligne cohomology class maps to* $\delta([Q]) \in H^3(M, \mathbb{Z}(1)) \overset{\sim}{\to} H^2(M, \mathbb{C}^*)$ *under the canonical map from* $H^3(M, \mathbb{Z}(2)_D^\infty) \otimes \mathbb{Z}(-1)$ *to* $H^3(M, \mathbb{Z}(1))$. *It maps to the 3-form* Ω *(the 3-curvature) under the canonical map* $H^3(M, \mathbb{Z}(2)_D^\infty) \otimes \mathbb{Z}(-1) \to A^3(M)_\mathbb{C}$

Proof. Recall from Chapter 1 the quasi-isomorphism of complexes of sheaves

$$
\begin{array}{ccccccc}
\mathbf{Z}(1) & \to & \underline{A}^0_{M,\mathbf{C}} & \to & \underline{A}^1_{M,\mathbf{C}} & \to & \underline{A}^2_{M,\mathbf{C}} \\
\downarrow & & \downarrow exp & & \downarrow Id & & \downarrow Id \\
0 & \to & \underline{\mathbf{C}}^*_M & \xrightarrow{d\ log} & \underline{A}^1_{M,\mathbf{C}} & \to & \underline{A}^2_{M,\mathbf{C}}
\end{array}.
$$

The first complex of sheaves is $\mathbf{Z}(2)^\infty_D \otimes \mathbf{Z}(-1)$, and the second one is the complex of sheaves (4-19), shifted by 1 to the right. Therefore $H^2(M, \underline{\mathbf{C}}^*_M \xrightarrow{d\ log} \underline{A}^1_{M,\mathbf{C}} \to \underline{A}^2_{M,\mathbf{C}})$ is isomorphic to $H^3(M, \mathbf{Z}(2)^\infty_D) \otimes \mathbf{Z}(-1)$, which justifies the first part of the statement.

The map $H^3(M, \mathbf{Z}(2)^\infty_D) \otimes \mathbf{Z}(-1) \to A^3(M)_{\mathbf{C}}$ corresponds to the map induced on degree 2 sheaf hypercohomology by the following morphism of complexes of sheaves

$$
\begin{array}{ccccc}
\underline{\mathbf{C}}^*_M & \xrightarrow{d\ log} & \underline{A}^1_{M,\mathbf{C}} & \to & \underline{A}^2_{M,\mathbf{C}} \\
& & & & \downarrow d \\
& & & & \underline{A}^3_{M,\mathbf{C}}
\end{array}
$$

The image of the Čech cocycle $(\underline{h}, -\underline{\alpha}, \underline{K})$ in $A^3(M)_{\mathbf{C}} = \Gamma(M, \underline{A}^3_{\mathbf{C}})$ is the 3-form which on U_i is equal to dK_i; thus it is equal to Ω. ∎

4.2.8. Corollary. *The image in $H^3(M, \mathbf{C})$ of the cohomology class $\delta([Q]) \in H^3(M, \mathbf{Z}(1))$ coincides with the cohomology class of Ω.*

We note the analogy with the treatment of line bundles with connection using Deligne cohomology presented in §2.2. Recall that Theorems 2.2.14 and 2.2.15 state that a closed 2-form is the curvature of some line bundle with connection if and only if its cohomology class, divided by $2\pi\sqrt{-1}$, is integral. From Corollary 4.2.8 we have an integrality condition for the 3-curvature of a B-bundle. We are not able to prove a converse to this result in this chapter, since the present theory, based on B-bundles, is not flexible enough for this purpose. Indeed, from the point of view adopted in Chapter 4, it is even completely unclear what a good choice of the Lie group B might be. In Chapter 5 we will use sheaves of groupoids to prove such a result for 3-forms.

4.3. Examples of projective Hilbert space bundles

In §2 we introduced a central extension $\tilde{B}$ of a Lie group B by $\mathbf{C}^*$ or $\mathbf{T}$, and we studied the differential geometry of principal B-bundles and their local liftings to $\tilde{B}$-bundles. We will give a few examples. The first example, already introduced in §1, comes from the cup-product of a class in $H^1(X, \mathbb{Z}(1))$ with a class in $H^2(X, \mathbb{Z})$. We have a canonical element of $H^1(\mathbb{T}, \mathbb{Z}(1))$ which corresponds to the universal covering map $\mathbb{R}(1) \to \mathbb{R}(1)/\mathbb{Z}(1) = \mathbb{T}$ with group of deck transformations equal to $\mathbb{Z}(1)$. We next consider an S^1-bundle $q : S \to X$, where X is a smooth manifold. This gives a class in $H^2(X, \mathbb{Z})$, using the isomorphism $H^1(X, \mathbb{R}/\mathbb{Z}) \xrightarrow{\sim} H^2(X, \mathbb{Z})$. Then over the product manifold $X \times \mathbb{T}$, we have both a $\mathbb{Z}(1)$-bundle and a S^1-bundle, pulled-back from S^1 and from X respectively. Following §1, we introduce a Heisenberg group H, which is a central extension of the product group $S^1 \times \mathbb{Z}(1)$ by $\mathbb{T}$ (this is slightly different from the Heisenberg group considered in §1). Note that H acts smoothly on the separable Hilbert space $E = L^2(S^1)$. So we are in the situation of §2, with $\tilde{B}$ replaced by H and B replaced by $S^1 \times \mathbb{Z}(1)$.

We will describe explicitly the corresponding projective space bundle $P \to X \times \mathbb{T}$. The fiber $P_{(x,z)}$ at $(x, z) \in X \times \mathbb{T}$ will be the projective Hilbert space $\mathbb{P}(E_{(x,z)})$, where $E_{(x,z)}$ depends on the choice of a universal cover $\tilde{C}_x$ of the fiber $C_x = q^{-1}(x)$, which is a circle without fixed base point. Let $T : \tilde{C}_x \to \tilde{C}_x$ be the automorphism corresponding to the generator of $H_1(S^1, \mathbb{Z})$. We have a Hilbert space

$$E_{x,z} = \{f \text{ measurable function on } \tilde{C}_x : f \circ T = z \cdot f \text{ and } |f| \text{ is } L^2 \text{ on } C_x\}. \tag{4-24}$$

The Hilbert space depends on the choice of $\tilde{C}_x$, which is only unique up to a non-unique diffeomorphism. The amount of non-uniqueness is exactly described by the cyclic group of deck transformations of $\tilde{C}_x$, which is generated by T. However, since T acts on $E_{(x,z)}$ by scalars, we find that the definition of the projective space $\mathbb{P}(E_{(x,z)})$ is indeed independent of the choice of $\tilde{C}_x$.

To describe the smooth bundle $P \to X \times \mathbb{T}$, we need to choose a class of smooth trivializations $\phi_{(x,z)} : \mathbb{P}(E) \xrightarrow{\sim} \mathbb{P}(E_{(x,z)})$ over open subsets of $X \times S^1$. We take an open set $U \subseteq X$ over which S is trivialized; this means that for $x \in U$ we can identify the fiber C_x with S^1 and take $\mathbb{R}$ as its universal cover. We have $T(t) = t + 1$, for t the coordinate on S^1. We also pick an open set $V \subset \mathbb{T}$ over which one can define a logarithm function u, i.e., a complex-valued function $u(z)$ such $e^u = z$. Then for $(x, z) \in U \times V$, we have an isomorphism $\phi_{(x,z)} : L^2(S^1) \xrightarrow{\sim} E_{(x,z)}$ given by

$$\phi_{(x,z)}(f)(t) = e^{t \cdot u} \cdot f(t). \tag{4-25}$$

The induced isomorphism of projective spaces is what we want. Notice that (4-25) also gives an isomorphism of $E^\infty = C^\infty(S^1)$ with the space $E_{(x,z)}^\infty$ of smooth functions in $E_{(x,z)}$.

Let us see how we can recover the $S^1 \times \mathbb{Z}(1)$-bundle $N \to X \times \mathbb{T}$. The fiber of N over (x, z) consists of all isomorphisms $\mathbf{P}(E) \xrightarrow{\sim} \mathbf{P}(E_{(x,z)})$ induced by some $\phi_{(x,z)}$ of the form (4-25), for suitable choices of a base point in C_x and of a logarithm u of z. Note that S^1 acts by translation on the set of trivializations of C_x, and that $\mathbf{Z}(1)$ acts by addition on the set of logarithms of z. We see again that N identifies with $S \times \mathbf{R}(1)$, which maps to $X \times \mathbf{T}$ by (q, exp), and the action of $S^1 \times \mathbf{Z}(1)$ on $S \times \mathbf{R}(1)$ is the natural one. In order to apply the differential-geometric methods of §2, we have to pick a connection on the $S^1 \times \mathbf{Z}(1)$-bundle $S \times \mathbf{R}(1) \to X \times \mathbf{T}$. For this we will need a connection A for $S \to X$, which is a 1-form with values in the Lie algebra of $S^1 = \mathbf{R}/\mathbf{Z}$. This Lie algebra has basis $\frac{d}{dt}$, for t the coordinate on S^1; it is often convenient to think of A as scalar-valued. The curvature $R = dA$ of A represents the class of S in $H^2(M, \mathbf{Z})$.

The Lie algebra of the structure group $S^1 \times \mathbf{Z}(1)$ is also $\mathbf{R} \cdot \frac{d}{dt}$. Our connection θ will be a $\mathbf{R} \cdot \frac{d}{dt}$-valued 1-form on $X \times S^1$, which is invariant under $S^1 \times \mathbf{Z}(1)$ and satisfies $\theta(\frac{d}{dt}) = \frac{d}{dt}$. We have to define $\theta(v)$ for v a tangent vector to $S \times \mathbf{R}(1)$. We set $\theta(v) = A(v) \cdot \frac{d}{dt}$ for tangent vectors v to S and $\theta(w) = 0$ for tangent vectors w to $\mathbf{R}(1)$. This defines our connection θ.

We need to be more explicit about our definition of the Heisenberg group H. So we define H to be the manifold $H = \mathbf{T} \times S^1 \times \mathbf{Z}(1)$, with product law

$$(z_1, t_1, u_1) \cdot (z_2, t_2, u_2) = (z_1 z_2 e^{t_1 u_2}, t_1 + t_2, , u_1 + u_2). \qquad (4 - 26)$$

Note here that for $t \in S^1$ and $u \in \mathbf{Z}(1)$, e^{tu} is a well-defined element of $\mathbf{T}$. The Lie algebra $\mathfrak{h}$ of H is a two-dimensional real abelian Lie algebra, with basis $(\sqrt{-1} \cdot Id, \frac{d}{dt})$. The adjoint action of H on $\mathfrak{h}$ factors through an action of the quotient group $S^1 \times \mathbf{Z}(1)$. S^1 acts trivially on $\mathfrak{h}$, but $u \in \mathbf{Z}(1)$ transforms $\frac{d}{dt}$ into $\frac{d}{dt} - u \cdot Id$. Thus the associated bundle $\mathcal{V}(\mathfrak{h})$ admits the following description: its sections over an open set U of $X \times \mathbf{T}$ are those $\mathfrak{h}$-valued functions $f(x, u)$ on $(Id, exp)^{-1}(U) \subseteq X \times \mathbf{R}(1)$ such that

$$f(x, u + v) = Ad(-v) \cdot f(x, u) \quad for \ v \in \mathbf{Z}(1). \qquad (4 - 27)$$

Writing $f(x, u) = g(x, u) \cdot Id + h(x, u)\frac{d}{dt}$, (4-27) means that $h(x, u + v) = h(x, u)$ and that $g(x, u + v) = g(x, u) + v \cdot h(x, u)$.

We have an exact sequence of associated vector bundles over $X \times \mathbf{T}$

$$0 \to \mathcal{V}(\mathbf{R}(1) \cdot Id) \to \mathcal{V}(\mathfrak{h}) \to \mathcal{V}(\mathbf{R} \cdot \frac{d}{dt}) \to 0 \qquad (4 - 28)$$

and, following §2, we want to find a splitting $l : \mathcal{V}(\mathfrak{h}) \to \mathbf{R}(1)$. We let $l(a, b) = f(x, u) \cdot Id$, where f is the function $f(x, u) = a(x, u) - u \cdot b(x, u)$. This is indeed periodic with respect to $\mathbf{Z}(1)$ in the variable $u \in \mathbf{R}(1)$. So l

does define a vector bundle map from $V(\mathfrak{h})$ to $\mathbf{R}(1)$, which maps $u \cdot Id$ to itself. We will use this splitting to define our scalar curvature.

With all these preparations behind us, we can study the scalar curvature of an H-bundle $\tilde{N} \to U$ over an open set U of $X \times \mathbf{T}$, such that $\tilde{N}/\mathbf{T}$ is isomorphic to $N_{/U}$. We may take U to be of the form $U = V \times W$, where $V \subseteq X$ is an open set such that $q^{-1}(V)$ is diffeomorphic to $V \times S^1$, and $W \subset \mathbf{T}$ an open set such that $exp^{-1}(W)$ is diffeomorphic to $W \times \mathbf{Z}(1)$. Then we can take $\tilde{N}$ to be $V \times W \times H$, with the obvious projection map to $V \times W \times (S^1 \times \mathbf{Z}(1))$. We define a connection $\tilde{\theta}$ on this H-bundle which is compatible with the connection θ on the $S^1 \times \mathbf{Z}(1)$-bundle $N \to X \times \mathbf{T}$. The connection A on $S \to X$ gives a 1-form B on V. Because of its required invariance property with respect to the action of $\mathbf{Z}(1)$, the connection $\tilde{\theta}$ will be determined uniquely by its restriction to $V \times W \times \mathbf{T} \times S^1$, where $\mathbf{T} \times S^1$ is a subset of H. We then set:

$$\tilde{\theta}(v) = \left\{ \begin{array}{ll} B(v) \cdot \dfrac{d}{dt} & \text{if } v \in TV \\ 0 & \text{if } v \in TW \\ v & \text{if } v \in \mathfrak{h} \end{array} \right\} \qquad (4-29)$$

Since $\mathfrak{h}$ is abelian, the curvature $\tilde{\Theta}$ of the connection $\tilde{\theta}$ is given by $\tilde{\Theta} = (dB) \otimes \frac{d}{dt} = R \otimes \frac{d}{dt}$, a 2-form with values in $\mathfrak{h}$. Applying the linear form l to get the scalar curvature $K(\tilde{\theta})$, and from the definition of l, we obtain

$$K(\tilde{\theta}) = R \otimes l(\frac{d}{dt}) = -u \cdot R = -Log(z) \cdot R,$$

where z is the coordinate on $W \subset \mathbf{T}$, and $u = Log(z)$ the corresponding coordinate on $exp^{-1}(W) \subset \mathbf{R}(1)$. Finally we obtain the formula for the 3-curvature:

4.3.1. Proposition. *The 3-curvature of the $S^1 \times \mathbf{Z}(1)$-bundle $N \to X \times \mathbf{T}$ is equal to*

$$-\frac{dz}{z} \wedge R. \qquad (4-30)$$

Its cohomology class is the opposite of the cup-product of the canonical class in $H^1(\mathbf{T}, \mathbf{Z}(1))$ and of the class $[S] \in H^2(X, \mathbf{Z})$ of the S^1-bundle $S \to X$.

This gives the differential-geometric counterpart for the constructions in §1 concerning the G-bundle over $X \times \mathbf{T}$; it is interesting to point out that we needed to restrict the structure group to $N \subset G$ in order for the differential geometry to make sense. The restriction of the structure group was already apparent in Theorem 4.1.4. We would like to point out the relation of Proposition 4.3.2 with the work of Raeburn and J. Rosenberg [R-R], which actually inspired the present approach. Assume X is compact. The space of continuous sections of the field of C^*-algebras associated to the G-bundle over $X \times \mathbb{T}$ identifies with the crossed-product algebra $C(S^1) \ltimes C(S)$ for the (free) action of S^1 on S. In fact, the center of this crossed-product algebra is $C(X \times S^1)$, and we get the standard continuous field of elementary C^*-algebras over $X \times S^1$. This assumes the usual identification of S^1 with $\mathbb{T}$ (by $t \mapsto e^{2\pi\sqrt{-1}t}$). The paper [R-R] discusses the case of the Hopf fibration $S^3 \to S^2$ in some detail.

There is another important class of examples, due to Gotay, Lashof, Sniatycki and Weinstein [G-L-S-W]. Let X be a smooth manifold, and let $p : Y \to X$ be a smooth *family of symplectic manifolds*. By this we mean the following:

(1) $p : Y \to X$ is a smooth submersion with connected finite-dimensional fibers Y_x such that $H^1(Y_x, \mathbb{R}) = 0$.

(2) We have a relative 2-form ω on Y such that, for each $x \in X$, the restriction of ω to the fiber $Y_x = p^{-1}(x)$ is a symplectic form.

In that situation, there is an obstruction to finding a closed 2-form β on Y such that the relative 2-form is equal to ω. However [G-L-S-W] show that this obstruction vanishes when the fibers are finite-dimensional and simply-connected.

We will make some further assumptions:

(3) The symplectic form ω_x on each fiber Y_x has *integral* cohomology class.

(4) p is a proper map.

It follows from (3) that the cohomology class of $[\omega_x]$ is in some sense constant and we have a bundle of projective Hilbert spaces over X defined as follows. For any x, we may choose a hermitian line bundle L on the fiber Y_x, with hermitian connection ∇, such that the curvature is exactly equal to $2\pi\sqrt{-1} \cdot \omega_x$ and we have the Hilbert space E_x of L^2-sections of L. Note that since Y_x is simply-connected, the pair (L, ∇) with curvature $2\pi\sqrt{-1} \cdot \omega_x$ is unique up to isomorphism.

To be more precise, if (L_1, ∇_1) and (L_2, ∇_2) are two such hermitian line bundles with hermitian connection, there exists an isomorphism $(L_1, \nabla_1) \xrightarrow{\sim} (L_2, \nabla_2)$, which is unique up to scalars. Hence the projective space $\mathbb{P}(E_x)$ is canonically defined. To define a fiber bundle $P \to X$ with fiber $\mathbb{P}(E_x)$, we need to give a class of local trivializations of E_x. Over a contractible neighborhood U of any $x \in X$, we may identify $p^{-1}(U)$ with $U \times Y_x$. Then we see ω_y (for $y \in U$) as a family of 2-forms on Y_x with the same cohomology class. Hence we can find a purely imaginary relative 1-form α such that $\omega_y = \omega_x + d\alpha_y$ for $y \in U$. Given the line bundle (L, ∇_x) on Y_x, we can take the line bundle on Y_y to be $(L, \nabla_x + \alpha_y)$. The spaces E_x and E_y are now identified since they are the spaces of sections of the same line bundle over Y_x.

This projective space bundle gives rise to a G-bundle $Q \to X$. However it is not clear how to do differential geometry for this bundle, in the sense of §2, and no Lie group B even presents itself. This is again a manifestation of the fact that the differential-geometric theory developed in §2 is neither very flexible nor general enough for many applications (as opposed to the general theory of sheaves of groupoids, to be presented in Chapter 5).

What we can do is discuss the obstruction to finding a closed 2-form β on Y restricting to ω on the fibers from the point of view of the Leray spectral sequence. We have no reason at this point to think we can take β to be closed, but we can take β to be closed up to a basic 3-form, as the next lemma will show. We will first need to recall from §1.6 the Leray spectral sequence for the mapping p. This is the spectral sequence associated to the filtered de Rham complex $A^\bullet(Y)$. The filtration is the Cartan filtration $\cdots \subset F^{p+1} \subset F^p \subset \cdots F^0 = A^\bullet(Y)$, where the degree n component of F^p consists of n-forms α such that $i(v_1) \cdots i(v_{n-p+1})\alpha = 0$, for any vertical vector fields $v_1, \cdots, v_{n-p+1}$. The spectral sequence for this filtered complex has E_1-term

$$E_1^{p,q} = A^p(X, \underline{H}^q(Y_x)) \qquad (4-31)$$

(p-forms on X with values in the local system with fiber $H^q(Y_x)$), and E_2-term

$$E_2^{p,q} = H^p(X, \underline{H}^q(Y_x)) \qquad (4-32)$$

4.3.2. Lemma. *With the assumptions (1), (2), (3), there exists a 2-form β on Y, which restricts to $[\omega]$ on the fibers of p, and satisfies $d\beta = p^*\nu$, for some 3-form ν on X.*

Proof. We look at the Leray spectral sequence for p. Since $[\omega_x]$ is integral for all x, the differential $d_1([\omega]) \in A^1(X, \underline{H}^2(Y_x))$ is 0, implying that $[\omega]$ lifts to a cohomology class in the quotient complex $A^\bullet(Y)/F^2$. The obstruction to lifting this to a cohomology class in $A^\bullet(Y)/F^3$ belongs to $H^3(F^2/F^3) = E_\infty^{2,1}$. But the group $E_2^{2,1} = H^2(X, \underline{H}^1(Y_x))$ is 0 because of assumption (1), hence we certainly have $E_\infty^{2,1} = 0$. Let now β' be an element of $A^3(Y)$ such that β' is closed modulo F^3 and the image of β' in $A^2(Y/X)$ is cohomologous to $[\omega]$. If $\alpha \in A^1(Y)$ is such that $d\alpha \equiv \omega - \beta'$ mod F^1, then $\beta = \beta' + d\alpha$ maps to the relative 2-form ω. Furthermore $d\beta = d\beta' \in F^3$ is a basic differential form, so it is of the form $p^*\nu$, for some 3-form ν on X. This completes the proof. ∎

The obstruction to finding a closed 2-form β restricting to ω on the fibers is exactly the cohomology class of ν in $H^3(X, \mathbb{R})$. Note that this class is equal to $d_3([\omega]) \in E_3^{3,0} = H^3(X, \mathbb{R})$, where d_3 is the differential of the Leray spectral sequence. We come to the theorem of Gotay, Lashof, Sniatycki and Weinstein.

4.3.3. Theorem. *Let $p : X \to Y$ be a smooth family of finite-dimensional symplectic manifolds satisfying (1), (2) and (3). Let $\omega \in A^2(Y/X)$ be the relative symplectic form. Then there exists a closed 2-form β on Y which restricts to ω on the fibers.*

Proof. Let $2n$ be the dimension of the fibers. We want to show that the cohomology class of the 3-form ν of Lemma 4.3.2 is trivial. We have $d\beta^{n+1} = (n+1)\beta^n \cup p^*\nu$. On the other hand, we can write $\beta^{n+1} \equiv \beta^n \wedge \alpha$ mod F^3 for a unique 2-form $\alpha \in F^2$ (i.e., α is horizontal). So we have

$$(n+1)\beta^n \cup p^*\nu \equiv \beta^n \wedge d\alpha + n\beta^{n-1} \wedge p^*(\nu) \wedge \alpha \text{ mod } dF^3.$$

The term $n\beta^{n-1} \wedge p^*\nu \wedge \alpha$ belongs to F^4. So we find that

$$\beta^n \wedge (p^*\nu - d(\frac{\alpha}{n+1})) \in F^4 + dF^3. \qquad (4-33)$$

Therefore $\beta^n \wedge (\nu - d(\frac{\alpha}{n+1}))$ gives the zero homology class in $H^{2n+3}(F^3/F^4) = A^3(X) \otimes H^{2n}(Y_x, \mathbb{R})$. Since $[\beta^n]$ is a generator of $H^{2n}(Y_x, \mathbb{R})$, we obtain $p^*\nu = d(\frac{\alpha}{n+1})$. The 2-form $\frac{\alpha}{n+1}$ is horizontal; since its exterior derivative is horizontal, this 2-form is basic, and so is of the type $p^*\gamma$ for some 2-form γ on X. Then $\nu = d(\gamma)$. ∎

We remark that in [G-L-S-W] this result is proved without the assumption that the cohomology class of ω_x is integral, but the authors then need to assume that X is simply-connected. We emphasize that only the fiber (not the base manifold) has to be finite-dimensional in order for this theorem to hold. The assumption that $H^1(Y_x, \mathbb{R}) = 0$ is however essential, as was shown in [G-L-S-W]. One way to understand Theorem 4.3.3 is to say that, in order to get a non-torsion class in $H^3(X, \mathbb{Z})$ from a family of symplectic manifolds over X, the fibers must be infinite-dimensional. We will describe such a family over a simply-connected compact Lie group in Chapter 5.

Chapter 5

Degree 3 Cohomology: Sheaves of Groupoids

We develop the theory of Dixmier-Douady sheaves of groupoids and relate it to degree 3 cohomology with integer coefficients. In §1 we explain the theory of descent for sheaves, based on local homeomorphisms. In §2, we introduce sheaves of groupoids (also called stacks) and gerbes. We relate gerbes on X with band a sheaf of abelian groups A with the cohomology group $H^2(X, A)$. A gerbe with band $\underline{\mathbb{C}}_X^*$ is called a Dixmier-Douady sheaf of groupoids. In §3 we introduce the notion of connective structure and curving for such sheaves of groupoids; we obtain the 3-curvature Ω, which is a closed 3-form such that the cohomology class of $\frac{\Omega}{2\pi\sqrt{-1}}$ is integral. We prove that any 3-form with these properties is the 3-curvature of some sheaf of groupoids, and relate this to the constructions of Chapter 4. In §4 we use the path-fibration to define a canonical sheaf of groupoids over a compact Lie group. In §5 we give other examples of sheaves of groupoids connected with Lie group actions on a smooth manifold and with sheaves of twisted differential operators.

5.1. Descent theory for sheaves

In Chapter 1 we introduced the notion of presheaves and sheaves over a space X. In this section we will present some aspects of the theory of descent for sheaves (in the sense of Grothendieck [Gr5]). The theory of Grothendieck is extremely general and applies in particular to sheaves over schemes. Here we are concerned with ordinary topological spaces, which leads to many simplifications. The presentation we give here has been influenced by some ideas in [K].

Recall that a *presheaf* F of sets on X consists of the data of a set $F(U)$ for each open subset U of X, and for each inclusion $V \subseteq U$ of open sets, a restriction mapping $\rho_{V,U} : F(U) \to F(V)$, satisfying a transitivity condition. We say that F is a sheaf if the following "glueing condition" holds. Given any open subset V of X, an open covering $(U_i)_{i \in I}$ of V by open subsets of V, and sections $s_i \in F(U_i)$ such that for each i and j, s_i and s_j have the same image in $F(U_{ij})$, there is a *unique* $s \in F(V)$ such that $\rho_{U_i,V}(s) = s_i$.

It is often useful to phrase this glueing property in terms of local homeo-morphisms. This has several advantages: first we obtain a formulation of the theory which does not (at least explicitly) involve indices. Also, many properties become more natural. In addition, the ground is prepared for the discussion of sheaves of groupoids in the next section. Instead of an open subset of X, we will more generally consider a continuous mapping $f : Y \to X$ of some space Y into X (in the terminology of Grothendieck, this is called a "space over X"). Recall from §1.4 that corresponding to a sheaf A over X, there is an inverse image sheaf $f^{-1}A$ over Y. Then we can define a set $A(Y \xrightarrow{f} X)$ (or simply $A(Y)$) to be the set $\Gamma(Y, f^{-1}(A))$ of global sections of the pull-back sheaf over Y. Now let $g : Z \to X$ be another continuous mapping. Assume there exists a continuous mapping $h : Z \to Y$ such that $g = f \circ h$. Then we have a pull-back map $h^{-1} : A(Y \xrightarrow{f} X) \to A(Z \xrightarrow{g} X)$.

Note that, given an open covering (U_i) of the open set $V \subseteq X$ as above, we have a continuous mapping $f : Y := \coprod_i U_i \to X$, where the restriction of f to each U_i is the inclusion map, yielding $A(Y \xrightarrow{f} X) = \prod_{i \in I} A(U_i)$. The glueing condition involves all the intersections U_{ij}, which are all open subsets of the space $Y \times_X Y$, the fiber product of $f : Y \to X$ with itself. The fiber product is the set of $(y_1, y_2) \in Y \times Y$ such that $f(y_1) = f(y_2)$. There are two projection maps $p_1, p_2 : Y \times_X Y \to Y$ such that $p_j(y_1, y_2) = y_j$ for $j = 1, 2$. However there is only one mapping $g : Y \times_X Y \to X$, given by $g(y_1, y_2) = f(y_1) = f(y_2)$. When $Y = \coprod_i U_i \to X$ is the disjoint union of the U_i and $f : Y \to X$ is the (local) inclusion, then $Y \times_X Y$ is the disjoint union of the $U_i \times_X U_j = U_{ij}$. The restriction of p_1 to U_{ij} is the inclusion of U_{ij} into the component U_i of Y, and the restriction of p_2 to U_{ij} is the inclusion of U_{ij} into U_j.

The glueing property therefore means that the following sequence

$$F(V) \xrightarrow{f^{-1}} F(Y) \underset{p_2^{-1}}{\overset{p_1^{-1}}{\rightrightarrows}} F(Y \times_X Y) \qquad (5-1)$$

is an exact sequence of sets, whenever V is an open subset of X, and $Y = \coprod_i U_i$, for (U_i) an open covering of V. What this means is that f^{-1} induces a bijection between $F(V)$ and the kernel $Ker(p_1^{-1}, p_2^{-1})$ of the pair (p_1^{-1}, p_2^{-1}), i.e., the set of elements $s \in F(Y)$ such that $p_1^{-1}(s) = p_2^{-1}(s)$. Of course f^{-1} induces a mapping from $F(V)$ to this kernel; surjectivity of the map implies the existence of a "glued together" section, and injectivity means that the glued together section is unique. Such a formulation of the glueing property is more elegant and compact, and it can be generalized to mappings other than the inclusion of a disjoint union of open sets.

5.1.1. Definition. *A mapping $f : Y \to X$ is said to be a local homeo-morphism if every $y \in Y$ has an open neighborhood U such that $f(U)$ is open* *in X, and the restriction of f to U gives a homeomorphism between U and $f(U)$.*

Note that the composition of two local homeomorphisms is again one, that the inclusion of a disjoint union of open sets is a local homeomorphism, and that a covering map is also a local homeomorphism (in fact, a very special one). If $f : Y \to X$ is a local homeomorphism, and $g : Z \to X$ is any continuous map, the map $Z \times_X Y \to Z$ is again a local homeomorphism. Given a local homeomorphism $f : Y \to X$, with X a smooth manifold, Y acquires the structure of a smooth manifold, with respect to which f is a local diffeomorphism.

5.1.2. Proposition. *If A is a sheaf on X, for every open set V of X and any <u>surjective</u> local homeomorphism $f : Y \to V$, the sequence of sets (5-1) is exact.*

Thus we get a new definition of a sheaf of sets A on X. In the new definition, such a sheaf A will associate to any local homeomorphism $f : Y \to X$ a set $A(Y \xrightarrow{f} X)$, and to a diagram $Z \xrightarrow{g} Y \xrightarrow{f} X$ of local homeomorphisms, a pull-back map $g^{-1} : A(Y \xrightarrow{f} X) \to A(Z \xrightarrow{fg} X)$. The pull-back map should be compatible with the composition of such diagrams, so that if we have local homeomorphisms $W \xrightarrow{h} Z \xrightarrow{g} Y \xrightarrow{f} X$, the pull-back map $(gh)^{-1} : A(Y \xrightarrow{f} X) \to A(W \xrightarrow{fgh} X)$ is the composite of $g^{-1} : A(Y \xrightarrow{f} X) \to A(Z \xrightarrow{fg} X)$ and of $h^{-1} : A(Z \xrightarrow{fg} X) \to A(W \xrightarrow{fgh} X)$. Then A is a sheaf if and only if, for any open set V of X and every surjective local homeomorphism $f : Y \to V$, the sequence of sets (5-1) is exact.

Note that, given a space X, we have a category C_X, whose objects are the local homeomorphisms $Y \xrightarrow{f} X$, and such that a morphism from $Z \xrightarrow{h} X$ to $Y \xrightarrow{f} X$ is a commutative diagram

$$
\begin{array}{ccc}
Z & \xrightarrow{\quad g \quad} & Y \\
 & \searrow h \quad \swarrow f & \\
 & X &
\end{array}
\qquad (5-2)
$$

where g is a local homeomorphism. A sheaf of sets F on X gives rise to a contravariant functor from C_X to the category *Sets* of sets. The category C_X gives a nice interpretation of the notion of a refinement $(V_j)_{j \in J}$ of an open covering $(U_i)_{i \in I}$ of X. Recall from Chapter 1 that this means that there exists a mapping $f : J \to I$ such that $V_j \subseteq U_{f(j)}$ for all $j \in J$. Observe that such a mapping gives rise to a local homeomorphism $h : \coprod_{j \in J} V_j \to \coprod_{i \in I} U_i$

which makes the following diagram commute

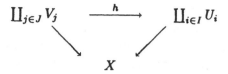

We now come to the glueing property. If $f : Y \to X$ is a surjective local homeomorphism, one can obtain a section of a sheaf A over X from a section s of A over $f : Y \to X$ satisfying the glueing condition $p_1^{-1}(s) = p_2^{-1}(s)$ in $A(Y \times_X Y)$. We apply this to morphisms of sheaves. Recall that given sheaves A and B over X, a morphism of sheaves $\phi : A \to B$ consists of mappings $\phi_U : A(U) \to B(U)$ for every open set U, such that, for $V \subseteq U$, we have a commutative diagram

$$
\begin{array}{ccc}
A(U) & \xrightarrow{\phi_U} & B(U) \\
\downarrow{\scriptstyle \rho_{V,U}} & & \downarrow{\scriptstyle \rho_{V,U}} \\
A(V) & \xrightarrow{\phi_V} & B(V)
\end{array}
$$

Similarly we have the notion of morphism of sheaves of groups. If A and B are two sheaves of groups over X, the set of morphisms from A to B is denoted by $Hom_X(A, B)$. For every open set U of X, there is a natural restriction map $Hom_X(A, B) \to Hom_U(A_{/U}, B_{/U})$. Thus we have a presheaf $U \mapsto Hom_U(A_{/U}, B_{/U})$, which is denoted by $\underline{Hom}(A, B)$. It is easy to see that this is a sheaf on X. Given a local homeomorphism $f : Y \to X$, the value of the sheaf $\underline{Hom}(A, B)$ on f is $Hom_Y(f^{-1}A, f^{-1}B)$. The fact that $\underline{Hom}(A, B)$ is a sheaf means that, given an open set V and a surjective local homeomorphism $f : Y \to V$, a morphism $\phi : A_{/V} \to B_{/V}$ of sheaves of groups over X is the same as a morphism $\psi : f^{-1}A \to f^{-1}B$ of the pull-back sheaves on Y such that we have $p_1^{-1}(\psi) = p_2^{-1}(\psi)$ as morphisms of sheaves on $Y \times_X Y$. This is usually phrased by saying that morphisms of sheaves satisfy *descent* with respect to local homeomorphisms.

More subtle is the fact that sheaves themselves can be descended. To explain this, we start with the more concrete case of principal G-bundles, for G a Lie group. Say X and Y are smooth manifolds, and $f : Y \to X$ is a smooth surjective local homeomorphism. Given a principal G-bundle $p : P \to X$, setting $Q = P \times_X Y$, we have the pull-back bundle $\pi : Q \to Y$. There is a natural isomorphism $\phi : p_1^* Q \xrightarrow{\sim} p_2^* Q$ of the two pull-back bundles over $Y \times_X Y$, since $f p_1 = f p_2$, which satisfies the following *cocycle condition*

$$ p_{13}^*(\phi) = p_{23}^*(\phi) \circ p_{12}^*(\phi), \tag{5-3} $$

an equality of morphisms $p_1^*Q \to p_3^*Q$ of G-bundles over the triple fiber product $Y \times_X Y \times_X Y$. In (5-3) $p_1, p_2, p_3 : Y \times_X Y \times_X Y \to Y$ are the projections on the three factors, and $p_{12}, p_{13}, p_{23} : Y \times_X Y \times_X Y \to Y \times_X Y$ are the projections on two of the three factors.

Conversely, consider a G-bundle $\pi : Q \to Y$, together with an isomorphism $\phi : p_1^*Q \xrightarrow{\sim} p_2^*Q$ of the two pull-back bundles over $Y \times_X Y$, which satisfies the cocycle condition (5-3). We recover a G-bundle $P \to X$, the pull-back of which to Y identifies with $Q \to Y$. Indeed we can construct the total space P as the quotient space of Q by some equivalence relation $\simeq$. The relation $\simeq$ identifies q_1 and q_2 if $y_1 = \pi(q_1)$ and $y_2 = \pi(q_2)$ satisfy $f(y_1) = f(y_2)$, and the isomorphism ϕ of G-bundles over $Y \times_X Y$, evaluated at the point (y_1, y_2) of $Y \times_X Y$, sends $(q_1; y_1, y_2)$ to $(q_2; y_1, y_2)$. Condition (5-3) means precisely that $\simeq$ is an equivalence relation. It is then easy to see that $P \to X$ is a G-bundle; the natural mapping $Q \to P$ now allows the G-bundle $Q \to Y$ to be identified with the pull-back of $P \to X$ to Y. The isomorphism ϕ is therefore the natural one between the two inverse images of P to Y, so that we have a one-to-one correspondence between the following kinds of objects:

(1) a principal G-bundle $p : P \to X$;

(2) a principal G-bundle $\pi : Q \to Y$, together with an isomorphism $\phi : p_1^*Q \xrightarrow{\sim} p_2^*Q$ of the two pull-back bundles over $Y \times_X Y$, which satisfies the cocycle condition (5-3).

The isomorphism ϕ in (2) is called a *descent datum* (or *descent isomorphism*) for the bundle Q with respect to the mapping $f : Y \to X$.

We not only have a one-to-one correspondence between isomorphism classes of objects of type (1) and (2), but we in fact have an equivalence of categories. Stated simply, we observe that there exists a category of G-bundles $p : P \to X$ over X. The objects are G-bundles. A morphism α from a G-bundle $p_1 : P_1 \to X$ to a G-bundle $p_2 : P_2 \to X$ is a bundle isomorphism $\alpha : P_1 \to P_2$ (so α is a diffeomorphism which commutes with the G-actions and induces the identity on the base space X). Note that this category is a *groupoid*, i.e., every morphism between objects is invertible.

The second category is associated to the surjective local homeomorphism $f : Y \to X$. The objects of this category are pairs $(\pi : Q \to Y, \phi)$ as in (2) above. A morphism from $(\pi_1 : Q_1 \to Y, \phi_1)$ to $(\pi_2 : Q_2 \to Y, \phi_2)$ is defined to be an isomorphism $\alpha : Q_1 \xrightarrow{\sim} Q_2$ of G-bundles over Y, such that the following diagram commutes

$$
\begin{array}{ccc}
p_1^*Q_1 & \xrightarrow{p_1^*\alpha} & p_1^*Q_2 \\
\downarrow{\phi_1} & & \downarrow{\phi_2} \\
p_2^*Q_1 & \xrightarrow{p_2^*\alpha} & p_2^*Q_2
\end{array}
$$

This category is also a groupoid. We now obtain

5.1.3. Proposition. *Let $f : Y \to X$ be a smooth surjective local home-omorphism. The pull-back functor f^* induces an equivalence of categories between the category of G-bundles over X and the category of G-bundles Q over Y, equipped with a descent isomorphism $\phi : p_1^* Q \tilde{\to} p_2^* Q$ satisfying the cocycle condition (5-3).*

The situation is entirely analogous for sheaves over X. We have the notion of descent datum for a sheaf A over Y, namely an isomorphism $\phi : p_1^{-1} A \tilde{\to} p_2^{-1} A$ of sheaves on Y, which satisfies the analog of (5-3) (with the pull-backs now denoted by p_{ij}^{-1} rather than p_{ij}^*). We define a category of such pairs (A, ϕ), and obtain

5.1.4. Proposition. *Let $f : Y \to X$ be a surjective local homeomorphism. The inverse image functor f^{-1} induces an equivalence of categories between the category of sheaves over X and the category of sheaves A over Y, equipped with a descent datum $\phi : p_1^{-1} A \tilde{\to} p_2^{-1} A$ satisfying the cocycle condition (5-3).*

Let $Sh(Y)$ be the category of sheaves over Y; we wish to study the functorial properties of $Sh(Y)$ as $f : Y \to X$ ranges over all local homeomorphisms. Given a diagram $Z \xrightarrow{g} Y \xrightarrow{f} X$ of local homeomorphisms, we have an inverse image functor $g^{-1} : Sh(Y) \to Sh(Z)$. This satisfies a certain transitivity condition, which looks complicated at first glance. Given a diagram $W \xrightarrow{h} Z \xrightarrow{g} Y \xrightarrow{f} X$ of local homeomorphisms, the two functors $(gh)^{-1}$ and $h^{-1} g^{-1}$ from $Sh(Y)$ to $Sh(W)$ are not exactly equal, but there is an invertible natural transformation $\theta_{g,h} : h^{-1} g^{-1} \tilde{\to} (gh)^{-1}$. The natural transformation satisfies a coherency condition relative to a diagram $T \xrightarrow{k} W \xrightarrow{h} Z \xrightarrow{g} Y \xrightarrow{f} X$ of local homeomorphisms, namely the commutativity of the diagram of functors and natural transformations

$$
\begin{array}{ccc}
k^{-1} h^{-1} g^{-1} & \xrightarrow{\theta_{g,h}} & k^{-1} (gh)^{-1} \\
\downarrow{\scriptstyle \theta_{h,k}} & & \downarrow{\scriptstyle \theta_{gh,k}} \\
(hk)^{-1} g^{-1} & \xrightarrow{\theta_{g,hk}} & (ghk)^{-1}
\end{array}
\tag{5-4}
$$

Two descent properties are satisfied. The first property implies that descent holds for morphisms of sheaves, while the second implies that descent holds for sheaves themselves.

(D1) Given $f : Y \to X$ and two sheaves A and B of groups on Y, the assignment $(g : Z \to Y) \to Hom_Z(g^{-1} A, g^{-1} B)$ defines a sheaf on Y, namely $\underline{Hom}(A, B)$.

Then the presheaf $(g : Z \to Y) \mapsto Isom_Z(g^{-1} A, g^{-1} B)$ is a subsheaf of

$\underline{Hom}(A, B)$, denoted by $\underline{Isom}(A, B)$.

(D2) Let V be an open subset of X, and $f : Y \to V$ a surjective local homeomorphism. Assume an object A of $Sh(Y)$ is given, together with an isomorphism $\phi : p_1^{-1}A \overset{\sim}{\to} p_2^{-1}A$ in $Sh(Y \times_X Y)$, such that the following diagram commutes

$$
\begin{array}{ccccc}
p_{23}^{-1}p_2^{-1}A & \overset{\phi}{\longrightarrow} & p_{23}^{-1}p_1^{-1}A & \overset{\theta_{p_1,p_{23}}}{\Longrightarrow} & p_2^{-1}A \\
\Big\downarrow{\scriptstyle\theta_{p_2,p_{23}}} & & & & \Big\downarrow{\scriptstyle\theta_{p_2,p_{12}}^{-1}} \\
p_3^{-1}A & & & & p_{12}^{-1}p_2^{-1}A \\
\Big\downarrow{\scriptstyle\theta_{p_2,p_{13}}^{-1}} & & & & \Big\downarrow{\scriptstyle\phi} \\
p_{13}^{-1}p_2^{-1}A & & & & p_{12}^{-1}p_1^{-1}A \\
\Big\downarrow{\scriptstyle\phi^{-1}} & & & & \Big\downarrow{\scriptstyle\theta_{p_1,p_{12}}} \\
p_{13}^{-1}p_1^{-1}A & \overset{\theta_{p_1,p_{13}}}{\longrightarrow} & & & p_1^{-1}A
\end{array}
\qquad (5-5)
$$

There exists an object $\mathcal{A}$ of $Sh(V)$, unique up to isomorphism, together with an isomorphism $\psi : f^{-1}\mathcal{A} \overset{\sim}{\to} A$ in $Sh(Y)$, such that the diagram in $Sh(Y \times_X Y)$

$$
\begin{array}{ccccc}
p_1^{-1}f^{-1}\mathcal{A} & \overset{\theta_{f,p_1}}{\longrightarrow} & g^{-1}\mathcal{A} & \overset{\theta_{f,p_2}^{-1}}{\longrightarrow} & p_2^{-1}f^{-1}\mathcal{A} \\
\Big\downarrow{\scriptstyle\psi} & & & & \Big\downarrow{\scriptstyle\psi} \\
p_1^{-1}A & & \overset{\phi}{\longrightarrow} & & p_2^{-1}A
\end{array}
$$

commutes (here g denotes the projection map $Y \times_X Y \to X$).

Furthermore, we have an equivalence of categories between $Sh(V)$ and the so-called *descent category* with respect to $f : Y \to V$, that is, the category whose objects are the pairs (A, ϕ) as above, and whose morphisms are the morphisms of sheaves compatible with the given isomorphisms over $Y \times_X Y$. A similar story could be told about the category of G-bundles over Y; everything works in exactly the same way.

We next discuss two constructions of sheaf theory in terms of local homeomorphisms. The first is that of the sheaf $\tilde{F}$ associated to a presheaf F. To begin, we need the notion of direct limit $\varinjlim_{Z \in C} F(Z)$ of a functor $F : C \to Sets$ from a category C to the category $Sets$. The direct limit is the quotient of the disjoint union $\coprod_{Z \in C} F(Z)$ by the equivalence relation which identifies $a \in F(Z)$ with $F(\phi)(a) \in F(W)$, for $Z \overset{\phi}{\to} W$ a morphism in C. To recover the direct limits indexed by an ordered set $(I, \le)$ as a special case, note that I leads to a category $C(I)$, with $Ob(C(I)) = I$ so that there is exactly one arrow from $i \in I$ to $i' \in I$ if $i \le i'$, and no arrow otherwise.

In the present context, we will take a presheaf F over X to mean a contravariant functor $(Y \overset{f}{\to} X) \mapsto F(Y)$ from the category C_X of local homeomorphisms $Y \xrightarrow{f} X$ to the category of sets. Given local homeomorphisms $Z \xrightarrow{g} Y \xrightarrow{f} X$, we define $Desc(Z \xrightarrow{g} Y)$ to be the subset of $F(Z)$ consisting of elements $a \in F(Z)$ such that $p_1^{-1}(a) = p_2^{-1}(a)$ in $F(Z \times_Y Z)$. Then we set

$$\tilde{F}(Y) = \lim_{Z \xrightarrow{g} Y} Desc(Z \xrightarrow{g} Y),$$

where the direct limit is indexed by the category C_Y of local homeomorphisms $Z \xrightarrow{g} Y$.

We next discuss the inverse image $f^{-1}(A)$ of a sheaf A under a continuous mapping $f : Y \to X$ (not necessarily a local homeomorphism), and start by describing a presheaf B on Y for which $f^{-1}(A)$ is the associated sheaf. Let $g : Z \to Y$ be a local homeomorphism. Before defining $B(Z \xrightarrow{g} Y)$, we introduce a certain ordered set S: An object of S is a commutative diagram

$$\begin{array}{ccc} Z & \xrightarrow{g} & Y \\ \downarrow{i} & & \downarrow{f} \\ W & \xrightarrow{h} & X \end{array} \qquad (5-6)$$

where h is a local homeomorphism such that the mapping $Z \to W \times_X Y$ is the inclusion of an open set.

The order on S is as follows: We say that the diagram

$$\begin{array}{ccc} Z & \xrightarrow{g} & Y \\ \downarrow{i} & & \downarrow{f} \\ W & \xrightarrow{h} & X \end{array}$$

is smaller than the diagram

$$\begin{array}{ccc} Z & \xrightarrow{g} & Y \\ \downarrow{i'} & & \downarrow{f} \\ W' & \xrightarrow{h'} & X \end{array}$$

if W is an open set of W', h is the restriction of h' to $W \subseteq W'$, and $i = i'$. To explain the motivation for this set S, note that if U is an open subset of Y and V an open subset of X such that $U \subseteq f^{-1}(V)$; then we have a diagram

$$\begin{array}{ccc} U & \longrightarrow & Y \\ \downarrow{f_{/U}} & & \downarrow{f} \\ V & \longrightarrow & X \end{array}$$

which belongs to S.

We define a presheaf B on Y by $B(Z\xrightarrow{g}Y) = \varinjlim_s A(W)$. The inverse image sheaf $f^{-1}A$ is the sheaf associated to the presheaf B, which is compatible with the construction of the inverse image of a sheaf given in Chapter 1.

A final topic is the relation between sheaves and principal bundles, which brings us to torsors. We can view a principal G-bundle $p : P \to X$ as giving rise to a sheaf with extra structure. Indeed, let $\underline{P}_X$ be the sheaf of smooth sections of p. Then the sheaf of groups $\underline{G}_X$ of smooth G-valued functions operates on $\underline{P}_X$ on the right. The action of $\underline{G}_X$ on $\underline{P}_X$ has the following property: if U is an open subset of X such that $\Gamma(U, \underline{P}_X) \neq \emptyset$, then an element s of $\Gamma(U, \underline{P}_X)$ allows us to identify the sheaves $\underline{P}_X$ and $\underline{G}_X$ over U. In other words, $\underline{P}_X$ is what is called a *torsor* under the sheaf of groups $\underline{G}_X$. Proposition 5.1.4 now applies to suitable categories of torsors, and hence contains Proposition 5.1.3 as a special case.

With respect to notation, a torsor under a sheaf of groups A is usually a sheaf which admits a left action of A rather than a right action as above. For future use, we now record the formal definition and some properties of torsors.

5.1.5. Definition and Proposition. (1) *Let H be a sheaf of groups on X. An H-torsor over X is a sheaf F, together with an action $H \times F \to F$ of the sheaf of groups H on F, such that every point has a neighborhood U with the property that for $V \subseteq U$ open, the set $F(V)$ is a principal homogeneous space under the group $H(V)$.*

(2) *For a Lie group G, a manifold X, and a principal G-bundle $p : P \to X$, the sheaf $\underline{P}_X$ of smooth sections of P is a torsor under the sheaf $\underline{G}_X$ of smooth G-valued functions.*

(3) *The set of isomorphism classes of H-torsors over X is in a natural bijection with the sheaf cohomology set $H^1(X, H)$.*

The trivial H-torsor is H itself, acting on itself by left translations. Any H-torsor is locally isomorphic to the trivial torsor H.

If we have two sheaves F_1 and F_2 on X which are locally isomorphic, the sheaf $\underline{Isom}(F_1, F_2)$ is a torsor under the sheaf of groups $\underline{Aut}(F_2)$. This can be applied to sheaves of sets, or groups, or vector spaces, and so on, so that consequently, a locally constant sheaf with fiber $\mathbb{C}^n$ defines a $GL(n, \mathbb{C})$-torsor.

We now describe the functorial properties of torsors. Given a sheaf of groups H over X and a continuous mapping $f : Y \to X$, the inverse image sheaf $f^{-1}H$ is a sheaf of groups on Y. Then given an H-torsor T over X, the inverse image $f^{-1}T$ is a torsor under $f^{-1}H$, which yields the

operation of inverse image of a torsor under f. Another operation associated
to a morphism $H \to K$ of sheaves of groups over the same space X can be
described as follows: Given a H-torsor T, the sheaf associated to the presheaf
$U \mapsto T(U) \times^H K = (T(U) \times K)/H$ is a torsor under K, denoted by $T \times^H K$.
In particular, if H is abelian, there is a notion of contracted product $F \times^H F'$
of H-torsors. To construct it, we first consider $F \times F'$, a torsor under the
sheaf of groups $H \times H$, and then using the product map $H \times H \to H$, we
get the H-torsor $F \times^H F' := (F \times F') \times^{H \times H} H$. We now obtain a structure
of an abelian group on the set of equivalence classes of H-torsors, and the
isomorphism of Proposition 5.1.5 (3) is a group isomorphism. The opposite
F^{-1} of a H-torsor F has a nice interpretation: its sections over $U \subseteq X$ are
torsor isomorphisms $F_{/U} \xrightarrow{\sim} H$.

Now let $f : Y \to X$ be a smooth mapping between smooth manifolds.
There is a natural inverse image map for differential forms, which gives a
morphism of sheaves of real vector spaces $f^{-1}\underline{A}^p_{X,\mathbb{C}} \to \underline{A}^p_{Y,\mathbb{C}}$, and also a
morphism of the complexified sheaves. Given a torsor T under $\underline{A}^1_{X,\mathbb{C}}$, the
inverse image $f^{-1}T$ is a torsor under $f^{-1}\underline{A}^1_{X,\mathbb{C}}$; then $f^{-1}T \times^{f^{-1}\underline{A}^1_{X,\mathbb{C}}} \underline{A}^1_{Y,\mathbb{C}}$ is
a torsor under $\underline{A}^1_{Y,\mathbb{C}}$, which will be denoted by f^*T. To summarize, given a
torsor T under $\underline{A}^1_{X,\mathbb{C}}$, f^*T is then a torsor under $\underline{A}^1_{Y,\mathbb{C}}$.

5.2. Sheaves of groupoids and gerbes

We introduce the notion of sheaf of categories over a space. This is not just
an abstract exercise, since all geometric constructs of a local nature lead
to a sheaf of categories. The precise formulation of the concept requires
a fair amount of care, as one needs to explicitly write down a number of
commutative diagrams. Our guide will be the study of the sheaf over X,
which to a local homeomorphism $Y \xrightarrow{f} X$, associates the category of sheaves
over Y.

5.2.1. Definition. (1) *A __presheaf of categories__ C over X consists of the
following data:*

(a) *for every local homeomorphism $f : Y \to X$, a category $C(Y \xrightarrow{f} X)$;*

(b) *for every diagram $Z \xrightarrow{g} Y \xrightarrow{f} X$ of local homeomorphisms, a functor
$g^{-1} : C(Y \xrightarrow{f} X) \to C(Z \xrightarrow{fg} X)$;*

(c) *for every diagram $W \xrightarrow{h} Z \xrightarrow{g} Y \xrightarrow{f} X$ of local homeomorphisms, an
invertible natural transformation $\theta_{g,h} : h^{-1}g^{-1} \xrightarrow{\sim} (gh)^{-1}$;*

We require the commutativity of diagrams of type (5-4).

(2) *The presheaf of categories C is said to be a __sheaf of categories__ (or*

stack in the terminology of Grothendieck) if the descent properties (D1) and (D2) (in section 1) are satisfied. We understand (D2) to mean that given a local homeomorphism $f : Y \to X$ and a surjective local homeomorphism $g : Z \to Y$, the natural functor gives an equivalence of categories between $C(Y \xrightarrow{f} X)$ and the descent category of pairs (A, ϕ) as in (D2).

We will often use the simpler notation $C(Y)$, instead of $C(Y \xrightarrow{f} X)$, when there is no risk of confusion. If $f : U \hookrightarrow X$ is the inclusion of an open set, we will denote the functor f^{-1} by $P \mapsto P_{/U}$.

It is worthwhile to explain what the descent conditions (D1) and (D2) mean concretely in the case of the surjective local homeomorphism $f : Y \to X$, where $Y = \coprod_i U_i$, for (U_i) an open covering of X. (D1) implies that, for P_1 and P_2, two objects of $C(X)$, a morphism $h : P_1 \to P_2$ in $C(X)$ amounts to the same thing as a family h_i of morphisms $h_i : (P_1)_{/U_i} \to (P_2)_{/U_i}$ in $C(U_i)$, which satisfy $(h_i)_{/U_{ij}} = (h_j)_{/U_{ij}}$. (D2) implies that given objects Q_i of $C(U_i)$, and isomorphisms $u_{ij} : (Q_j)_{/U_{ij}} \xrightarrow{\sim} (Q_i)_{/U_{ij}}$ in $C(U_{ij})$, which satisfy the glueing condition $u_{ik} = u_{ij} u_{jk}$ over U_{ijk}, there is a unique way to glue the Q_i to a global object Q of $C(X)$.

From §1, we have a few examples of sheaves of categories, for instance, $C(Y) = Sh(Y)$, or the category of sheaves of groups (resp. abelian groups, and so on) on Y. Another case is $C(Y) = Bund_G(Y)$, the category of G-bundles over Y. This latter example is a sheaf of groupoids (i.e., every category $C(Y)$ is a groupoid).

Recall from Chapter 1 that, to a given presheaf F (of sets) on a space X, there corresponds an associated sheaf $\tilde{F}$, together with a mapping $F \mapsto \tilde{F}$ of presheaves. There is a similar construction, which to a presheaf of categories C over X, associates a sheaf of categories $\tilde{C}$. To explain this we have to introduce the notion of the direct limit $\varinjlim_{i \in I} C_i$ of a family of categories C_i indexed by the set of objects of a category I (the indexing category). It is assumed of course that, for an arrow $\alpha : i \to i'$ in I, there is a corresponding functor $\alpha_* : C_i \to C_{i'}$ (we use the notation α_*, reminiscent of a direct image functor). However, given a diagram $i_1 \xrightarrow{\alpha_1} i_2 \xrightarrow{\alpha_2} i_3$ in I, we do not assume that the functor $(\alpha_1 \alpha_2)_*$ coincides with the composite $\alpha_{1,*} \alpha_{2,*}$ (since this would not be true in any interesting geometric situation), but rather that there is given an invertible natural transformation

$$\theta_{\alpha_1, \alpha_2} : \alpha_{1,*} \alpha_{2,*} \xrightarrow{\sim} (\alpha_1 \alpha_2)_*.$$

We then require the commutativity of diagrams analogous to (5-4).

In that situation the set of objects of the direct limit category $\varinjlim_{i \in I} C_i$ is just the disjoint union $\coprod_{i \in I} Ob(C_i)$. We need to define the notion of morphism from $P_1 \in Ob(C_{i_1})$ to $P_2 \in Ob(C_{i_2})$. For any diagram D of the type

$$
\begin{array}{ccc}
i_1 & \xrightarrow{\alpha_1} & j \\
 & & \downarrow{Id} \\
i_2 & \xrightarrow{\alpha_2} & j
\end{array}
$$

we consider the set S_D of morphisms $\beta : \alpha_{1,*}(P_1) \to \alpha_{2,*}(P_2)$ in C_j, and the following equivalence relation on the disjoint union $\coprod_D S_D$. Given D as above and a similar diagram D'

$$
\begin{array}{ccc}
i_1 & \xrightarrow{\alpha_1'} & j' \\
 & & \downarrow{Id} \\
i_2 & \xrightarrow{\alpha_2'} & j'
\end{array}
$$

we say that $\beta : \alpha_{1,*}(P_1) \to \alpha_{2,*}(P_2)$ is equivalent to $\beta' : \alpha'_{1,*}(P_1) \to \alpha'_{2,*}(P_2)$ if and only if there exists a diagram

$$
\begin{array}{ccc}
j & \xrightarrow{\gamma} & k \\
 & & \downarrow{Id} \\
j' & \xrightarrow{\gamma'} & k
\end{array}
$$

such that $\gamma\alpha_j = \gamma'\alpha'_j$ for $j = 1, 2$, and such that there is a commutative diagram

$$
\begin{array}{ccccc}
(\gamma\alpha_1)_*(P_1) & \xrightarrow{\theta_{\gamma,\alpha_1}^{-1}} & \gamma_*\alpha_{1,*}(P_1) & \xrightarrow{\beta} & \gamma_*\alpha_{2,*}(P_2) \\
\downarrow{\theta_{\gamma',\alpha_1'}^{-1}} & & & & \downarrow{\theta_{\gamma,\alpha_2}} \\
\gamma'_*\alpha'_{1,*}(P_1) & \xrightarrow{\beta'} & \gamma'_*\alpha'_{2,*}(P_2) & \xrightarrow{\theta_{\gamma',\alpha_2'}} & (\gamma\alpha_2)_*(P_2)
\end{array}
$$

In practice, the categories C_i are indexed by the opposite of a category I (where I could be the category of local homeomorphisms whose target is a fixed space X). Given an arrow $i' \xrightarrow{\alpha} i$, we now have a functor, denoted by α^{-1} (like an inverse image), from C_i to $C_{i'}$.

We return to the presheaf of categories C over X. Given a diagram $Z \xrightarrow{g} Y \xrightarrow{f} X$ of local homeomorphisms, we have a descent category $Desc(C, g)$. The objects of this category are descent data for $Z \xrightarrow{g} Y$, in the sense of §1. For a commutative diagram

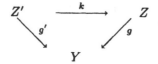

we have an inverse image functor $k^{-1} : Desc(\mathcal{C}, g) \to Desc(\mathcal{C}, g')$. There are natural transformations $\theta_{k_1,k_2} : k_2^{-1} k_1^{-1} \tilde{\to} (k_1 k_2)^{-1}$. We can now take the direct limit category $\lim\limits_{Z \xrightarrow{g} Y} Desc(\mathcal{C}, g)$.

5.2.2. Definition and Proposition. *The assignment* $Y \mapsto \lim\limits_{Z \xrightarrow{g} Y} Desc(\mathcal{C}, g)$ *defines a sheaf of categories on* X, *called the sheaf of categories associated to the presheaf of categories* $\mathcal{C}$.

The above construction of an associated sheaf of categories is very handy in defining the inverse image $f^{-1}\mathcal{C}$ of a stack $\mathcal{C}$ on X for a continuous mapping $f : Y \to X$. We proceed in two steps, and recall the notations at the end of §1 where we described the inverse image of a sheaf on X. First we define a presheaf of categories $\mathcal{B}$ on Y. Let $Z \xrightarrow{g} Y$ be a local homeomorphism. Recall the ordered set S introduced at the end of §1. Then we set

$$\mathcal{B}(Z \xrightarrow{g} Y) = \lim\limits_{\overrightarrow{S}} \mathcal{C}(W \to X)$$

(a direct limit category). The pull-back functors are obtained from those of $\mathcal{C}$, given the fact that a local homeomorphism $Z \to Z'$ between spaces over Y induces a mapping from the set S' for Z' to the set S for Z. Thus we get a presheaf of categories $\mathcal{B}$, and $f^{-1}\mathcal{C}$ is defined to be the sheaf of categories associated to $\mathcal{B}$.

There is no difficulty in defining the direct image $f_*\mathcal{C}$ of a sheaf of categories $\mathcal{C}$ over Y for $f : Y \to X$ a continuous mapping. Given a local homeomorphism $g : Z \to X$, we set $(f_*\mathcal{C})(Z \xrightarrow{g} X) = \mathcal{C}(Z \times_X Y \to Y)$. The pull-back functors are easy to define.

Let us now study a fundamental example of a sheaf of categories, which is closely related to the material in Chapter 4. Let $1 \to A \to C \to B \to 1$ be an exact sequence of (possibly infinite-dimensional) Lie groups, such that the projection $C \to B$ is a locally trivial A-bundle. Assume we are given a smooth B-bundle $p : P \to X$ over a manifold X. We wish to analyze the problem of finding a C-bundle $q : Q \to X$ such that the associated B-bundle $Q/A \to X$ is isomorphic to $P \to X$. The analysis may take place over every open set of X, or more generally, in the spirit of our discussion of the descent properties of sheaves in §1, over a space Y together with a local homeomorphism $f : Y \to X$. We then have a category $\mathcal{C}(Y \xrightarrow{f} X)$, whose objects are pairs $(Q \xrightarrow{q} Y, \alpha)$, where $Q \xrightarrow{q} Y$ is a C-bundle, and $\phi : Q/A \tilde{\to} P \times_X Y$ is an isomorphism of B-bundles. Note that the quotient Q/A is indeed a B-bundle over Y. The bundle $P \times_X Y \to Y$ is the pull-back of $p : P \to X$ under $f : Y \to X$. A morphism $\psi : (Q_1 \xrightarrow{q_1} Y, \alpha_1) \to (Q_2 \xrightarrow{q_2} Y, \alpha_2)$ is an isomorphism $\psi : Q_1 \tilde{\to} Q_2$ of C-bundles over Y such that the following

diagram commutes

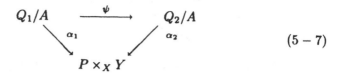

$$(5-7)$$

The pull-back functors for C will be denoted by f^*, g^*, and so on, instead of f^{-1}, g^{-1}, and so on. Given a diagram $Z \xrightarrow{g} Y \xrightarrow{f} X$, there is a natural pull-back functor $g^* : C(Y \xrightarrow{f} X) \to C(Z \xrightarrow{fg} X)$, defined on objects by $g^*(Q, \phi) = (Q \times_Y Z, g^*\phi) = (g^*Q, g^*\phi)$, where we denote by $g^*\phi$ the isomorphism $g^*(\phi) : (Q \times_Y Z)/A \tilde\to P \times_X Z$ of B-bundles over Z obtained from ϕ by pull-back. In the situation of a diagram $W \xrightarrow{h} Z \xrightarrow{g} Y \xrightarrow{f} X$, there is an obvious natural transformation of functors $\theta_{g,h} : h^*g^* \to (gh)^*$ from $C(Y \xrightarrow{f} X)$ to $C(W \xrightarrow{fgh} X)$, which is induced by the corresponding natural transformation for the categories of C-bundles. More precisely, for an object $(Q \xrightarrow{q} Y, \alpha)$ of $C(Y \xrightarrow{f} X)$, $\theta_{g,h}(Q \xrightarrow{q} Y, \alpha)$ is the isomorphism $\theta_{g,h} : (h^*g^*Q, h^*g^*\alpha) \tilde\to ((gh)^*Q, (gh)^*\alpha)$, where $\theta_{g,h}$ is the natural isomorphism of C-bundles discussed earlier in this section, which leads to a commutative diagram

$$
\begin{array}{ccc}
h^*g^*Q/A & \xrightarrow{\theta_{g,h}} & (gh)^*Q/A \\
\downarrow{h^*g^*\alpha} & & \downarrow{(gh)^*\alpha} \\
(P \times_X Y) \times_Y Z & \xrightarrow{\simeq} & P \times_X Z
\end{array}
$$

We then have

5.2.3. Proposition. *The above constructions give a sheaf of groupoids on X associated to the principal B-bundle $p : P \to X$ and the exact sequence $1 \to A \to C \to B \to 1$ of Lie groups.*

In §4.2, we considered a central extension of Lie groups

$$1 \to \mathbb{C}^* \to \tilde B \to B \to 1$$

and showed in effect that a B-bundle over X leads to a class in $H^2(X, \underline{\mathbb{C}}_X^*)$. The construction of this class was in fact implicitly based on the above sheaf of categories C consisting of local liftings of the structure group to $\tilde B$. We wish to relate the more general sheaves of groupoids associated to a central extension $1 \to A \to C \to B \to 1$ to the degree 2 sheaf cohomology $H^2(X, \underline{A}_X)$. We start by formulating some properties of the sheaf of groupoids of Proposition 5.2.3.

(G1) Given any object $(Q \xrightarrow{q} Y, \alpha)$ (or simply Q) of $\mathcal{C}(Y \xrightarrow{f} X)$, the sheaf $\underline{Aut}(Q)$ of automorphisms of this object is a sheaf of groups on Y, which is locally isomorphic to the sheaf $\underline{A}_Y$ of smooth A-valued functions. Such a local isomorphism $\alpha : \underline{Aut}(Q) \xrightarrow{\sim} \underline{A}_Y$ is unique up to inner automorphisms of A.

(G2) Given two objects Q_1 and Q_2 of $\mathcal{C}(Y \xrightarrow{f} X)$, there exists a surjective local homeomorphism $g : Z \to Y$ such that $g^{-1}Q_1$ and $g^{-1}Q_2$ are isomorphic (this means that Q_1 and Q_2 are locally isomorphic).

(G3) There exists a surjective local homeomorphism $f : Y \to X$ such that the category $\mathcal{C}(Y \xrightarrow{f} X)$ is non-empty.

Let us justify these statements briefly. Given an object $(Q \xrightarrow{q} Y, \alpha)$ as in (G1), an automorphism of this object is a bundle automorphism ψ of Q which is compatible with α. The compatibility with α means that ψ induces the identity mapping on Q/A. In other words, ψ is a bundle automorphism for the A-bundle $Q \to Q/A$. Now locally on Y, Q is isomorphic to the trivial C-bundle $Q = C \times Y \xrightarrow{p_2} Y$. Then a bundle automorphism of Q is just given by the left action of a smooth A-valued function on $C \times Y$ (such a section s acts by $(c, y) \mapsto (s(y) \cdot c, y)$. (G1) is proved.

To establish (G2), it is enough to show that any point $y \in Y$ has a neighborhood $U = U_y$ such that the restrictions of (Q_1, α_1) and (Q_2, α_2) to U are isomorphic. Indeed we can then take $Z = \coprod_{y \in Y} U_y \to Y$. First we assume that the restriction of Q_1 and Q_2 to U are trivial. We also assume that the restriction of f^*P to U is trivial (after shrinking U if necessary). Moreover, α_1 and α_2 are given by the left actions of smooth functions $g_1, g_2 : U \to B$. Now we use the fact that the projection mapping $\pi : C \to B$ has local smooth sections. After shrinking U if necessary, we can find a smooth function $\gamma : U \to C$ which projects to $g_2^{-1} g_1$. Then (Id, γ) yields an isomorphism $(Q_1, \alpha_1) \xrightarrow{\sim} (Q_2, \alpha_2)$ over U.

(G3) is easily proved by the same method.

5.2.4. Definition. (*Giraud* [Gi]) *Let X be a manifold, and let A be a Lie group. A gerbe with band $\underline{A}_X$ over X is a sheaf of groupoids over X satisfying properties (G1), (G2) and (G3).*

The point of gerbes with band $\underline{A}_X$, for a given Lie group A, is that they give rise to degree 2 cohomology classes in $H^2(X, \underline{A}_X)$. In the same fashion, one can define a gerbe over X with band H for some sheaf of groups H over X (one replaces $\underline{A}_X$ with H in axiom (G1)). This is developed in the treatise of Giraud [Gi], who in fact uses gerbes as an approach to non-abelian degree 2 cohomology. We will be content here with studying the abelian case, which means that we assume that the Lie group A is *commutative*.

We now begin the study of the objects of a gerbe on X over various local homeomorphisms $Y \xrightarrow{f} X$. Given a gerbe C with band equal to the commutative sheaf of groups H, a local homeomorphism $f : Y \to X$, and two objects P_1 and P_2 of $C(Y)$, there is a corresponding H-torsor, namely the sheaf $\underline{Isom}(P_1, P_2)$. Indeed, the sheaf of groups $H = \underline{Aut}(P_2)$ acts on $\underline{Isom}(P_1, P_2)$ on the left, and this action is locally simply transitive because property (G2) says that the objects P_1 and P_2 are locally isomorphic. There is a very important converse to this construction. Let P be an object of $C(Y)$, and let I be an H-torsor over Y. Then there is a corresponding twisted object $P \times^H I$, unique up to isomorphism, such that $\underline{Isom}(P, P \times^H I) \xrightarrow{\sim} I$ as H-torsors. To construct this object, let $g : Z \to Y$ be a surjective local homeomorphism such that there exists a section s of $g^{-1}I$. Thus over $Z \times_Y Z$ we have two sections of the H-torsor $p_1^{-1}g^{-1}I \simeq p_2^{-1}g^{-1}I$, namely, $p_1^{-1}(s)$ and $p_2^{-1}(s)$.

The difference $u = p_2^{-1}(s) - p_1^{-1}(s)$ is a section of H over $Z \times_Y Z$ which by construction satisfies the cocycle condition $p_{13}^{-1}(u) = p_{12}^{-1}(u) + p_{23}^{-1}(u)$ over $Z \times_Y Z$. We may view u as an automorphism of the object $p_1^{-1}g^{-1}P$ of $C(Z \times_Y Z)$, or as an isomorphism $u : p_1^{-1}g^{-1}P \xrightarrow{\sim} p_2^{-1}g^{-1}P$, satisfying the cocycle condition (5-3). By the descent property (D2), this gives an object Q of $C(Y)$, the pull-back of which to Z is identified with the pull-back of P to Z. There is an isomorphism $\omega : I \xrightarrow{\sim} \underline{Isom}(P, Q)$; it is characterized by the fact that the pull-back of ω to Z sends the section s to the identity map (note that this makes sense, since the pull-back of Q to Z is identical to the pull-back of P to Z). The object Q of $C(Y)$ was constructed from a section u of the H-torsor $g^{-1}I$ over Z. Other choices will lead to an isomorphic object Q'. In fact, the isomorphism $\beta : Q \xrightarrow{\sim} Q'$ is the unique one which makes the following diagram commutative

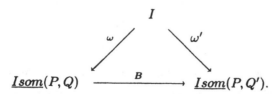

Therefore Q is well-defined up to a unique isomorphism. We summarize this discussion of twisting in

5.2.5. Proposition. (1) *Let H be a sheaf of commutative groups over a space X. Given a gerbe C on X with band H, $f : Y \to X$ a local homeomorphism, and two objects P_1 and P_2 of $C(Y)$, the sheaf $\underline{Isom}(P_1, P_2)$ is an H-torsor.*

(2) *Given an object P of $C(Y)$ and an H-torsor I over Y, there exists an object Q of $C(Y)$ such that $I \xrightarrow{\sim} \underline{Isom}(P, Q)$ as H-torsors. The object Q is*

unique up to a unique isomorphism. It is denoted $P \times^H I$ and is called the object obtained by twisting P by the H-torsor I.

(3) *If $C(Y) \neq \emptyset$, the set of isomorphism classes of objects of $C(Y)$ is in a natural bijection with $H^1(Y, H)$.*

Proof. We have already proved (1) and (2). The constructions of (1) and (2) are inverse to each other in a natural way. More precisely, given an object P of $C(Y)$, the construction $I \mapsto P \times^H I$ establishes a bijection between the set of isomorphism classes of H-torsors over Y and the set of isomorphism classes of objects of $C(Y)$. By Proposition 5.1.5, the latter set is in bijection with $H^1(Y, H)$. ∎

We describe various operations on gerbes. First we note

5.2.6. Proposition. *Let $f : Y \to X$ be a continuous mapping, let H be a sheaf of groups over X, and let C be a gerbe over X with band H. Then the inverse image sheaf of groupoids $f^{-1}C$ is a gerbe over Y whose band is the sheaf $f^{-1}H$.*

Proof. We need to verify (G1), (G2) and (G3) for $f^{-1}C$. Let $Z \xrightarrow{g} Y$ be a local homeomorphism, and let P be an object of the category $(f^{-1}C)(Z)$. From the description of $(f^{-1}C)(Z)$ as a direct limit of categories, it follows that P is locally represented by an object Q of $C(W)$ for some commutative diagram

$$
\begin{array}{ccc}
Z & \xrightarrow{g} & Y \\
\downarrow{i} & & \downarrow{f} \\
W & \xrightarrow{h} & X
\end{array}
$$

as in §1. We may as well assume that i is onto. Then an automorphism of P is the same as an automorphism of Q, i.e., an element of $\Gamma(W, H)$, which gives an element of $\Gamma(Z, i^{-1}H) = \Gamma(Z, f^{-1}H)$. The last equality follows from the fact that Z is an open subset of $W \times_X Y$. This proves (G1).

If we are given two objects P_1 and P_2 of $(f^{-1}C)(Z)$, they are represented by objects Q_1, Q_2 of $C(W_1), C(W_2)$, respectively, for commutative diagrams

$$
\begin{array}{ccc}
Z & \xrightarrow{g} & Y \\
\downarrow{i_j} & & \downarrow{f} \\
W_j & \xrightarrow{h_j} & X
\end{array}
$$

(for $j = 1, 2$). Let $W = W_1 \times_X W_2$, which leads to a similar diagram. We obtain by pull-back two objects of $C(W)$, which we will still denote by Q_1, Q_2. There exists a surjective local homeomorphism $W' \to W$ such that the pull-backs of Q_1 and Q_2 to W' are isomorphic. Then the pull-backs of P_1 and P_2 to $Z \times_W W'$ are isomorphic, proving (G2).

(G3) is proved in the same way as (G2). ∎

The direct image of a gerbe is not a gerbe in general. We have the following result, although it will not be used in this book.

5.2.7. Proposition. *Let* $f : Y \to X$ *be a continuous mapping, let* H *be a sheaf of abelian groups over* Y, *and let* C *be a gerbe over* Y *with band* H. *Assume that the direct image sheaves* $R^1 f_* H$ *and* $R^2 f_* H$ *are both* 0. *Then the direct image sheaf of categories* $f_* C$ *is a gerbe with band* $f_* H$.

We come to the notion of product of gerbes. Let H_1, H_2 be sheaves of groups over a space X and let C_j ($j = 1, 2$) be gerbes over X with band H_j. Then the product gerbe $C_1 \times C_2$ is defined by $(C_1 \times C_2)(Y) = C_1(Y) \times C_2(Y)$. This is a gerbe with band $H_1 \times H_2$. Moreover, there is an external product of gerbes. If C_1 (resp. C_2) is a gerbe over X_1 (resp. X_2), with band H_1 (resp. H_2), we can form $C_1 \boxtimes C_2 := p_1^{-1} C_1 \times p_2^{-1} C_2$; this is a gerbe over $X_1 \times X_2$ with band the sheaf $p_1^{-1} H_1 \times p_2^{-1} H_2$.

The last construction to be described has to do with a morphism $\phi : H \to K$ of sheaves of abelian groups over a space X: To a given gerbe C over X with band H, it associates a gerbe over X with band K. This new gerbe is the sheaf of categories associated to the presheaf $(Y \xrightarrow{f} X) \mapsto C(Y) \times^{H(Y)} K(Y)$, where $C(Y) \times^{H(Y)} K(Y)$ is the following category: Its objects are the same as those of $C(Y)$, and the morphisms are given by

$$Hom_{C(Y) \times^{H(Y)} K(Y)}(P_1, P_2) = Hom_{C(Y)}(P_1, P_2) \times^{H(Y)} K(Y). \qquad (5-8)$$

We will need the notion of *equivalence* between two gerbes C and D with band H on a manifold X. Such an equivalence Φ will consist of the following data:

(a) For every local homeomorphism $f : Y \to X$, an equivalence of categories $\Phi(Y \xrightarrow{f} X) = \Phi(Y)$ from $C(Y \xrightarrow{f} X)$ to $D(Y \xrightarrow{f} X)$. For every object P of $C(Y \xrightarrow{f} X)$, we should have a commutative diagram

$$\begin{array}{ccc}
Aut_{C(Y)}(P) & \xrightarrow{\Phi(Y)} & Aut_{D(Y)}(\Phi(P)) \\
 & \searrow^{\alpha_C} \qquad \swarrow_{\alpha_D} & \\
 & H &
\end{array}$$

where the isomorphisms α_C and α_D are those given by the property (G1) of the gerbes C and D.

(b) For every diagram $Z \xrightarrow{g} Y \xrightarrow{f} X$ of local homeomorphisms, an invertible natural transformation $\beta_g : g_D^{-1}\Phi(Y) \to \Phi(Z)g_C^{-1}$, where g_C^{-1} and g_D^{-1} are the pull-back functors of the gerbes C and D, respectively. We require that, given a diagram $W \xrightarrow{h} Z \xrightarrow{g} Y \xrightarrow{f} X$ of local homeomorphisms, the following diagram is commutative

$$
\begin{array}{ccc}
h_D^{-1}g_D^{-1}\Phi(Y) & \xrightarrow{\beta_g} \ h_D^{-1}\Phi(Z)g_C^{-1} \ \xrightarrow{\beta_h} & \Phi(W)h_C^{-1}g_C^{-1} \\
\downarrow{\scriptstyle \theta_{g,h}} & & \downarrow{\scriptstyle \theta_{g,h}} \\
(gh)_D^{-1}\Phi(Y) & \xrightarrow{\beta_{gh}} & \Phi(W)(gh)_C^{-1}.
\end{array}
\qquad (5-9)
$$

Equivalence between gerbes over X with given commutative band H is an equivalence relation. Now to construct the notion of a contracted product $C_1 \times^H C_2$ of gerbes with band H, take the product gerbe $C_1 \times C_2$, with band $H \times H$. Then, using the product map $H \times H \to H$, the gerbe $C_1 \times^H C_2$ with band H is obtained. We now give the following direct description of the contracted product. For a local homeomorphism $f : Y \to X$, an object of $(C_1 \times^H C_2)(Y)$ is a quadruple $(Z \xrightarrow{g} Y, P_1, P_2, \alpha)$, where g is a surjective local homeomorphism, P_1 and P_2 are objects of $C_1(Z)$ and $C_2(Z)$, and α is a section over $Z \times_Y Z$ of the quotient of the sheaf

$$
\underline{Isom}_{C_1}(p_1^{-1}P_1, p_2^{-1}P_1) \times \underline{Isom}_{C_2}(p_1^{-1}P_2, p_2^{-1}P_2)
$$

by the action of H given by $a \cdot (\phi_1, \phi_2) = (a \cdot \phi_1, a^{-1} \cdot \phi_2)$ such that s satisfies the cocycle condition (5-3) modulo H. A morphism $(Z \xrightarrow{g} Y, P_1, P_2, \alpha) \to (Z' \xrightarrow{g} Y, P_1', P_2', \alpha')$ is induced by a local homeomorphism $W \to Z \times_Y Z'$, and isomorphisms $p_1^{-1}P_1 \xrightarrow{\sim} p_2^{-1}P_1'$ in $C_1(W)$ and $p_1^{-1}P_2 \xrightarrow{\sim} p_2^{-1}P_2'$ in $C_2(W)$, which are compatible with the descent data α and α' (the reader will surely not write down the corresponding commutative diagram).

Thus the set of equivalence classes of gerbes over X with band H is an abelian group. The identity element corresponds to the *trivial gerbe* with band H, which associates to $Y \xrightarrow{f} X$ the category of H-torsors over Y; its pull-back functors are the natural ones. We now come to the relation with degree 2 sheaf cohomology.

5.2.8. Theorem. (*Giraud* [Gi]) *Let X be a smooth manifold, and let H be a sheaf of abelian groups over X. For a gerbe C over X with band H, construct a cohomology class in $H^2(X, H)$ as follows. Let $(U_i)_{i \in I}$ be a "good" open covering of X, and let P_i be an object of the category $C(U_i)$. Then there exists an isomorphism $u_{ij} : (P_j)_{/U_{ij}} \xrightarrow{\sim} (P_i)_{/U_{ij}}$ in the category $C(U_{ij})$. We* *define a section h_{ijk} of H over U_{ijk} by*

$$h_{ijk} = u_{ik}^{-1} u_{ij} u_{jk} \in Aut(P_k) \simeq H, \qquad (5-10)$$

so that $\underline{h} = (h_{ijk})$ is an H-valued Čech 2-cocycle. The corresponding class in $H^2(X,H)$ is independent of all the choices. In this way, we obtain an isomorphism between the group of equivalence classes of gerbes over X with band H, and the group $H^2(X,H)$.

Proof. It is immediate to see that $\underline{h}$ is a Čech 2-cocycle. The construction of this 2-cocycle is clearly compatible with replacing the open covering by a refinement. By Proposition 5.1.9, all objects of $\mathcal{C}(U_i)$ are isomorphic so that the choice of these objects is irrelevant. We now use the fact that all morphisms in $\mathcal{C}(Y)$ commute with the action of H (by property (G1)). If we replace the isomorphism u_{ij} by $a_{ij}u_{ij}$ for $a_{ij} \in \Gamma(U_{ij}, H)$, we also replace h_{ijk} by $h_{ijk}a_{ik}^{-1}a_{ij}a_{jk}$, which differs from h_{ijk} by a Čech coboundary. We have proved that the class of $\underline{h}$ in $H^2(X,H)$ is independent of all choices.

It is then easy to see that we have a group homomorphism from the group of equivalence classes of gerbes with band H to $H^2(X,H)$. To show this homomorphism is injective, assume that h_{ijk} is a Čech coboundary. After modifying the isomorphisms u_{ij}, we arrange to have the glueing condition $u_{ik} = u_{ij}u_{jk}$ satisfied. Given an object $P' = (P_i) \in \mathcal{C}(\coprod_i U_i)$ and an isomorphism $p_1^{-1}P' \xrightarrow{\sim} p_2^{-1}P'$ given by the (u_{ij}), by the descent property we now obtain an object P of $\mathcal{C}(X)$ such that $P_{/U_i}$ is identified with P_i. We thus have an equivalence of gerbes from the trivial gerbe to $\mathcal{C}$, which sends an H-torsor I to $P \times^H I$ (cf. Proposition 5.1.5).

To prove surjectivity, we construct a gerbe $\mathcal{C}$ associated to a Čech 2-cocycle $\underline{h} = (h_{ijk})$ with values in H. Let $f : Y \to X$ be a local homeomorphism. Then an object of $\mathcal{C}(Y)$ consists of a family (P_i) of H-torsors over $f^{-1}(U_i) \subseteq Y$, together with isomorphisms $u_{ij} : (P_j)_{/f^{-1}(U_{ij})} \xrightarrow{\sim} (P_i)_{/f^{-1}(U_{ij})}$ which satisfy the glueing condition

$$u_{ik}^{-1} u_{ij} u_{jk} = h_{ijk} \in H(f^{-1}(U_{ijk})).$$

A morphism between two such objects consists of a family of morphisms between the H-torsors over $f^{-1}(U_i)$, which commutes with the glueing isomorphisms over $f^{-1}(U_{ij})$. The pull-back functors are the natural ones. It is clear that we have a sheaf of groupoids over X. To show $\mathcal{C}$ is a gerbe, we need to verify properties (G1) to (G3). For (G1), note that an automorphism of the object $((P_i), (u_{ij}))$ as above consists of a family $a_i \in \Gamma(U_i, H)$ such that a_i and a_j have the same restriction to $f^{-1}(U_{ij})$. Since H is a sheaf, this is the same as a section of H over Y. For (G2) and (G3), it will be enough to show that, for any $i_0 \in I$, the restriction of the sheaf of categories $\mathcal{C}$ to U_{i_0} is equivalent to the trivial gerbe with band H for which (G2) and (G3) are true. The equivalence is obtained by mapping the object $((P_i), (U_{ij}))$ to P_{i_0}. To show that this too is an equivalence, we observe that because h_{ijk} is a Čech 2-cocycle, we can find u_{ij} over $U_{i_0} \cap U_{ij}$ such that $h_{ijk} = u_{ik}^{-1} u_{ij} u_{jk}$ over $U_{i_0} \cap U_{ijk}$ (set $u_{ij} = h_{i_0ij}$). We now verify that the gerbe $\mathcal{C}$ gives rise to the 2-cocycle $\underline{h}$. By the above method, we have found an object Q_{i_0} of $\mathcal{C}(U_{i_0})$, for which all the torsors P_i over U_{i_0i} are trivial, and $u_{ij} = h_{i_0ij}$. Given another element i_1 of I, we have $h_{i_0ij} h_{i_1ij}^{-1} = h_{i_0i_1i} h_{i_0i_1j}^{-1}$ by the cocycle relation. So the family $(h_{i_0i_1i})_{i \in I}$ gives an isomorphism $v_{i_0i_1} : Q_{i_1/U_{i_0i_1}} \xrightarrow{\sim} Q_{i_0/U_{i_0i_1}}$. The corresponding Čech 2-cocycle is, over $U_{i_0i_1i_2}$,

$$v_{i_0i_2}^{-1} v_{i_0i_1} v_{i_1i_2} = h_{i_0i_2i}^{-1} h_{i_0i_1i} h_{i_1i_2i}$$
$$= h_{i_0i_1i_2}.$$

Surjectivity is proved. ∎

We note a compatibility with inverse image of gerbes. With the notations of Theorem 5.2.8, if $f : Y \to X$ is a continuous mapping, and $\mathcal{C}$ is a gerbe over X with band H, then the class in $H^2(Y, f^{-1}H)$ of the gerbe $f^{-1}\mathcal{C}$ is the image of $[\mathcal{C}] \in H^2(X, H)$ under the pull-back map $H^2(X, H) \to H^2(Y, f^{-1}H)$. The pull-back map on cohomology is obtained from a morphism of complexes of sheaves $f^{-1}I^\bullet \to J^\bullet$, where $I^\bullet$ is an injective resolution of H and $J^\bullet$ is an injective resolution of $f^{-1}H$.

5.2.9. Theorem. *Let $1 \to A \to C \to B \to 1$ be an exact sequence of Lie groups, with A central in C, such that the map $C \to B$ has local sections. Let $P \to X$ be a principal B-bundle, with class $[P] \in H^1(X, \underline{B}_X)$. Let $\mathcal{G}$ be the corresponding gerbe on X with band $\underline{A}_X$. Then the class $[\mathcal{G}] \in H^2(X, \underline{A}_X)$ is equal to $\delta_1([P])$, where $\delta_1 : H^1(X, \underline{B}_X) \to H^2(X, \underline{A}_X)$ is the connecting homomorphism as in Theorem 4.1.3.*

Proof. Let (U_i) be an open covering of X such that the restriction of P to U_i has a smooth section s_i. Let $b_{ij} : U_{ij} \to B$ be the transition function, such that $s_j = s_i \cdot b_{ij}$. Then we have an object Q_i of $\mathcal{G}(U_i)$, given by the pair $(C \times U_i, s_i)$, where s_i is the isomorphism $B \times U_i \xrightarrow{\sim} P_{/U_i}$ induced by s_i. If U_{ij} is contractible, we can find a smooth function $c_{ij} : U_{ij} \to B$ which projects to b_{ij}, enabling the use of the c_{ij} to give an isomorphism $u_{ij} : (Q_j)_{/U_{ij}} \xrightarrow{\sim} (Q_i)_{/U_{ij}}$. The corresponding $\underline{A}_X$-valued 2-cocycle is $u_{ik}^{-1} u_{ij} u_{jk} = c_{ik}^{-1} c_{ij} c_{jk}$, which is equal to $\delta_1(\underline{b})_{ijk}$. ∎

Of special significance in this book are gerbes over a smooth manifold X with band the sheaf $\underline{\mathbb{C}}_X^*$ of smooth $\mathbb{C}^*$-valued functions.

5.2.10. Definition and Proposition. *Let X be a smooth manifold. A Dixmier-Douady sheaf of groupoids over X is a gerbe over X with band $\underline{\mathbb{C}}_X^*$. The group of equivalence classes of gerbes over X with band $\underline{\mathbb{C}}_X^*$ is canonically isomorphic to $H^2(X, \underline{\mathbb{C}}_X^*) \tilde{\to} H^3(X, \mathbb{Z}(1))$.*

A unitary Dixmier-Douady sheaf of groupoids is a gerbe with band $\underline{\mathbb{T}}_X$.

Thus the Dixmier-Douady sheaves of groupoids give the geometric interpretation of the group $H^3(X, \mathbb{Z}(1)) = 2\pi\sqrt{-1} \cdot H^3(X, \mathbb{Z})$. Given a smooth mapping $f : Y \to X$ between manifolds, we have the notion of pull-back f^*C of a Dixmier-Douady sheaf of groupoids over X. First $f^{-1}C$ is a gerbe over Y with band $f^{-1}\underline{\mathbb{C}}_X^*$. Then since we have a morphism $f^{-1}\underline{\mathbb{C}}_X^* \to \underline{\mathbb{C}}_Y^*$ of sheaves of groups over Y, we obtain a gerbe with band $\underline{\mathbb{C}}_Y^*$, denoted by f^*C. This corresponds to the pull-back map $H^3(X, \mathbb{Z}(1)) \to H^3(Y, \mathbb{Z}(1))$ on degree 3 cohomology.

In the rest of this chapter, we will try to convince the reader that Dixmier-Douady sheaves of groupoids are ubiquitous. So far we only have given one type of example, namely, the one arising from a central extension of Lie groups. There is another general type of example, for which we consider a smooth locally trivial fibration $p : Y \to X$, where the manifolds X and Y satisfy the assumptions of Proposition 1.4.12, so that their de Rham cohomology identifies with their Čech cohomology. We then introduce a complex-valued relative 2-form ω. The following conditions should be verified:

(F1) The restriction of ω to any fiber $p^{-1}(x)$ is a closed 2-form with integral cohomology class, and each fiber $p^{-1}(x)$ is connected.

(F2) Each fiber $p^{-1}(x)$ satisfies: $H^1(p^{-1}(x), \mathbb{C}^*) = 0$ (this will certainly hold if the fibers are simply-connected).

We need the following notion

5.2.11. Definition. *Let $p : Y \to X$ be a smooth fibration, and let L be a line bundle over Y. A fiberwise connection ∇ on L is a rule which assigns to every open subset U of Y and to every section s of L over U, an element $\nabla(s) \in (\underline{A}_{Y/X}^1 \otimes L)(U)$, i.e., a relative 1-form on U with values in L, so that the following properties hold:*

(1) The construction $s \mapsto \nabla(s)$ is compatible with restriction to smaller open sets.

(2) For any given U, the map $s \mapsto \nabla(s)$ is $\mathbb{C}$-linear.

(3) *The Leibniz rule holds, that is,*

$$\nabla(f \cdot s) = f \cdot \nabla(s) + df \otimes s.$$

Note that this differs from the notion of connection (cf. Definition 2.2.1) only in that the 1-forms are replaced by relative 1-forms. A connection $\tilde{\nabla}$ on L induces, in an obvious way, a relative connection.

We now define a sheaf of groupoids $\mathcal{C}$ over X as follows. Let $Z \xrightarrow{g} X$ be a local homeomorphism; the pull-back fibration $p : Y \times_X Z \to Z$ has the same properties as p. We now define a category $\mathcal{C}(Z \xrightarrow{g} X)$; an object of this category is a pair (L, ∇), where L is a line bundle over $Y \times_X Z$, and ∇ is a fiberwise connection for L, relative to the fibration $p : Z \times_X Y \to Z$. The connection ∇ must be such that its (fiberwise) curvature is equal to the pull-back of $2\pi\sqrt{-1} \cdot \omega$ to Z. A morphism from (L_1, ∇_1) to (L_2, ∇_2) is an isomorphism $\phi : L_1 \tilde{\to} L_2$ of line bundles over $Z \times_X Y$ such that $\phi_*(\nabla_1) = \nabla_2$. Given a diagram $W \xrightarrow{h} Z \xrightarrow{g} X$ of local homeomorphisms, there is an obvious pull-back functor $h^* : \mathcal{C}(Z \xrightarrow{g} X) \to \mathcal{C}(W \xrightarrow{gh} X)$. The transitivity isomorphisms are also obvious.

5.2.12. Theorem. *The above presheaf of categories $\mathcal{C}$ is a Dixmier-Douady sheaf of groupoids over X.*

Proof. It is easy to see that $\mathcal{C}$ is indeed a sheaf of groupoids. Let $Z \xrightarrow{g} X$, and let (L, ∇) be an object of $\mathcal{C}(Z \xrightarrow{g} X)$. We examine the sheaf $\underline{Aut}(L, \nabla)$. A section ϕ of this sheaf is a bundle automorphism of L which preserves ∇. A bundle automorphism of L is given by a smooth function $f : Z \times_X Y \to \mathbb{C}^*$; the compatibility with ∇ means that the vertical derivative of f is 0, i.e., that g is locally constant along the fibers. Since the fibers are connected, this is equivalent to saying that f is the pull-back of a smooth function on the base space Z. So we find that $\Gamma(Z, \underline{Aut}(L, \nabla)) = \Gamma(Z, \underline{\mathbb{C}}_Z^*) = \Gamma(Z, \underline{\mathbb{C}}_X^*)$. (G1) is proved.

Now let (L_1, ∇_1) and (L_2, ∇_2) be two objects of $\mathcal{C}(Z \xrightarrow{g} X)$. The line bundle $L_2 \otimes L_1^{\otimes -1}$ over $Y \times_X Z$ is equipped with the difference fiberwise connection $\nabla_2 - \nabla_1$. The fiberwise curvature of this connection is 0. The restriction of $L_2 \otimes L_1^{\otimes -1}$ to a fiber $p^{-1}(z)$ (for $z \in Z$) is then a flat line bundle. Having $H^1(p^{-1}(z), \mathbb{C}^*) = 0$ by assumption, the flat line bundle is trivial (cf. Proposition 2.1.12 (2)), and therefore, there exists a horizontal section of it on any fiber of p. It follows that, locally on Z, we can find a nowhere vanishing section of $L_2 \otimes L_1^{\otimes -1}$ which restricts to a horizontal section on any fiber of $p : Y \times_X Z \to X$. Such a section produces an isomorphism between (L_1, ∇_1) and (L_2, ∇_2). This proves (G2).

To prove (G3), let $x \in X$ and let U be a contractible neighborhood of x. Recalling the methods used in the proof of Lemma 4.3.2, we observe that there exists a closed 2-form β on $p^{-1}(U) \subseteq Y$ such that the restriction of β to any fiber is equal to ω. Since we have $H^2(p^{-1}(U), \mathbf{R}) \xrightarrow{\sim} H^2(p^{-1}(x), \mathbf{R})$ for any $x \in U$, the cohomology class of β is integral. By the theorem of Kostant-Weil (cf. Theorem 2.2.15), there exists a line bundle with connection (L, D) over $p^{-1}(U)$ with curvature $2\pi\sqrt{-1} \cdot \beta$. Then if ∇ is the fiberwise connection induced by D, (L, ∇) is an object of $\mathcal{C}(U)$. This establishes property (G3). ∎

To get a unitary Dixmier-Douady sheaf of groupoids, we assume that ω is a real 2-form, and we define $\mathcal{C}(Z \xrightarrow{g} X)$ to consist of triples (L, h, ∇), where (L, ∇) is, as before, and h is a hermitian structure on L which is compatible with ∇.

Note that for a local homeomorphism $g : Z \to X$, any object P of $\mathcal{C}(Z \xrightarrow{g} X)$ gives rise to an object of $(f^*\mathcal{C})(Z \times_X Y \to Y)$, which will be denoted by f^*P.

The theory of gerbes (with non-commutative band) is intimately related to tannakian categories and to motives; we refer the reader to the book [D-M-O-S].

5.3. Differential geometry of gerbes

In the last section we introduced gerbes over a space and the notion of a Dixmier-Douady sheaf of groupoids over a manifold X, which means a gerbe over X with band, the sheaf of groups $\underline{\mathbf{C}}_X^*$. Proposition 5.2.10 states that Dixmier-Douady sheaves of groupoids represent the degree 3 cohomology group $H^3(X, \mathbf{Z}(1))$. In this section we develop the differential geometry relevant to Dixmier-Douady sheaves of groupoids and, in particular, construct a degree 3 closed differential form, the 3-*curvature*, which represents the cohomology class of such a sheaf of groupoids. These ideas are completely analogous with the well-known notions of connection and curvature for line bundles, as already developed in Chapter 2. To obtain the degree 3 differential form, we must consider two nested structures. The first one is called *connective structure*, and the second one is *curving*. The idea of these structures was given to the author by Pierre Deligne.

In the present section, "sheaf of groupoids" will always mean "Dixmier-Douady sheaf of groupoids," unless otherwise specified.

Let $\mathcal{G}$ be a sheaf of groupoids over a smooth manifold X. We wish to define some notion of connections for objects of $\mathcal{G}$. If $\mathcal{G}$ is the trivial gerbe, such that the objects of $\mathcal{G}(Y \xrightarrow{f} X)$ are the $\underline{\mathbf{C}}_Y^*$-torsors over Y, we have the notion of connection for any such torsor P. In fact, such a torsor is equivalent to a principal $\mathbb{C}^*$-bundle over Y. As was shown in Corollary 2.2.3, the set of all connections on P forms a principal homogeneous space under the vector space $A^1(Y) \otimes \mathbb{C}$. The set of connections over various open sets actually forms a sheaf over Y, denoted by $Co(P)$, which is then a torsor under the sheaf $\underline{A}^1_{Y,\mathbb{C}}$. Let us observe some formal properties of the assignment $P \mapsto Co(P)$ of such an $\underline{A}^1_{Y,\mathbb{C}}$-torsor to any $\underline{\mathbb{C}}^*_Y$-torsor.

Recall from §1 that, if $f : Y \to X$ is a smooth mapping between manifolds, then for a torsor T under $\underline{A}^1_{X,\mathbb{C}}$, we can define its inverse image f^*T, which is a torsor under $\underline{A}^1_{Y,\mathbb{C}}$.

(C1) (invariance under base change) Given a diagram $Z \xrightarrow{g} Y \xrightarrow{f} X$ of local homeomorphisms, and an object P of $\mathcal{G}(Y)$, there is a natural isomorphism $\alpha_g : g^* \, Co(P) \tilde{\to} Co(g^*P)$ of $\underline{A}^1_{Z,\mathbb{C}}$-torsors. This is compatible with the composition of pull-backs in the sense that, for a diagram $W \xrightarrow{h} Z \xrightarrow{g} Y \xrightarrow{f} X$ of local homeomorphisms, we have a commutative diagram

$$
\begin{array}{ccccc}
h^*g^*Co(P) & \xrightarrow{\alpha_g} & h^*Co(g^*P) & \xrightarrow{\alpha_h} & Co(h^*g^*P) \\
\downarrow{\scriptstyle can} & & & & \downarrow{\scriptstyle (\theta_{g,h})_*} \\
(gh)^*Co(P) & & \xrightarrow{\alpha_{gh}} & & Co((gh)^*P)
\end{array}
$$

where can is the canonical isomorphism of torsors, and $(\theta_{g,h})_*$ is as in (C2).

(C2) (invariance under isomorphisms) Given a local homeomorphism $f : Y \to X$, any isomorphism $\phi : P_1 \tilde{\to} P_2$ between $\underline{\mathbb{C}}^*_Y$-torsors induces an isomorphism $\phi_* : Co(P_1) \tilde{\to} Co(P_2)$ of the associated $\underline{A}^1_{Y,\mathbb{C}}$-torsors. This is compatible with the composition of isomorphisms in the sense that $(\psi\phi)_* = \psi_* \circ \phi_*$. The isomorphisms ϕ_* must satisfy two types of requirements:

(R1) (behavior under gauge transformations) Let $g : Y \to \mathbb{C}^*$ be a smooth function, which gives an automorphism g of the $\underline{\mathbb{C}}^*_Y$-torsor $P \to Y$. Then we have $g_*(\nabla) = \nabla - g^{-1}dg$ for $\nabla \in Co(P)$.

(R2) (compatibility with pull-backs) Let $f : Y \to X$ be a local homeomorphism, and let $\phi : P_1 \tilde{\to} P_2$ be an isomorphism of $\underline{\mathbb{C}}^*_Y$-torsors. Let $g : Z \to Y$ be a local homeomorphism. Then we have the commutative diagram

$$
\begin{array}{ccc}
g^*Co(P_1) & \xrightarrow{g^*(\phi_*)} & g^*Co(P_2) \\
\downarrow{\scriptstyle \alpha_g} & & \downarrow{\scriptstyle \alpha_g} \\
Co(g^*P_1) & \xrightarrow{(g^*\phi)_*} & Co(g^*P_2).
\end{array}
$$

We can now define the notion of connective structure on a Dixmier-Douady sheaf of groupoids.

5.3.1.   Definition. *Let C be a Dixmier-Douady sheaf of groupoids over* *a smooth manifold* X. *A* <u>*connective structure*</u> *for* $\mathcal{C}$ *is an assignment of an* $\underline{A}^1_{Y,\mathbb{C}}$-*torsor over* $f : Y \to X$ *to any object* P *of* $\mathcal{C}(Y) = \mathcal{C}(Y \xrightarrow{f} X)$, *together with the following data:*

(1) *for a diagram* $Z \xrightarrow{g} Y \xrightarrow{f} X$ *of local homeomorphisms, and an object* P *of* $\mathcal{C}(Y)$, *an isomorphism* $\alpha_g : g^* Co(P) \xrightarrow{\sim} Co(g^* P)$ *of* $\underline{A}^1_{Z,\mathbb{C}}$-*torsors. This isomorphism must satisfy the conditions of (C1) above.*

(2) *Any isomorphism* $\phi : P_1 \xrightarrow{\sim} P_2$ *between objects of* $\mathcal{C}(Y)$ *induces an isomorphism* $\phi_* : Co(P_1) \xrightarrow{\sim} Co(P_2)$ *of the associated* $\underline{A}^1_{Y,\mathbb{C}}$-*torsors. This isomorphism must satisfy the properties in (C2) above.*

It might not be obvious at first sight that a connective structure exists on any given sheaf of groupoids $\mathcal{C}$, but we will later prove that this is indeed the case. We first give an example already encountered in §4.2. Let B be a Lie group, and let $\tilde{B}$ be a central extension of B by $\mathbb{C}^*$. Let $p : Q \to X$ be a principal B-bundle. Then we have a sheaf of groupoids $\mathcal{C}$, as defined in §2, such that for a local homeomorphism $f : Y \to X$, the category $\mathcal{C}(Y)$ has as objects the pairs $(\tilde{Q}, \phi)$, where $\tilde{Q} \to Y$ is a $\tilde{B}$-bundle, and $\phi : \tilde{Q}/\mathbb{C}^* \xrightarrow{\sim} f^* Q$ is an isomorphism of B-bundles. If we pick a connection θ on the B-bundle Q, we have the following connective structure. Recall that for $(\tilde{Q}, \phi)$ as above, a connection $\tilde{\theta}$ on $\tilde{Q} \to Y$ is said to be compatible with θ if the $\mathfrak{b}$-valued 1-form $\phi^* \theta$ on $\tilde{Q}$ is equal to the projection of the $\tilde{\mathfrak{b}}$-valued 1-form $\tilde{\theta}$. By Lemma 4.2.1, the set of connections on $\tilde{Q} \to Y$, which are compatible with θ, forms an affine space under the vector space $A^1(Y) \otimes \mathbb{C}$. We thus obtain a sheaf $Co(\tilde{Q}, \phi)$ consisting of connections $\tilde{\theta}$ compatible with θ. This is a torsor under $\underline{A}^1_{Y,\mathbb{C}}$. The base change isomorphism α_g of (C1) is obtained by pulling-back compatible connections. The isomorphisms ϕ_* of (C2) are also obvious. We have now obtained a connective structure on the sheaf of groupoids $\mathcal{G}$.

Note that if we have two sheaves of groupoids $\mathcal{G}$ and $\mathcal{G}'$ with band $\underline{\mathbb{C}}^*_X$, and if we have an equivalence of gerbes $\Phi : \mathcal{G} \to \mathcal{G}'$, then a connective structure $P \mapsto Co'(P)$ on $\mathcal{G}'$ induces a connective structure Co on $\mathcal{G}$ such that $Co(P) = Co'(\Phi(P))$. We easily obtain the structures of (C1) and (C2).

5.3.2. Proposition. *Let* $\mathcal{G}$ *be a Dixmier-Douady sheaf of groupoids over a manifold* X *(always assumed to be paracompact). There exists a connective structure for* $\mathcal{G}$.

Proof. Let $\mathcal{U} = (U_i)_{i \in I}$ be a good open covering of X. Let P_i be an object of $\mathcal{G}(U_i)$. Choose an isomorphism $u_{ij} : (P_j)_{/U_{ij}} \xrightarrow{\sim} (P_i)_{U_{ij}}$. Let $h_{ijk} = u_{ik}^{-1} u_{ij} u_{jk}$ be the corresponding $\underline{\mathbb{C}}^*_X$-valued Čech 2-cocycle. Then $d\log(h_{ijk})$ is an $\underline{A}^1_{X,\mathbb{C}}$-valued 2-cocycle. Since the group $\check{H}^2(\mathcal{U}, \underline{A}^1_{X,\mathbb{C}})$ is 0, there exist 1-forms α_{ij}

on U_{ij} such that $\alpha_{jk} - \alpha_{ik} + \alpha_{ij} = -d \, \log(h_{ijk})$ on U_{ijk}.

We can now construct a connective structure. Let $f : Y \to X$ be a local homeomorphism, and let $Q \in \mathcal{G}(Y)$. We will construct an $\underline{A}^1_{Y,\mathbf{C}}$-torsor as follows. Over $f^{-1}(U_i) \subseteq Y$, we have the $\underline{\mathbf{C}}^*_Y$-torsor $\underline{Isom}(Q, P_i)$. Consider the corresponding $\underline{A}^1_{Y,\mathbf{C}}$-torsor $R_i = \underline{Isom}(Q, P_i) \times^{\underline{\mathbf{C}}^*_Y} \underline{A}^1_{Y,\mathbf{C}}$. The isomorphism u_{ij} induces an isomorphism $\underline{Isom}(Q, P_j) \overset{\sim}{\to} \underline{Isom}(Q, P_i)$, hence $d \log(u_{ij})$ gives an isomorphism $(R_j)_{/f^{-1}U_{ij}} \overset{\sim}{\to} (R_i)_{/f^{-1}U_{ij}}$; then $\psi_{ij} = \alpha_{ij} \cdot d \log(u_{ij})$ also gives such an isomorphism, and the cocycle condition $\psi_{ik} = \psi_{ij}\psi_{jk}$ is satisfied. By the glueing property for torsors, we obtain a uniquely defined $\underline{A}^1_{Y,\mathbf{C}}$-torsor, denoted by $Co(Q)$. It is clear that an isomorphism of objects of $\mathcal{G}(Y)$ induces an isomorphism of the corresponding $\underline{A}^1_{Y,\mathbf{C}}$-torsors, and that the automorphism g of Q induces the automorphism $\alpha \mapsto \alpha - g^{-1}dg$ on $Co(Q)$ (this is because $\underline{Isom}(Q, P_i)$ is a contravariant functor of Q). The pull-back isomorphisms are easy to construct. So $Q \mapsto Co(Q)$ is a connective structure on $\mathcal{G}$. ∎

There is a very nice way of putting a connective structure on the Dixmier-Douady sheaf of groupoids associated to a smooth fibration $p : Y \to X$ and a closed relative 2-form ω, which satisfies condition (F1) of §2.

5.3.3. Lemma. *Under the assumption (F1) of §2 for the closed relative 2-form ω, there exists a 2-form β on Y such that*

(1) *The restriction of β to the fibers of p is equal to ω.*

(2) *$d\beta \in F^2$, where $F^\bullet$ denotes the Cartan filtration of $A^\bullet(Y)$, as in §1.5.*

Proof. This is proved in the same way as Lemma 4.3.2. ∎

Given a 2-form β as in Lemma 5.3.3, we obtain a connective structure on the gerbe $\mathcal{C}$ associated to $p : Y \to X$ and to ω (cf. Theorem 5.2.12). Recall that, for a local homeomorphism $Z \overset{g}{\to} X$, the category $\mathcal{C}(Z \overset{g}{\to} X)$ has as objects the pairs (L, ∇), where L is a line bundle on $Y \times_X Z$, and ∇ is a fiberwise connection on L, with fiberwise curvature equal to $2\pi\sqrt{-1} \cdot \omega$.

5.3.4. Lemma. *In the situation of the preceding paragraph, there exists a connection $\tilde{\nabla}$ on L such that*

(1) *$\tilde{\nabla}$ is compatible with the fiberwise connection ∇.*

(2) *The curvature K of $\tilde{\nabla}$ satisfies $K - 2\pi\sqrt{-1} \cdot \beta \in F^2$.*

The set of connections on L satisfying (1) and (2) is an affine space under $A^1(Z)_{\mathbf{C}}$.

Proof. We first show that such a connection exists locally on Z. Assume that the line bundle L is trivial, so that θ is given by a relative 1-form α, satisfying $d\alpha = 2\pi\sqrt{-1} \cdot \omega \in A^2(Y/X)_{\mathbb{C}}$. We then look at the short exact sequence

$$H^1(A^\bullet(Y)_{\mathbb{C}}/F^2) \to H^1(A^\bullet(Y/X)_{\mathbb{C}}) \to H^2(F^1/F^2).$$

The group $H^2(F^1/F^2)$ is equal to $A^1(X, H^1(Y_x, \mathbb{C})) = 0$. Hence we can find a 1-form $\tilde{\alpha}$, such that $d\tilde{\alpha} \in F^2$, and its cohomology class $[\tilde{\alpha}]$ maps to $[\alpha] \in H^1(A^\bullet(Y/X)_{\mathbb{C}})$. Modifying $\tilde{\alpha}$ by a boundary, we arrange that the restriction of $\tilde{\alpha}$ to the fibers is exactly equal to α. Now, by assumption, the 2-form β is closed modulo F^2. We have seen that the group $H^2(F^1/F^2)$ is 0. The 2-form $2\pi\sqrt{-1} \cdot \beta - d\tilde{\alpha}$ is closed modulo F^2 and has zero restriction to the fibers, giving a closed element of F^1/F^2; hence it is a boundary.

Let $\gamma \in F^1$ be such that $2\pi\sqrt{-1} \cdot \beta - d\tilde{\alpha} \equiv d\gamma \pmod{F^2}$. The connection given by the 1-form $\tilde{\alpha} + \gamma$ satisfies (1) and (2), and so locally, a connection $\tilde{\nabla}$ on L with properties (1) and (2) exists. Another connection of this type is of the form $\tilde{\nabla} + \gamma$, where γ is some 1-form on $Y \times_X Z$. Property (1) means that γ has zero restriction to the fibers, and so belongs to F^1. Property (2) means that $d\gamma \in F^2$; this holds if and only if the 1-form γ is basic with respect to the fibration $Y \times_X Z \to Z$, in other words, is the pull-back of a 1-form on the base Z. Thus (locally) we have shown that the set of connections satisfying (1) and (2) is an affine space under $A^1(Z)_{\mathbb{C}}$. The existence of a global connection satisfying (1) and (2) then follows by using partitions of unity. Alternatively, we can form a sheaf of such connections, which is a torsor under $\underline{A}^1_{Z,\mathbb{C}}$. Such a torsor has a global section since $H^1(Z, \underline{A}^1_{Z,\mathbb{C}}) = 0$. ∎

We are now ready to define the connective structure. The torsor $Co(L, \nabla)$ consists of the connections $\tilde{\nabla}$ satisfying properties (1) and (2) in Lemma 5.3.4. Conditions (R1) and (R2) are easily verified so that we obtain

5.3.5. Proposition. *By assigning to $(L, \nabla) \in \mathcal{C}(Z \xrightarrow{g} X)$ the above $\underline{A}^1_{X,\mathbb{C}}$-torsor $Co(L, \nabla)$, we obtain a connective structure on the sheaf of groupoids $\mathcal{C}$.*

Notice that this connective structure only depends on the class of β modulo F^2.

In the case of a line bundle L over a manifold X, the set of connections on L is an affine space under $A^1(X)_{\mathbb{C}}$. There is an analogous statement for the case of connective structures over a Dixmier-Douady sheaf of groupoids.

If we have a connective structure $P \mapsto Co(P)$ on a sheaf of groupoids $\mathcal{C}$ over X, and some $\underline{A}^1_{X,\mathcal{C}}$-torsor Q, then we have a new connective structure $P \mapsto Co(P) \times^{\underline{A}^1_{X,\mathcal{C}}} Q$. There is a converse to this.

5.3.6. Proposition. *Let $\mathcal{C}$ be a Dixmier-Douady sheaf of groupoids over the manifold X. Let $P \mapsto Co(P)$ be a connective structure on $\mathcal{C}$. If $P \mapsto Co'(P)$ is another connective structure, there is a well-defined $\underline{A}^1_{X,\mathcal{C}}$-torsor Q over X such that $Co'(P) \xrightarrow{\sim} Co(P) \times^{\underline{A}^1_{X,\mathcal{C}}} Q$.*

Proof. We first show that, for a local homeomorphism $Y \xrightarrow{f} X$, and P_1 and P_2, two objects of $\mathcal{C}(Y \xrightarrow{f} X)$, the $\underline{A}^1_{Y,\mathcal{C}}$-torsors $Co'(P_1) \times^{\underline{A}^1_{Y,\mathcal{C}}} Co(P_1)^{-1}$ and $Co'(P_2) \times^{\underline{A}^1_{Y,\mathcal{C}}} Co(P_2)^{-1}$ are canonically isomorphic. Let us construct an isomorphism using descent. Let $Z \xrightarrow{g} Y$ be a surjective local homeomorphism such that there exists an isomorphism $\phi : g^{-1}P_1 \xrightarrow{\sim} g^{-1}(P_2)$. Then we have an induced isomorphism $\phi_* : Co(g^{-1}(P_1)) \xrightarrow{\sim} Co(g^{-1}(P_2))$ of torsors, and similarly, an isomorphism $\phi'_* : Co'(g^{-1}(P_1)) \xrightarrow{\sim} Co'(g^{-1}(P_2))$. Hence we get an isomorphism $\tilde{\phi}_*$ between $Co'(P_1) \times^{\underline{A}^1_{Y,\mathcal{C}}} Co(P_1)^{-1}$ and $Co'(P_2) \times^{\underline{A}^1_{Y,\mathcal{C}}} Co(P_2)^{-1}$. This isomorphism is independent of the choice of ϕ; indeed, ϕ is unique up to a section u of the sheaf $\underline{\mathbb{C}}^*_Z$ (acting by automorphisms on g^*P_1). Furthermore, such an automorphism u acts on both $Co(g^*(P_1))$ and $Co'(g^*(P_1))$ by $-u^{-1}du$, according to property (R1). The uniqueness result implies that the isomorphism $\tilde{\phi}_*$ satisfies the descent condition $p_1^* \tilde{\phi}_* = p_2^* \tilde{\phi}_*$ over $Z \times_Y Z$, hence it comes from an isomorphism $Co'(P_1) \times^{\underline{A}^1_{Y,\mathcal{C}}} Co(P_1)^{-1} \xrightarrow{\sim} Co'(P_2) \times^{\underline{A}^1_{Y,\mathcal{C}}} Co(P_2)^{-1}$ over Y, which is canonical. Now given a third object P_3 of $\mathcal{C}(Y \xrightarrow{f} X)$, the isomorphism $Co'(P_1) \times^{\underline{A}^1_{Y,\mathcal{C}}} Co(P_1)^{-1} \xrightarrow{\sim} Co'(P_3) \times^{\underline{A}^1_{Y,\mathcal{C}}} Co(P_3)^{-1}$ is the composition of the isomorphisms $Co'(P_1) \times^{\underline{A}^1_{Y,\mathcal{C}}} Co(P_1)^{-1} \xrightarrow{\sim} Co'(P_2) \times^{\underline{A}^1_{Y,\mathcal{C}}} Co(P_2)^{-1}$ and $Co'(P_2) \times^{\underline{A}^1_{Y,\mathcal{C}}} Co(P_2)^{-1} \xrightarrow{\sim} Co'(P_3) \times^{\underline{A}^1_{Y,\mathcal{C}}} Co(P_3)^{-1}$. We now construct by descent a torsor Q under $\underline{A}^1_{X,\mathcal{C}}$. Take a surjective local homeomorphism $f : Y \to X$ such that $\mathcal{C}(Y)$ has an object P. Then the torsor $Co'(P) \times^{\underline{A}^1_{Y,\mathcal{C}}} Co(P)^{-1}$ is equipped with a descent isomorphism, and so comes from an $\underline{A}^1_{X,\mathcal{C}}$-torsor Q over X. Up to a canonical isomorphism, the torsor Q is independent of $Y \xrightarrow{f} X$ and of P. We then have a canonical isomorphism $Co'(P) \xrightarrow{\sim} Co(P) \times^{\underline{A}^1_{X,\mathcal{C}}} Q$ of torsors. ∎

Let $\mathcal{C}$ be a sheaf of groupoids over X and let Co be a connective structure on $\mathcal{C}$. Let $f : Y \to X$ be a local homeomorphism, let $P \in \mathcal{C}(Y)$, and let Q be some $\underline{\mathbb{C}}^*_Y$-torsor. We can construct the twisted object $P \times^{\underline{\mathbb{C}}^*_Y} Q$ of $\mathcal{C}(Y)$, as in Proposition 5.2.5. Then we have the $\underline{A}^1_{Y,\mathcal{C}}$-torsor $Conn(Q)$ of connections on Q. There is a canonical isomorphism of $\underline{A}^1_{Y,\mathbb{C}}$-torsors

$$Co(P \times^{\underline{\mathbb{C}}^*_Y} Q) \simeq Co(P) \times^{\underline{A}^1_{Y,\mathbb{C}}} Conn(Q). \qquad (5-11)$$

The existence of this isomorphism is a consequence of property (C2) of Co and of the fact that $Conn(Q)$ identifies with the $\underline{A}^1_{Y,\mathbb{C}}$-torsor obtained from Q by the sheaf homomorphism $-d \log : \underline{\mathbb{C}}^*_Y \to \underline{A}^1_{Y,\mathbb{C}}$.

There is a natural notion of equivalence $\Phi : (\mathcal{C}_1, Co_1) \to (\mathcal{C}_2, Co_2)$ between sheaves of groupoids equipped with connective structures. This consists of an equivalence of gerbes in the sense of §2, and of isomorphisms of $\underline{A}^1_{X,\mathbb{C}}$-torsors $Co_1(P) \xrightarrow{\sim} Co_2(\Phi(P))$, which are functorial and compatible with pull-back isomorphisms. We leave the details to the reader.

For unitary sheaves of groupoids, we have a similar notion of unitary connective structure; in this case we associate to an object P, a torsor under the sheaf $\sqrt{-1} \cdot \underline{A}^1_X$ of purely imaginary 1-forms. In the case of a smooth fibration $p : Y \to X$ and a real relative 2-form ω, there results a unitary sheaf of groupoids consisting of hermitian line bundles with hermitian relative connections. If in Lemma 5.3.4 we restrict further to hermitian connections, we obtain a unitary connective structure.

Given a smooth map $f : Y \to X$ between manifolds, we have the notion of the pull-back $f^*\mathcal{C}$ of a sheaf of groupoids $\mathcal{C}$ over X. For a local homeomorphism $Z \xrightarrow{g} Y$, an object P of $(f^*\mathcal{C})(Z)$ is locally of the form i^*Q, where $Q \in \mathcal{C}(W)$, and we have a commutative diagram of type (5-6)

$$\begin{array}{ccc} Z & \xrightarrow{g} & Y \\ \downarrow{\scriptstyle i} & & \downarrow{\scriptstyle f} \\ W & \xrightarrow{h} & X \end{array}$$

in which $Z \to W \times_X Y$ is the inclusion of an open set. Further, we have the $\underline{A}^1_{W,\mathbb{C}}$-torsor $Co(Q)$. We then introduce a connective structure on $f^*\mathcal{C}$ by defining $Co(P)$ to be equal to $i^*Co(Q)$. The details are left to the reader.

To pursue the analogy between sheaves of groupoids and line bundles, we have to define the curvature of an element ∇ of the torsor $Co(P)$ associated to an object of $\mathcal{C}(Y \xrightarrow{f} X)$.

5.3.7.   Definition. *Let $\mathcal{C}$ be a sheaf of groupoids over the manifold X, equipped with a connective structure $P \mapsto Co(P)$. A $\underline{curving}$ of the connective structure is a rule which assigns to any object $P \in \mathcal{C}(Y \xrightarrow{f} X)$ and to any section ∇ of the $\underline{A}^1_{Y,\mathbb{C}}$-torsor $Co(P)$ a complex-valued 2-form $K(\nabla)$ on Y (called the $\underline{curvature}$ of ∇), so that the following properties are verified:*

(1) *(compatibility with pull-backs).   Let $P \in \mathcal{C}(Y \xrightarrow{f} X)$ and $\nabla \in \Gamma(Y, Co(P))$.   Given a local homeomorphism $g : Z \to Y$, the curvature*

$K(\alpha_g(g^*\nabla))$ of the section $\alpha_g(g^*\nabla)$ of $Co(g^*P)$ is equal to $g^*K(\nabla)$.

(2) *(invariance under isomorphism).* Let (P,∇) be as in *(1),* and let $\phi: P \tilde{\to} P'$ be an isomorphism with another object P' of $C(Y)$. Let $\phi_*(\nabla)$ be the corresponding object of $Co(P')$. Then we have $K(\nabla) = K(\phi_*(\nabla))$.

(3) *Let α be a complex-valued 1-form on Y. Then we have*

$$K(\nabla + \alpha) = K(\nabla) + d\alpha.$$

As our first example, we explain how to find a curving for the connective structures previously discussed. In the case of the sheaf of groupoids associated with a B-bundle $p : Q \to X$ and a central extension $\tilde{B}$ of B by $\mathbb{C}^*$, the work was already done in §4.2. Recall the construction in Definition 4.2.3, and fix a connection θ on the bundle. Then for $\tilde{Q} \in C(Y \xrightarrow{f} X)$, we have a principal $\tilde{B}$-bundle $q : \tilde{Q} \to Y$, together with an isomorphism $\tilde{Q}/\mathbb{C}^* \tilde{\to} Q \times_X Y$ of B-bundles. The torsor $Co(\tilde{Q})$ consists of those connections $\tilde{\theta}$ on $\tilde{Q}$ which are compatible with the connection θ on the B-bundle $Q \times_X Y \to Y$. For the exact sequence of vector bundles on X,

$$0 \to \mathcal{V}(\mathbb{C}) \to \mathcal{V}(\tilde{\mathfrak{b}}) \to \mathcal{V}(\mathfrak{b}) \to 0,$$

associated to representations of B, we choose a splitting $l : \mathcal{V}(\tilde{\mathfrak{b}}) \to \mathbb{C}$ so that the scalar curvature of $\tilde{\theta}$ is defined by $K(\tilde{\theta}) = l(\tilde{\Theta})$, where $\tilde{\Theta}$ (a $\tilde{\mathfrak{b}}$-valued 2-form) is the curvature of $\tilde{\theta}$. Finally, the assignment $\tilde{\theta} \mapsto K(\tilde{\theta})$ is a curving of the connective structure.

The second example comes from a smooth fibration $p : Y \to X$ and from a closed relative 2-form ω satisfying conditions (F1) and (F2) of §2. Lemma 4.3.2 states that there exists a 2-form β on Y which restricts to ω on the fibers and satisfies $d\beta \in F^3$ so that we have the connective structure associated to the class of β in $A^2(Y)_{\mathbb{C}}/F^2$. Let $(L,\nabla) \in C(Z \xrightarrow{g} X)$ and let $\tilde{\nabla}$ be an element of $Co(P)$, i.e., a connection on L satisfying the conditions of Lemma 5.3.4. We can now observe that the 2-form $K - 2\pi\sqrt{-1}\cdot\beta$ is basic, because it belongs to F^2 and its exterior derivative $-2\pi\sqrt{-1}\cdot d\beta$ belongs to F^3. Therefore it is the pull-back of a 2-form on X. So we set

$$K(\tilde{\nabla}) = K - 2\pi\sqrt{-1}\cdot\beta \in A^2(Z)_{\mathbb{C}}. \qquad (5-12)$$

5.3.8. Proposition. *The assignment $(L,\tilde{\nabla}) \mapsto K(\tilde{\nabla})$ as in (5-12) defines a curving of the connective structure on X.*

We note that in the case of a smooth fibration $p : Y \to X$ equipped with a relative 2-form, the notions of connective structure and curving give a geometric interpretation of some of the differentials of the Leray spectral sequence.

In the case of a unitary connective structure, we have the notion of unitary curving, for which it is further imposed that the curvature $K(\nabla)$ be a purely imaginary 2-form.

5.3.9. Proposition. (1) *Given a sheaf C of groupoids with connective structure $P \mapsto Co(P)$ over a manifold X, there exists a curving of Co.*

(2) *Given a curving $\nabla \mapsto K(\nabla)$, any other curving is of the form $\nabla \mapsto K(\nabla) + \beta$ for some complex-valued 2-form β.*

Proof. First we show that a curving exists locally. We may as well assume that the sheaf of groupoids is the sheaf of groupoids of all $\underline{\mathbb{C}}_X^*$-torsors over X (in other words, the trivial Dixmier-Douady sheaf of groupoids). According to Proposition 5.3.6, there exists some $\underline{A}_{X,\mathbb{C}}^1$-torsor Q such that the connective structure is given by $Co(P) = Conn(P) \times^{\underline{A}_{X,\mathbb{C}}^1} Q$, where P is a $\underline{\mathbb{C}}_X^*$-torsor, and $Conn(P)$ is the $\underline{A}_{X,\mathbb{C}}^1$-torsor of connections on P (which is the torsor of connections on the corresponding $\mathbb{C}^*$-bundle). Pick a global section D of Q. Then for ∇ a section of $Co(P)$, $\nabla - D$ is a section of $Conn(P)$, i.e., a connection on P. Setting $K(\nabla)$ as the curvature of the connection $\nabla - D$ gives a curving.

Now if we have a curving $\nabla \mapsto K(\nabla)$ and a 2-form β, then $\nabla \mapsto K(\nabla) + \beta$ is another curving. Conversely, if $\nabla \mapsto K'(\nabla)$ is another curving, then the difference $\beta(P, \nabla) = K'(\nabla) - K(\nabla)$ does not depend on the section ∇ of $Co(P)$, by property (3) of curvings. So, for given $P \in \mathcal{C}(Y \xrightarrow{f} X)$, we have a 2-form $\beta(P)$ on Y which is well-behaved under pull-backs by property (1) of curvings, and is independent of the object P. Indeed, since this is a local question, we have only to consider two isomorphic objects P_1 and P_2. Then $\beta(P_1) = \beta(P_2)$ by property (2) of curvings, and the existence of the 2-form β on X now follows by descent. Finally, the existence of a global curving follows by a partition of the unity argument. ∎

Let $f : Y \to X$ be a smooth mapping between manifolds, and let $(\mathcal{G}, Co)$ be a sheaf of groupoids on X with connective structure. Let $f^*\mathcal{G}$ be the pull-back sheaf of groupoids on Y, together with the pull-back connective structure f^*Co. If we have a curving $\nabla \mapsto K(\nabla)$ on $(\mathcal{G}, Co)$, there is a pull-back curving f^*K on f^*Co, described as follows: Given a commutative diagram

$$
\begin{array}{ccc}
Z & \xrightarrow{\;g\;} & Y \\
\downarrow{\scriptstyle i} & & \downarrow{\scriptstyle f} \\
W & \xrightarrow{\;h\;} & X
\end{array}
$$

in which $Z \to W \times_X Y$ is the inclusion of an open set, and given an object Q of $\mathcal{G}(W)$ and a section of ∇ of the torsor $Co(Q)$, we have

$$(f^*K)(i^*\nabla) = i^*K(\nabla) \in A^2(Z)_\mathbf{C}.$$

Here $i^*\nabla$ is the induced section of the $\underline{A}^1_{Z,\mathbf{C}}$-torsor $Co(i^*Q)$.

There is no reason for the 2-form $K(\nabla)$ associated to $\nabla \in Co(P)$ to be closed. Instead we have

5.3.10. Theorem and Definition. *Let X be a manifold, and let $\mathcal{C}$ be a Dixmier-Douady sheaf of groupoids over X. Let $P \mapsto Co(P)$ be a connective structure on $\mathcal{C}$, and let $\nabla \mapsto K(\nabla)$ be a curving of this connective structure. Then there exists a unique closed complex-valued 3-form Ω on X such that for any $P \in \mathcal{C}(Y \xrightarrow{f} X)$ and for any $\nabla \in \Gamma(Y, Co(P))$, we have*

$$f^*\Omega = dK(\nabla). \tag{5 - 13}$$

The 3-form Ω is called the 3-curvature *of the curving. The construction of the 3-curvature is compatible with pull-back of sheaves of groupoids with connective structure and curving.*

Proof. If $\mathcal{C}(X)$ is not empty, then (5-13) defines a 3-form Ω on X. The fact that this defines Ω unambiguously follows because of rule (3) for curvings, and because any 2 objects of $\mathcal{C}(X)$ are locally isomorphic. Now if $\mathcal{C}(X)$ is empty, there exists a surjective local homeomorphism $f : Y \to X$ such that $\mathcal{C}(Y \xrightarrow{f} X) \neq \emptyset$. Then we have a 3-form Ω on Y. Because we have $p_1^*\Omega = p_2^*\Omega$ on $Y \times_X Y$, Ω is the pull-back of a 3-form on X, which must be closed, since it is locally exact. ∎

Let us give some examples. Let $p : Y \to X$ be a smooth fibration, and let ω be a closed relative 2-form, satisfying (F1) and (F2). Let β be a 2-form on Y satisfying the conclusion of Lemma 4.3.3. The curving is then defined by formula (5-12). Since K is closed, the 3-curvature Ω of this curving has pull-back to Y equal to $-2\pi\sqrt{-1} \cdot d\beta$.

The 3-curvature of a unitary curving is of course purely imaginary.

Note that Proposition 4.2.5, which has to do with the sheaf of groupoids associated to a B-bundle and a central extension $\tilde{B}$ of B by $\mathbb{C}^*$, in effect gives the calculation of the 3-curvature for the curving corresponding to a splitting of the exact sequence of bundles (5-11). Corollary 4.2.8 identifies the cohomology class of the 3-curvature, and shows that it belongs to the image of $H^3(X, \mathbb{Z}(1))$.

We wish to pursue the analogy between sheaves of groupoids and line bundles as far as possible, and, in particular, to prove an analog for 3-forms of the Kostant-Weil theorem. In a spirit very similar to §2.2 (2.2.10-2.2.15), we will therefore relate the group of equivalence classes of sheaves of groupoids equipped with connective structure and curving to smooth Deligne cohomology. We have already defined the notion of an equivalence $\Phi : (\mathcal{C}_1, Co_1) \to (\mathcal{C}_2, Co_2)$ between sheaves of groupoids with a connective structure. If the connective structures are equipped with curvings K_1 and K_2, we require that, for $\nabla \in Co_1(P)$, the equality

$$K_1(\nabla) = K_2(\Phi(\nabla))$$

holds. Then we say that Φ is an equivalence of sheaves of groupoids equipped with connective structure and curving.

Let $\mathcal{C}$ be a sheaf of groupoids. As in the proof of Theorem 5.2.8, let (U_i) be an open covering of X such that there exists an object P_i of $\mathcal{C}(U_i)$, and let $u_{ij} : (P_j)_{/U_{ij}} \xrightarrow{\sim} (P_i)_{/U_{ij}}$ be an isomorphism. Now let $\mathcal{C}$ be equipped with a connective structure Co. Choosing an object ∇_i of the torsor $Co(P_i)$, we define a 1-form α_{ij} on U_{ij} by

$$\alpha_{ij} = \nabla_i - (u_{ij})_*(\nabla_j), \qquad (5-14)$$

and using the notation $h_{ijk} = u_{ik}^{-1} u_{ij} u_{jk}$, as in Theorem 5.2.8, we have

$$\alpha_{ij} + \alpha_{jk} - \alpha_{ik} = \nabla_i - (u_{ij} u_{jk} u_{ki})_*(\nabla_i) = h_{ijk}^{-1} dh_{ijk}.$$

Therefore $(\underline{h}, -\underline{\alpha})$ forms a Čech 2-cocycle with coefficients in the complex of sheaves

$$\underline{\mathbb{C}}_X^* \xrightarrow{d \, \log} \underline{A}_{X,\mathbb{C}}^1. \qquad (5-15)$$

Now assume that we have a curving $\nabla \mapsto K(\nabla)$ of the connective structure. Let $K_i = K(\nabla_i)$, a complex-valued 2-form on U_i. We have $K_i - K_j = d\alpha_{ij}$ on U_{ij}. Therefore $(\underline{h}, -\underline{\alpha}, \underline{K})$ forms a Čech 2-cocycle with values in the complex of sheaves

$$\underline{C}^*_X \xrightarrow{d \, \log} \underline{A}^1_{X,C} \xrightarrow{d} \underline{A}^2_{X,C} \tag{5-16}$$

The following theorem is a generalization of Proposition 4.2.9.

5.3.11. Theorem. *The cohomology class of the cocycle* $(\underline{h}, -\underline{\alpha}, K)$ *in the hypercohomology group* $\mathcal{A} = H^2(X, \underline{C}^*_X \xrightarrow{d \, \log} \underline{A}^1_{X,C} \xrightarrow{d} \underline{A}^2_{X,C})$ *is independent of all the choices. In this fashion, we identify* $\mathcal{A}$ *with the group of equivalence classes of sheaves of groupoids over* X *with connective structure and curving. Furthermore,* $\mathcal{A}$ *identifies in turn with the smooth Deligne cohomology group* $(2\pi\sqrt{-1})^{-2} \cdot H^3(X, \mathbb{Z}(3)^\infty_D)$.

We then have a cohomological interpretation of the 3-curvature, involving the morphism of complexes of sheaves

$$
\begin{array}{ccccc}
\underline{C}^*_X & \xrightarrow{d \, \log} & \underline{A}^1_{X,C} & \xrightarrow{d} & \underline{A}^2_{X,C} \\
\downarrow & & \downarrow & & \downarrow{\scriptstyle d} \\
0 & \longrightarrow & 0 & \longrightarrow & \underline{A}^3_{X,C}
\end{array}
\tag{5-17}
$$

5.3.12. Theorem. *Let* $(C, Co, K(-))$ *be a sheaf of groupoids over* X *with connective structure and curving. Then the 3-curvature* Ω *is obtained by applying the morphism of complexes of sheaves (5-17) to the class of* $(C, Co, K(-))$ *in* $H^2(X, \underline{C}^*_X \xrightarrow{d \, \log} \underline{A}^1_{X,C} \xrightarrow{d} \underline{A}^2_{X,C})$. *Furthermore, the cohomology class of* $\frac{\Omega}{2\pi\sqrt{-1}}$ *belongs to the image of* $H^3(X, \mathbb{Z}) \to H^3(X, \mathbb{C})$.

Proof. Since the 3-curvature Ω is equal to dK_i on U_i, it is indeed obtained by applying the morphism of complexes (5-17) to the Čech 2-cocycle $(\underline{h}, -\underline{\alpha}, K)$. The integrality of $[\frac{\Omega}{2\pi\sqrt{-1}}]$ follows from the exact sequence of complexes of sheaves

$$
\begin{array}{ccccc}
0 & 0 & 0 \\
\downarrow & \downarrow & \downarrow \\
\underline{C}^*_X & \longrightarrow & \underline{A}^1_{X,C} & \longrightarrow & \underline{A}^{2}_{X,C}{}^{cl} \\
\downarrow{\scriptstyle Id} & & \downarrow{\scriptstyle Id} & & \downarrow{\scriptstyle incl} \\
\underline{C}^*_X & \longrightarrow & \underline{A}^1_{X,C} & \longrightarrow & \underline{A}^2_{X,C} \\
\downarrow & & \downarrow & & \downarrow{\scriptstyle d} \\
0 & \longrightarrow & 0 & \longrightarrow & \underline{A}^{3}_{X,C}{}^{cl} \\
\downarrow & & \downarrow & & \downarrow \\
0 & & 0 & & 0
\end{array}
\tag{5-18}
$$

where $\underline{A}_{X,\mathbb{C}}^{p}{}^{cl} \subseteq \underline{A}_{X,\mathbb{C}}^{p}$ is the subsheaf of closed p-forms. The complex of sheaves $\underline{\mathbb{C}}_X^* \to \underline{A}_{X,\mathbb{C}}^1 \to \underline{A}_{X,\mathbb{C}}^2{}^{cl}$ is quasi-isomorphic to the constant sheaf $\mathbb{C}_X^*$. Therefore we get an exact sequence

$$0 \to H^2(X,\mathbb{C}^*) \to H^2(X,\underline{\mathbb{C}}_X^* \xrightarrow{d \log} \underline{A}_{X,\mathbb{C}}^1 \xrightarrow{d} \underline{A}_{X,\mathbb{C}}^2) \to A^3(X)_{\mathbb{C}}^{cl} \to H^3(X,\mathbb{C}^*).$$
$$(5-19)$$

It follows that $[\Omega]$ belongs to the kernel of the exponential map exp : $H^3(X,\mathbb{C}) \to H^3(X,\mathbb{C}^*)$, hence to the image of $H^3(X,\mathbb{Z}(1)) \to H^3(X,\mathbb{C})$. ∎

We now observe that the 3-curvature of a sheaf of groupoids is "quantized." The converse result is

5.3.13.   Theorem. *Let X be a manifold, and let Ω be a closed complex-valued 3-form such that $[\Omega]$ belongs to the image of $H^3(X,\mathbb{Z}(1)) \to H^3(X,\mathbb{C})$. Then there exists a sheaf of groupoids over X, with connective structure and curving, having 3-curvature Ω.*

Proof. This follows immediately from Theorem 5.3.11 and from the exact sequence (5-19). ∎

Note that the combination of Theorem 5.3.12 and Theorem 5.3.13 yields the theorem we wanted about the quantization of 3-forms, and that it is entirely analogous to the Kostant-Weil theory for 2-forms. Hence this section goes further than §4.3. Such complete results could not have been obtained from bundles of projective Hilbert spaces, so that our use of the abstract theory of sheaves of groupoids was essential.

It is interesting to study the sheaves of groupoids, with connective structure and curving, having zero 3-curvature. According to (5-19), these sheaves of groupoids are classified by the group $H^2(X,\mathbb{C}^*)$. We now introduce the notion of a flat sheaf of groupoids.

5.3.14.   Definition. *Let X be a manifold. A flat (Dixmier-Douady) sheaf of groupoids is a gerbe over X with band, the constant sheaf $\mathbb{C}_X^*$.*

Note that a flat sheaf of groupoids $\mathcal{G}$ over X gives rise, via the morphism of sheaves $\mathbb{C}_X^* \to \underline{\mathbb{C}}_X^*$, to a usual Dixmier-Douady sheaf of groupoids, say $\mathcal{G}'$, which has a natural connective structure and curving. The connective structure is characterized by the fact that $Co(P)$ is the trivial torsor $\underline{A}_{X,\mathbb{C}}^1$, for any object P of the flat sheaf of groupoids $\mathcal{G}$. The curving is characterized by the fact that for such P, and for $\alpha \in \underline{A}_{X,\mathbb{C}}^1$, the curvature $K(\alpha)$ is just $d\alpha$. So we have

5.3.15. Proposition. *A flat sheaf of groupoids is the same as a sheaf of groupoids with connective structure and curving whose 3-curvature is 0.*

Proof. We explain how to find a flat sheaf of groupoids $\mathcal{G}$ corresponding to a sheaf of groupoids $\mathcal{G}'$ with connective structure and curving whose 3-curvature is 0. For a local homeomorphism $f : Y \to X$, let $\mathcal{G}(Y \xrightarrow{f} X)$ consist of pairs (P, ∇), where $P \in \mathcal{G}'(Y \xrightarrow{f} X)$ and $\nabla \in Co(P)$ satisfy $K(\nabla) = 0$. Then it is easy to see that $\mathcal{G}$ is a flat Dixmier-Douady sheaf of groupoids. ∎

There is a theory of sheaves of groupoids in the holomorphic case, for which we briefly describe the main ideas and results.

5.3.16.   Definition. *Let X be a (finite-dimensional) complex manifold. Then a <u>holomorphic Dixmier-Douady sheaf of groupoids</u> over X is a gerbe over X with band the sheaf of groups $\mathcal{O}_X^*$.*

To define a holomorphic connective structure for a holomorphic sheaf of groupoids $\mathcal{G}$, we assign to an object $P \in \mathcal{G}(Y \xrightarrow{f} X)$ a torsor $Co(P)$ under Ω_X^1, together with the analogs of the structures (C1) and (C2), satisfying (R1) and (R2). A holomorphic curving of a holomorphic connective structure associates to $\nabla \in Co(P)$ a holomorphic 2-form $K(\nabla)$, such that the conditions of Definition 5.3.7 are satisfied; in condition (3), only holomorphic 1-forms are allowed. We then obtain

5.3.17. Theorem. *Let X be a (finite-dimensional) complex manifold. We then have*

(1) *The group of equivalence classes of holomorphic sheaves of groupoids over X is isomorphic to $H^2(X, \mathcal{O}_X^*)$.*

(2) *The group of equivalence classes of holomorphic sheaves of groupoids over X with holomorphic connective structure is isomorphic to the hyper-cohomology group $H^2(X, \mathcal{O}_X^* \xrightarrow{d \, \log} \Omega_X^1)$, hence to the Deligne cohomology group $(2\pi\sqrt{-1})^{-1} \cdot H^3(X, \mathbb{Z}(2)_D)$.*

(3) *The group of equivalence classes of holomorphic sheaves of groupoids over X with holomorphic connective structure and curving is isomorphic to the hypercohomology group $H^2(X, \mathcal{O}_X^* \xrightarrow{d \, \log} \Omega_X^1 \xrightarrow{d} \Omega_X^2)$, hence to the Deligne cohomology group $(2\pi\sqrt{-1})^{-2} \cdot H^3(X, \mathbb{Z}(3)_D)$.*

(4) *The 3-curvature of a holomorphic curving is a closed holomorphic 3-form.*

We will present an interesting example of a holomorphic sheaf of groupoids with holomorphic connective structure in §4. We now wish to point out that various differentials of the Hodge to de Rham spectral sequence can be naturally interpreted as obstructions to finding a holomorphic connective structure or curving on a holomorphic sheaf of groupoids. Note that, according to the monomorphism $\mathcal{O}_X^* \hookrightarrow \mathbb{C}_X^*$ of sheaves, a holomorphic sheaf of groupoids $\mathcal{G}$ gives rise to a sheaf of groupoids $\mathcal{G}^\infty$. The class $[\mathcal{G}] \in H^2(X, \mathcal{O}_X^*)$ maps to $[\mathcal{G}^\infty] \in H^3(X, \mathbb{Z}(1))$ under the boundary map associated to the exponential sequence of sheaves $0 \to \mathbb{Z}(1) \to \mathcal{O}_X \xrightarrow{\exp} \mathcal{O}_X^* \to 0$. Therefore the class $[\mathcal{G}^\infty] \in H^3(X, \mathbb{Z}(1))$ is in the kernel of the map $H^3(X, \mathbb{Z}(1)) \to H^3(X, \mathcal{O}_X)$, so that $[\mathcal{G}^\infty]$ gives rise to an element $a_{1,2}$ in the term $E_\infty^{1,2}$ of the Hodge to de Rham spectral sequence. An analysis of this spectral sequence shows that $E_\infty^{1,2} = E_4^{1,2}$ is a subspace of

$$E_2^{1,2} = \left[Ker(d : H^2(X, \Omega^1) \to H^2(X, \Omega_X^2)) \right] / d \, H^2(X, \mathcal{O}_X).$$

The element $a_{1,2} \in H^2(X, \Omega_X^1)/d \, H^2(X, \mathcal{O}_X)$ is essentially the obstruction to putting a holomorphic connective structure on $\mathcal{G}$. Assume now that such a holomorphic connective structure is given. For the discussion of holomorphic curvings, we will assume for clarity that X is a compact Kähler manifold, so that, by Hodge's theorem, the spectral sequence degenerates at E_1. $[\mathcal{G}^\infty]$ then gives rise to an element $a_{2,1} \in E_\infty^{2,1} = E_1^{2,1} = H^1(X, \Omega_X^2)$, which is the obstruction to finding a holomorphic curving. If this obstruction is zero, we are left with an element $a_{3,0} \in H^0(X, \Omega_X^3)$, which is of course the 3-curvature of the sheaf of groupoids.

5.4. The canonical sheaf of groupoids on a compact Lie group

In §5.3 we developed the notions of connective structure and curving for a (Dixmier-Douady) sheaf of groupoids over a manifold M. In particular, we showed that these notions are very natural in the context of a smooth mapping $f : Y \to M$ between manifolds, and a closed relative 2-form ω on Y which satisfies assumptions (F1) and (F2) of §2; we repeat these assumptions here for convenience:

(F1) The restriction of ω to each fiber Y_x has integral cohomology class, and Y_x is connected.

(F2) The fibers Y_x satisfy $H^1(Y_x, \mathbb{C}^*) = 0$.

We begin with a 3-form ν on a finite-dimensional manifold M and construct such a geometric setup, assuming that $2\pi\sqrt{-1} \cdot \nu$ satisfies the condition required of the 3-curvature of a sheaf of groupoids, namely, that the cohomology class of ν is integral. A topological assumption is required here:

(A) M is 2-connected, that is to say, M is connected and $\pi_1(M) = \pi_2(M) = 0$.

Further, choose a base point of M, denoted by 1, and following the pattern of Serre's thesis [Ser], consider the *based path space* P_1M, defined as the space of smooth maps $\gamma : [0,1] \to M$, such that $\gamma(0) = 1 \in M$. We put on P_1M the topology of uniform convergence of the path and all its derivatives.

5.4.1. Lemma. (1) P_1M *is a smooth manifold modeled on the topological vector space* $P_0\mathbb{R}^n$, *where* $n = \dim(M)$. P_1M *is contractible.*

(2) *The* <u>*evaluation map*</u> $ev : P_1M \to M$ *given by* $ev(\gamma) = \gamma(1)$ *is a smooth, locally trivial fibration, with fiber the submanifold* $\Omega M = \{\gamma \in P_1M : \gamma(1) = 1\}$ *of* P_1M.

Note that, what we denote by ΩM, is not the same as the space L_1M of based smooth loops with values in M. However, the inclusion $L_1M \hookrightarrow \Omega M$ is a homotopy equivalence.

Proof. (1) is proved in the same fashion as Proposition 3.1.2. Let $r : [0,1] \times P_1M \to P_1M$ be the map $r(t,\gamma) = \gamma_t$, where γ_t is the path $\gamma_t(x) = \gamma(tx)$. Then we have $r(1,\gamma) = \gamma$, $r(0,\gamma) = 1$ (where 1 is the constant path); hence P_1M is contractible. To show (2), let U be a contractible open set of M and let $a \in U$. Let $\rho : [0,1] \times U \to U$ be a smooth map such that $\rho(0,y) = a$ and $\rho(1,y) = y$. It is easy to show that there exists a smooth mapping $F : U \to Diff(M)$, where $Diff(M)$ is the group of diffeomorphisms, such that $F(y) \cdot a = y$ for any $y \in U$, and $F(a) = Id$. Then we have a diffeomorphism $\phi : ev^{-1}(a) \times U \tilde{\to} ev^{-1}(U)$ so that $\phi(\gamma, y)$ is the path

$$\phi(\gamma, y)(x) = F(\rho(x, y)) \cdot \gamma(x).$$

The inverse diffeomorphism is $\phi^{-1}(\gamma)(x) = F(\rho(x, ev(\gamma)))^{-1} \cdot \gamma(x)$, showing that ev is a smooth fibration. ∎

We now use the method of §3.5 to transgress the 3-form ν on M to a 2-form β on P_1M. More precisely, we define β by

$$\beta_\gamma(u, v) = \int_0^1 \nu_{\gamma(x)}(\frac{d\gamma}{dx}, u(x), v(x))dx, \qquad (5-19)$$

which is completely analogous to (3-1). In fact, the same formula gives a 2-form on the manifold of all smooth maps $[0,1] \to M$.

5.4.2. Proposition. (1) *The 2-form β on P_1M satisfies*

$$d\beta = ev^*\nu. \qquad (5-20)$$

Hence β restricts to a closed 2-form β_x on any fiber $ev^{-1}(x)$.

(2) *For any $x \in M$, the closed 2-form β_x on $ev^{-1}(x)$ has an integral cohomology class.*

Proof. We prove the more general formula

$$d\beta = ev^*\nu - T(d\nu) \tag{$*$}$$

for a 3-form ν which does not have to be closed, where $T(d\nu)$ denotes the transgression of $d\nu$, defined by a formula similar to (5-19). To prove ($*$), we note that $d\beta(u,v,w) = \int_0^1 d(i(\frac{d\gamma}{dx}) \cdot \nu)(u,v,w)dx$. On the other hand, we have $T(d\nu)(u,v,w) = \int_0^1 i(\frac{d\gamma}{dx})\nu(u,v,w)dx$. Therefore we get

$$d\beta(u,v,w) + T(d\nu)(u,v,w) = \int_0^1 \mathcal{L}(\frac{d\gamma}{dx}) \cdot \nu(u,v,w) \cdot dx$$
$$= \nu_{\gamma(1)}(u(1),v(1),w(1)) = \nu_{ev(\gamma)}(d\ ev(u), d\ ev(v), d\ ev(w))$$

This proves (1).

To prove (2), we first show that β_1 has an integral cohomology class. We deduce this from Proposition 3.5.2. Note that ΩM contains the submanifold $L_1 M$ of smooth loops with origin 1, and that the inclusion $L_1 M \hookrightarrow \Omega M$ is a homotopy equivalence. Therefore it will suffice to show that the restriction of β_1 to $L_1 M$ has an integral cohomology class. But $L_1 M$ is also a submanifold of LM, and we can apply Proposition 3.5.2 (iii). Next, $x \in M \mapsto [\beta_x]$ gives a smooth function on M with values in the local system $H^2(ev^{-1}(x), \mathbb{C})$, so that it will be enough to show that this function is constant. The derivative of this function is the element of $A^1(M, \underline{H}^2(ev^{-1}(x), \mathbb{C}))$ obtained by applying the differential d_1 of the Leray spectral sequence to the relative 2-form ω induced from β. Since ω lifts to a cycle β in the complex $A^\bullet(P_1 M)_{\mathbb{C}}/F^2$, it is killed by d_1. This establishes (2). ∎

Notice that we have $\pi_1(\Omega M) = \pi_2(M) = 0$ by (A). Hence the smooth fibration $ev : P_1 M \to M$ and the relative 2-form induced by β satisfy the assumptions of Theorem 5.2.11; we therefore have a sheaf of groupoids. Furthermore, β satisfies the assumptions of Proposition 5.3.9 so that we have a connective structure and curving. To summarize:

5.4.3. Theorem. *Let M be a 2-connected finite-dimensional manifold. Let ν be a closed complex-valued 3-form on M such that the cohomology class of ν is integral, and let $1 \in M$ be a base point. Then*

(1) *There exists a sheaf of groupoids $\mathcal{C}$ on M such that, for a local homeomorphism $f : Y \to M$, $\mathcal{C}(Y)$ consists of pairs (L, ∇), where L is a line bundle on $P_1 M \times_M Y$, and ∇ is a fiberwise connection on L, with fiberwise curvature equal to $2\pi\sqrt{-1} \cdot \beta$.*

(2) *There is a connective structure on $\mathcal{C}$, which assigns to a pair (L, ∇) as in (1), the $\underline{A}^1_{Y,\mathbb{C}}$-torsor of connections $\tilde{\nabla}$ on L which are compatible with ∇ and whose curvature K satisfies*

$$K - 2\pi\sqrt{-1} \cdot \beta \in F^2.$$

(3) *The connective structure admits a curving such that $K(\tilde{\nabla}) = K - 2\pi\sqrt{-1} \cdot \beta$ for $\tilde{\nabla}$, as in (2). The 3-curvature of this curving is equal to $-2\pi\sqrt{-1} \cdot \nu$.*

(4) *Any sheaf of groupoids with connective structure and curving whose 3-curvature is $-2\pi\sqrt{-1}\nu$ is equivalent to $\mathcal{C}$.*

Proof. We only have to prove (4). The assumptions on M imply that $H^2(M, \mathbb{C}^*) = 0$. The exact sequence (5-19) shows that the group of equivalence classes of sheaves of groupoids with connective structure and curving injects into $A^3(M)^{cl}_{\mathbb{C}}$; in other words, if two such sheaves of groupoids have the same 3-curvature, they are equivalent. ∎

Note that the sheaf of groupoids, connective structure and curving described above only depend on M, the 3-form ν, and the base point $1 \in M$.

A very interesting example is that of a finite-dimensional Lie group G. Let $\mathfrak{g}$ be the Lie algebra of G. For E a vector space, let E^* be its dual. We first recall that the complex $A^\bullet(G)^G$ of left-invariant differential forms on G identifies with the *standard complex* of Lie algebra cohomology, i.e., the complex $(\wedge^\bullet(\mathfrak{g})^*, d)$, with differential d given by

$$(d\alpha)(\xi_1, \ldots, \xi_{p+1}) = \sum_{1 \leq i < j \leq p+1} (-1)^{i+j}\alpha([\xi_i, \xi_j], \xi_1, \ldots, \hat{\xi}_i, \ldots, \hat{\xi}_j, \ldots, \xi_{p+1})$$

$$(5-21)$$

where α is a skew-symmetric p-linear functional on $\mathfrak{g}$. The cohomology of this complex is the *Lie algebra cohomology* $H^\bullet(\mathfrak{g})$; see [C-E] for instance. The inclusion of complexes $A^\bullet(G)^G \hookrightarrow A^\bullet(G)$ induces homomorphisms $H^p(\mathfrak{g}) \to H^p(G, \mathbb{R})$. For G compact connected, these homomorphisms are bijective. Recall that the Lie algebra $\mathfrak{g}$ of G is equipped with a G-invariant symmetric bilinear form, the *Killing form B* given by

$$B(\xi, \eta) = Tr_{\mathfrak{g}}(ad(\xi) \cdot ad(\eta)). \qquad (5-22)$$

When $\mathfrak{g}$ is semisimple, B is non-degenerate.

5.4.4. Lemma. *Let $\langle\ ,\ \rangle$ be an invariant symmetric bilinear form on the Lie algebra $\mathfrak{g}$. Then the trilinear form ν on $\mathfrak{g}$ defined by*

$$\nu(\xi,\eta,\zeta) = \langle[\xi,\eta],\zeta\rangle \tag{5-23}$$

is skew-symmetric and belongs to the kernel of $d : \wedge^3(\mathfrak{g})^* \to \wedge^4(\mathfrak{g})^*$.

Proof. First we show that ν is skew-symmetric. Clearly we have $\nu(\eta,\xi,\zeta) = -\nu(\xi,\eta,\zeta)$. Because $\langle\ ,\ \rangle$ is invariant, we have

$$\langle[\xi,\eta],\zeta\rangle = -\langle\eta,[\xi,\zeta]\rangle = -\langle[\xi,\zeta],\eta\rangle,$$

hence $\nu(\xi,\eta,\zeta) = -\nu(\xi,\zeta,\eta)$. So ν is skew-symmetric. Now we have

$$(d\nu)(\xi_1,\xi_2,\xi_3,\xi_4) = -a(1,2) + a(1,3) - a(1,4) - a(2,3) + a(2,4) - a(3,4)$$

where we put $a(1,2) = \langle[[\xi_1,\xi_2],\xi_3],\xi_4\rangle$, and so on. We have $-a(1,2) + a(1,3) + a(2,3) = 0$ by the Jacobi identity. We also have

$$a(1,2) = \langle[\xi_1,\xi_2],[\xi_3,\xi_4]\rangle = a(3,4),$$

and similarly $a(1,4) = a(2,3)$ and $a(2,4) = a(1,3)$. Therefore $-a(1,4) + a(2,4) - a(3,4) = -a(2,3) + a(1,3) - a(1,2) = 0$. ∎

We next examine the conditions which ensure that the cohomology class of the 3-form ν is integral. We will restrict ourselves to compact Lie groups and begin by stating some well-known facts.

5.4.5. Proposition. *($E.$ Cartan [Car]) Let G be a compact simple simply-connected Lie group (so $\mathfrak{g}$ is a simple Lie algebra). Then we have*

(1) $\pi_2(G) = 0$.

(2) $H_3(G,\mathbb{Z}) = \mathbb{Z}$.

(3) *The space of bi-invariant 3-forms on G is the span of the 3-form ν of (5-23), where $\langle\ ,\ \rangle$ is the Killing form.*

It is worthwhile to expand a little bit on the important fact that $H_3(G,\mathbb{Z})$ is *canonically* isomorphic to $\mathbb{Z}$ (not just up to sign). For the simplest of such groups, the group $SU(2) = S^3$, we choose the natural orientation. Note that every automorphism of $SU(2)$ is orientation-preserving: this is obvious for inner automorphisms but it is also true for the outer automorphism $g \mapsto \bar{g}$. Further, for arbitrary G as in Proposition 5.4.5, the generator of $H_3(G, \mathbb{Z})$ is represented by a subgroup $SU(2)$ of G corresponding to a "long root" of G (see below); this is a theorem of Bott [**Bott2**]. The long roots are all conjugate, giving a unique generator γ of $H_3(G, \mathbb{Z})$. Then there is a unique bi-invariant 3-form ν on G, called the *canonical 3-form on G*, such that $\int_\gamma \nu = 1$. For a detailed discussion, see [**P-S**]. For our purposes we simply quote one characterization of it. Recall that if $\mathfrak{h} \subset \mathfrak{g}$ is a *Cartan subalgebra* (i.e., a maximal abelian Lie subalgebra), the complexified Lie algebra $\mathfrak{g}_\mathbb{C}$ decomposes under the action of $\mathfrak{h}$ as

$$\mathfrak{g}_\mathbb{C} = \mathfrak{h}_\mathbb{C} \oplus \sum_{\alpha \in R} \mathfrak{g}_\alpha,$$

where R is a finite subset of $\sqrt{-1} \cdot \mathfrak{h}^* \setminus \{0\}$, called the *root system*. Each root space $\mathfrak{g}_\alpha$ is one-dimensional. For each $\alpha \in R$, the root spaces $\mathfrak{g}_\alpha$ and $\mathfrak{g}_{-\alpha}$ generate a 3-dimensional Lie subalgebra of $\mathfrak{g}_\mathbb{C}$, isomorphic to $\mathfrak{sl}(2, \mathbb{C})$. The *coroot* $\check{\alpha}$ is the element of $\mathfrak{g}_\mathbb{C}$ which corresponds to $\begin{pmatrix} 1 & 0 \\ 0 & -1 \end{pmatrix} \in \mathfrak{sl}(2, \mathbb{C})$ under this isomorphism. The opposite of the Killing form has a positive-definite restriction to $\mathfrak{h}$. Accordingly, we may define the length of a root. There are at most two root lengths; a root of maximal length is called a *long root*.

5.4.6. Proposition. [**P-S**] *Let* $\langle \ , \ \rangle$ *be a symmetric invariant bilinear form on* $\mathfrak{g}$, *such that* $\langle \check{\alpha}, \check{\alpha} \rangle = 4\pi$ *for each (or for one) long root* α. *Then the 3-form* ν *given by (5-23) is the canonical 3-form on G.*

Let us illustrate this for $SU(n)$, with the Lie algebra as the space $\mathfrak{su}(n) = \{M \in \mathfrak{gl}(n, \mathbb{C}) : M^* = -M\}$ of skew-hermitian matrices. Consider the symmetric invariant bilinear form $\langle \xi, \eta \rangle = Tr_{\mathbb{C}^n}(\xi\eta)$. All roots have the same length. A particular coroot is $H = \begin{pmatrix} 1 & 0 & 0 & \cdots \\ 0 & -1 & 0 & \cdots \\ 0 & 0 & 0 & \cdots \\ \vdots & \vdots & \vdots & \ddots \end{pmatrix}$. We have $\langle H, H \rangle = 2$; therefore the canonical 3-form on $SU(n)$ is

$$\nu(\xi, \eta, \zeta) = \frac{1}{8\pi} \cdot Tr_{\mathbb{C}^n}([\xi, \eta] \cdot \zeta). \tag{5-24}$$

From Theorem 5.4.3 we obtain immediately

5.4.7. Theorem. *Let G be a simply-connected simple compact Lie group. Let ν be the canonical 3-form on G. There exists a canonical sheaf of* *groupoids C, with connective structure and curving, whose curvature is equal to* $-2\pi\sqrt{-1}\cdot\nu$. *This sheaf of groupoids with connective structure and curving is unique up to equivalence.*

Note that P_1G and ΩG are Lie groups (under pointwise product of paths). The fibration $ev : P_1G \to G$ is actually a homomorphism of Lie groups, and is therefore a principal fiber bundle with structure group ΩG. Let $\langle\ ,\ \rangle$ be the invariant bilinear form on $\mathfrak{g}$ such that the 3-form (5-23) is equal to the canonical 3-form ν. We have the following formula for β:

$$\beta_\gamma(v_1, v_2) = \int_0^1 \langle \gamma(x)^{-1} \cdot \frac{d\gamma}{dx}, [\xi_1(x), \xi_2(x)]\rangle dx, \qquad (5-25)$$

where we put $\xi_j(x) = \gamma(x)^{-1} \cdot v_j(x) \in \mathfrak{g}$.

Let $g \in G$ and let $l_g : G \to G$ denote left translation by g, i.e., $l_g(h) = g \cdot h$. Then it follows from Theorem 5.4.7 and the left-invariance of ν that the sheaf of groupoids C (with connective structure and curving) is equivalent to its pull-back l_g^*C. There is, however, no canonical equivalence $C \to l_g^*C$. We can write down such an equivalence, but it will depend on a choice of a smooth path δ from $1 \in G$ to g. The main point will be to compare the 2-form β on P_1G, obtained by transgression, with the manifestly left-invariant 2-form ω on P_1G defined by

$$\omega_\gamma(v_1, v_2) = \int_0^1 \langle \xi_1'(x), \xi_2(x)\rangle dx - \int_0^1 \langle \xi_1(x), \xi_2'(x)\rangle dx. \qquad (5-26)$$

Note that the restriction of β to ΩG is equal to the 2-form considered in [P-S]. The following lemma was communicated to me by Bill Richter.

5.4.8. Lemma. *Let α be the 1-form on P_1G defined by*

$$\alpha_\gamma(v) = \int_0^1 \langle \gamma(x)^{-1} \cdot \gamma'(x), \gamma(x)^{-1} \cdot v(x)\rangle\ dx. \qquad (5-27)$$

Then we have $d\alpha = \omega - \beta$.

Now since ω is left-invariant, we will consider the difference $\mu_\delta = l_\delta^*\alpha - \alpha$, for $\delta \in P_1G$, which is a 1-form on $P_1(G)$. To express it in a nice form, we extend the invariant bilinear form $\langle\ ,\ \rangle$ on $\mathfrak{g}$ to a bilinear form on the Lie algebra $P_0\mathfrak{g}$, denoted by the same symbol. We set

$$\langle v_1, v_2\rangle = \int_0^1 \langle v_1(x), v_2(x)\rangle \cdot dx.$$

Using this symmetric bilinear form, we have an injection from $P_0\mathfrak{g}$ to its dual $(P_0\mathfrak{g})^*$. Then we have the following description of μ_δ, which we include here because it is very simple.

5.4.9. Lemma. *For $\delta \in P_1G$, the 1-form μ_δ is the right-invariant 1-form corresponding to the element $\delta^{-1}(x) \cdot \delta'(x)$ of $P_0(\mathfrak{g})$.*

Now for $\delta \in P_1G$, we have $l_\delta^*\beta - \beta = d\mu_\delta$. We are ready to construct an equivalence of sheaves of groupoids $\Phi_g : C \to l_g^*C$, using the path δ from 1 to g. Let $f : Y \to G$ be a local homeomorphism. An object of $C(Y)$ is a pair (L, ∇), where L is a line bundle over $P_1G \times_G Y$, and ∇ is a fiberwise connection on L, with fiberwise curvature equal to $2\pi\sqrt{-1} \cdot \beta$. Now the connection $l_\delta^*\nabla - 2\pi\sqrt{-1} \cdot \mu_\delta$ on the pull-back line bundle l_δ^*L has fiberwise curvature $2\pi\sqrt{-1} \cdot \beta$; hence the pair $(l_\delta^*L, l_\delta^*\nabla - 2\pi\sqrt{-1} \cdot \mu_\delta)$ is an object of l_δ^*C. Thus $\Phi(L, \nabla) = (l_\delta^*L, l_\delta^*\nabla - 2\pi\sqrt{-1} \cdot \mu_\delta)$ defines an equivalence of sheaves of groupoids. To get an equivalence of sheaves of groupoids with connective structures, we observe that, if $\tilde{\nabla}$ is a connection on L as above, such that its curvature K satisfies $K - 2\pi\sqrt{-1} \cdot \beta \in F^2$, then the curvature K' of $l_\delta^*\tilde{\nabla} - 2\pi\sqrt{-1} \cdot \mu_\delta$ satisfies $K' - 2\pi\sqrt{-1} \cdot l_\delta^*\beta \in F^2$. We now have an equivalence of sheaves of groupoids with connective structure and curving by setting

$$\Phi_g(L, \tilde{\nabla}) = (l_\delta^*L, l_\delta^*\tilde{\nabla} - 2\pi\sqrt{-1} \cdot \mu_\delta).$$

Turning to the holomorphic sheaves of groupoids over a complex simple simply-connected group $G_{\mathbb{C}}$, we note that if $G \subset G_{\mathbb{C}}$ is a maximal compact subgroup, the symmetric space $G_{\mathbb{C}}/G$ is contractible so that $G_{\mathbb{C}}$ is homotopy equivalent to G. Therefore the topological facts in Proposition 5.4.5 also apply to $G_{\mathbb{C}}$. Any left-invariant differential form on G gives rise to a holomorphic differential form on $G_{\mathbb{C}}$, and the canonical 3-form ν on G gives rise therefore to a closed bi-invariant holomorphic 3-form on $G_{\mathbb{C}}$, again denoted by ν.

5.4.10. Theorem.   *Let $G_{\mathbb{C}}$ be a simply-connected simple complex Lie group. Let ν be the canonical bi-invariant holomorphic 3-form on $G_{\mathbb{C}}$. Then there exists a holomorphic sheaf of groupoids over $G_{\mathbb{C}}$, with holomorphic connective structure and curving whose 3-curvature is equal to ν. Such a sheaf of groupoids is unique up to equivalence of sheaves of groupoids with connective structure and curving.*

Proof. According to Theorem 5.3.8, the group of equivalence classes of holomorphic sheaves of groupoids over $G_{\mathbb{C}}$ equipped with holomorphic connective structures and curvings is isomorphic to the hypercohomology group $H^2(G_{\mathbb{C}}, \mathcal{O}_{G_{\mathbb{C}}} \to \Omega^1_{G_{\mathbb{C}}} \to \Omega^2_{G_{\mathbb{C}}})$. The holomorphic analog of the exact sequence (4-19) is

$$H^2(G_{\mathbb{C}}, \mathbb{C}^*) \to H^2(G_{\mathbb{C}}, \mathcal{O}_{G_{\mathbb{C}}} \to \Omega^1_{G_{\mathbb{C}}} \to \Omega^2_{G_{\mathbb{C}}}) \to \Omega^3(G_{\mathbb{C}})^{cl} \to H^3(G_{\mathbb{C}}, \mathbb{C}^*).$$

The group $H^2(G_{\mathbb{C}}, \mathbb{C}^*)$ is 0 because $G_{\mathbb{C}}$ is 2-connected. Therefore $H^2(G_{\mathbb{C}}, \mathcal{O}_{G_{\mathbb{C}}} \to \Omega^1_{G_{\mathbb{C}}} \to \Omega^2_{G_{\mathbb{C}}})$ injects into $\Omega^3(G_{\mathbb{C}})_{cl}$, which means that a holomorphic sheaf of groupoids (with connective structure and curving) is determined by its 3-curvature. ∎

An explicit construction of a sheaf of groupoids over $G_{\mathbb{C}}$ with 3-curvature ν would involve, in one way or the other, algebraic K-theory. We can at least give an explicit construction of a holomorphic sheaf of groupoids with connective structure and curving on the group $SL(2, \mathbb{C})$. Since we have the holomorphic mapping $p : SL(2, \mathbb{C}) \to \mathbb{C}^2 \setminus \{0\}$, given by $p(\begin{pmatrix} x & y \\ z & w \end{pmatrix}) = (x, y)$, it is enough to construct such a holomorphic sheaf of groupoids over $\mathbb{C}^2 \setminus \{0\}$. We have $\mathbb{C}^2 \setminus \{0\} = U_1 \cup U_2$, where U_1 is the open set $x \neq 0$ and U_2 the open set $y \neq 0$. Over the intersection $U_{12} = \mathbb{C}^* \times \mathbb{C}^*$, we have the invertible functions x and y; from the construction of §2.2, we obtain a holomorphic line bundle $\{x, y\}$ over U_{12}, equipped with a connection ∇ of curvature $K = \dfrac{1}{2\pi\sqrt{-1}} \cdot \dfrac{dx}{x} \wedge \dfrac{dy}{y}$. We can then define a holomorphic sheaf of groupoids $\mathcal{C}$ over $\mathbb{C}^2 \setminus \{0\}$ as follows. For $f : Y \to \mathbb{C}^2 \setminus \{0\}$ a local homeomorphism, an object of $\mathcal{C}(Y)$ is a triple (L_1, L_2, ϕ), where
  (1) L_1 is a holomorphic line bundle over $f^{-1}(U_1) \subseteq Y$.
  (2) L_2 is a holomorphic line bundle over $f^{-1}(U_2) \subseteq Y$.
  (3) $\phi : L_1 \xrightarrow{\sim} L_2 \otimes \{x, y\}$ is a holomorphic isomorphism of line bundles over $f^{-1}(U_{12})$.
We define a holomorphic connective structure on $\mathcal{C}$, for which $Co(L_1, L_2, \phi)$ consists of pairs (∇_1, ∇_2), where ∇_1 (resp. ∇_2) is a holomorphic connection on L_1 (resp. L_2), and ϕ is an isomorphism of line bundles with connection, when $L_2 \otimes \{x, y\}$ is equipped with the connection $\nabla_2 + \nabla$. This connective structure does not have a curving. However after pulling back to $SL(2, \mathbb{C})$, there is a curving. To describe it, choose a holomorphic 2-form β_1 on $p^{-1}(U_1)$ (resp. β_2 on $p^{-1}(U_2)$), such that $\beta_2 - \beta_1 = \dfrac{1}{2\pi\sqrt{-1}} \cdot \dfrac{dx}{x} \wedge \dfrac{dy}{y}$ on $p^{-1}(U_{12})$. Then, given (∇_1, ∇_2) as above, let K_1 (resp. K_2) be the curvature of ∇_1 (resp. ∇_2). We can now assign to (∇_1, ∇_2) the 2-form which is equal to $K_j + \beta_j$ on the inverse image of U_j $(j = 1, 2)$, which defines a curving of the pull-back gerbe.

5.5. Examples of sheaves of groupoids

We consider two types of examples: the first is associated to a compact Lie group acting on a manifold, and the second is associated to a symplectic manifold.

In the first example, we deal with a type of space which, in general, is not a manifold. We consider the quotient space M/G, where the compact Lie group G acts smoothly on the smooth manifold M. The space M/G is Hausdorff. The projection map $M \to M/G$ will be denoted by q. In this case we can still define the notion of smooth mapping $f : V \to X$, where V is an open subset of M/G and X is some manifold; we say that f is smooth if $f \circ q$ is a smooth mapping from $q^{-1}(V)$ to X. We may thus define a sheaf $\underline{X}_{M/G}$ of a smooth mappings with values in X. The existence of smooth G-invariant partitions of unity shows that the sheaf $\underline{\mathbb{C}}_{M/G}$ is fine. Now a line bundle over M/G is defined by using transition functions which are smooth in the above sense, permitting the group $Pic^\infty(M/G)$ of isomorphism classes of line bundles over M/G to be identified with $H^1(M/G, \underline{\mathbb{C}}^*_{M/G}) \simeq H^2(M/G, \mathbb{Z}(1))$. Similarly, the group of equivalence classes of Dixmier-Douady sheaves of groupoids over M/G is isomorphic to $H^2(M/G, \underline{\mathbb{C}}^*_{M/G}) \xrightarrow{\sim} H^3(M/G, \mathbb{Z}(1))$.

5.5.1. Definition. *A G-equivariant line bundle L on M is a line bundle $p : L \to M$ together with a smooth action of G on L by bundle automorphisms (cf. Definition 2.4.2) which lifts the action of G on M. An isomorphism between equivariant line bundles is an isomorphism of line bundles which commutes with the G-actions.*

There is an obvious notion of tensor product of G-equivariant line bundles on M, and the dual of an equivariant line bundle is equivariant in a natural way. Thus we may define the *equivariant Picard group* $Pic_G^\infty(M)$ as the group of isomorphism classes of G-equivariant line bundles over M.

5.5.2. Examples.

Let $M = G/H$, for H a closed Lie subgroup of G. Let $1 \in G/H$ be the base point. If L is a G-equivariant line bundle over G/H, the fiber L_1 is a one-dimensional representation of the stabilizer H, on which H acts by some character $\chi : H \to \mathbb{C}^*$. Conversely, given such a character χ, we have the 1-dimensional representation $\mathbb{C}_\chi$ of H, and we construct the associated line bundle $L = G \times^H \mathbb{C}_\chi$, which is the quotient of $G \times \mathbb{C}_\chi$ by the H-action $h \cdot (g, \lambda) = (gh^{-1}, \chi(h) \cdot \lambda)$. This is a line bundle over G/H, by the projection map $p(g, \lambda) = g \cdot H$. G acts on $G \times \mathbb{C}_\chi$ by left multiplication on the first factor, and this action descends to an action on L. Thus L is a G-equivariant line bundle. The constructions $L \mapsto L_1$ and $E \mapsto G \times^H E$ are inverse to one another, so that a G-equivariant line bundle is really the same thing as a one-dimensional character of H. Hence we have $Pic_G^\infty(G/H) = X(H)$, the group of characters $H \to \mathbb{C}^*$.

We will determine $Pic_G^\infty(M)$. First we note that if $\mathcal{L}$ is a line bundle over M/G, the pull-back line bundle $q^*\mathcal{L}$ is defined as the fiber product $M \times_{M/G} \mathcal{L}$. The natural action of G on this fiber product makes $q^*\mathcal{L}$ into a G-equivariant line bundle over M.

5.5.3. Lemma. *The pull-back map* $Pic^\infty(M/G) \to Pic_G^\infty(M)$ *is injective.*

Proof. It is enough to show that if $\mathcal{L}$ is a line bundle over M/G, and if $q^*\mathcal{L}$ is trivial as an equivariant line bundle, then $\mathcal{L}$ is trivial. But if $q^*\mathcal{L}$ is isomorphic to the trivial equivariant line bundle, it has a nowhere vanishing G-invariant section, giving a nowhere vanishing section of $\mathcal{L}$. ■

However, not every G-equivariant line bundle over M is obtained by this process. To analyze this, we need to introduce a sheaf $\mathcal{X}$ on G/H. The stalk of $\mathcal{X}$ at $\bar{x} = q(x)$ will be equal to the character group $X(G_x)$, where G_x is the stabilizer of x. The group $X(G_x)$ depends only on the orbit of x; in fact, another point y of the orbit is of the form $y = g \cdot x$, and conjugation by g gives an isomorphism $X(G_x) \xrightarrow{\sim} X(G_y)$. Furthermore, this isomorphism is independent of the choice of g satisfying $g \cdot x = y$, because an inner automorphism of G_x acts trivially on $X(G_x)$. Thus $X(G_x)$ is intrinsically associated to the point $\bar{x}$ of M/G. To combine these stalks into a sheaf, we have to describe the local section of $\mathcal{X}$ defined by an element of $X(G_x)$. For this we need the slice theorem.

5.5.4. Theorem. *(Palais, see* [Bo]) *For $x \in M$, there exists a G-invariant neighborhood U of x and a G-equivariant smooth projection map $f : U \to G \cdot x$, such that $f(x) = x$.*

For any $y \in U$, with $f(y) = g \cdot x$, we have the inclusion $G_y \subseteq G_{g \cdot x} = g \cdot G_x \cdot g^{-1}$. The restriction map then gives a group homomorphism $X(G_x) = X(G_{g \cdot x}) \to X(G_y)$, which is independent of the choice of g and of the map $f : U \to G \cdot x$, as in 5.5.4. We use this to define a section of $\mathcal{X}$ over U/G associated to a character $\chi \in X(G_x)$.

Now, given a line bundle L over a G-invariant open set U of M, we have a section $\kappa(L)$ of the sheaf $\mathcal{X}$ over $U/G \subseteq M/G$. This section $\kappa(L)$ is called the *associated character* of L; its value at $q(x)$ is the character by which the stabilizer group G_x acts on the fiber L_x.

5.5.5. Lemma. *Let L be a G-equivariant line bundle over M such that the associated character is trivial. Then L identifies with the pull-back of a line* *bundle $\mathcal{L}$ on M/G.*

Proof. Define $\mathcal{L}^+ = L^+/G$, with $\bar{p} : \mathcal{L}^+ \to M/G$ as the projection map. There is a natural action of $\mathbb{C}^*$ on $\mathcal{L}^+$, and the assumption implies that $\bar{p}$ is a principal $\mathbb{C}^*$-fibration. In fact, the fiber $\bar{p}^{-1}(\bar{x})$ identifies with the quotient of L_x^+ by G_x; but G_x acts on this fiber by the character $\kappa(L)_x$, which is trivial. Hence the fiber is a principal homogeneous space under $\mathbb{C}^*$. It is clear that the pull-back $q^*(\mathcal{L}^+)$ identifies with L^+, so that L is the pull-back of the line bundle corresponding to $\mathcal{L}^+$. ∎

Conversely, a line bundle $\mathcal{L}$ over M/G, by pull-back, gives a G-equivariant line bundle over M, with trivial associated character.

At this point, we have an exact sequence

$$0 \to Pic^\infty(M/G) \to Pic_G^\infty(M) \to \Gamma(M/G, \mathcal{X}).$$

We want to extend this to the right and therefore ask the question: When does a section of $\mathcal{X}$ come from an equivariant line bundle? There is no local obstruction, as is shown by the following lemma.

5.5.6. Lemma. *Given $x \in M$ and $\chi \in X(G_x)$, there exists a G-equivariant open set U containing x and a G-equivariant line bundle L over U, such that $\kappa(L)_{q(x)} = \chi$.*

Proof. Choose U and f as in Theorem 5.5.4. Let L' be the line bundle over $G \cdot x \simeq G/G_x$ corresponding to the character χ, and let $L = f^*(L')$, a G-equivariant line bundle over U. Because $f(x) = x$, we have $\kappa(L)_{q(x)} = \kappa(L')_{q(x)} = \chi$. ∎

Now let η be a global section of $\mathcal{X}$. We define a sheaf of groupoids $\mathcal{C}$ over M/G as follows. For a local homeomorphism $f : Y \to M/G$, the category $\mathcal{C}(Y \xrightarrow{f} M/G)$ has for objects the equivariant line bundles over $M \times_{M/G} Y$ with associated character η, and for morphisms the equivariant isomorphisms of line bundles. Then we have

5.5.7. Proposition. *$\mathcal{C}$ is a Dixmier-Douady sheaf of groupoids over M/G.*

Proof. It is easy to see that $\mathcal{C}$ is a sheaf of groupoids. We next establish properties (G1), (G2) and (G3). First let $f : Y \to M/G$, and let L be an object of $\mathcal{C}(Y \xrightarrow{f} M/G)$. An automorphism of L is a smooth $\mathbb{C}^*$-valued function on $M \times_{M/G} Y$ which is G-invariant. This is equivalent to a smooth function from Y to $\mathbb{C}^*$. (G1) is proved. Let L_1, L_2 be two objects of $\mathcal{C}(Y \xrightarrow{f} M/G)$. The equivariant line bundle $L_2 \otimes L_1^{\otimes -1}$ on $M \times_{M/G} Y$ has trivial associated character, so by Lemma 5.5.5 it descends to a line bundle on Y. Locally on Y, there exists a nowhere vanishing section of this line bundle, which provides a G-equivariant isomorphism $L_1 \xrightarrow{\sim} L_2$. This proves (G2). Property (G3) follows from Lemma 5.5.6. ∎

Therefore we obtain

5.5.8. Proposition. *Given a section s of $\mathcal{X}$ over M/G, there is a well-defined element of $H^2(M/G, \mathbb{C}^*_{M/G})$ which is trivial if and only if s comes from an element of $Pic^\infty_G(M)$.*

This immediately yields

5.5.9. Theorem. *There is an exact sequence:*

$$0 \to H^2(M/G, \mathbb{Z}(1)) \to Pic^\infty_G(M) \to \Gamma(M/G, \mathcal{X})$$
$$\to H^2(M/G, \mathbb{C}^*_{M/G}) = H^3(M/G, \mathbb{Z}(1)).$$

Example: Let S^1 act on the 2-sphere by rotations. Then S^2/S^1 is the closed segment $[0, 1]$. The sheaf $\mathcal{X}$ has stalk at 0 and 1 equal to $\mathbb{Z}$, and the stalk at any other point is equal to 0. As S^2/S^1 is contractible, the exact sequence of Theorem 5.5.9 simplifies to give $Pic^\infty_{S^1}(S^2) = \mathbb{Z} \oplus \mathbb{Z}$.

In algebraic geometry, there is a result of Knop, Kraft and Vust [K-K-V] which is analogous to Theorem 5.5.9, in that it gives a three term exact sequence

$$0 \to Pic(M//G) \to Pic^G(M) \to \Gamma(M//G, \mathcal{X})$$

(here M is a quasi-projective algebraic variety, G a reductive algebraic group which acts on M, and $M//G$ is the Mumford quotient variety). It is not clear whether this exact sequence could be extended to the right in general, by replacing $H^3(M/G, \mathbb{Z}(1))$ with the Brauer group of $M//G$.

In our second example, we will associate a flat sheaf of groupoids to a manifold X equipped with a closed complex-valued 2-form ω. Recall from §2.2 the theorem of Weil and Kostant, which says that there exists a line bundle L over X, with connection L whose curvature is equal to $2\pi\sqrt{-1} \cdot \omega$, if and only if $[\omega] \in H^2(X, \mathbb{C})$ belongs to the image of $H^2(X, \mathbb{Z}) \to H^2(X, \mathbb{C})$. This is equivalent to saying that $[\omega]$ belongs to the kernel of the exponential map $H^2(X, \mathbb{C}) \to H^2(X, \mathbb{C}^*)$.

We introduce a sheaf of groupoids $\mathcal{G}$ over X as follows. For a local homeomorphism $f : Y \to X$, the category $\mathcal{G}(Y \xrightarrow{f} X)$ has for objects the pairs (L, ∇), where L is a line bundle over Y, and ∇ is a connection on L with curvature $2\pi\sqrt{-1}\cdot\omega$. Morphisms are isomorphisms of line bundles with connections. The pull-back functors are the natural ones.

5.5.10. Proposition. $\mathcal{G}$ *is a flat Dixmier-Douady sheaf of groupoids over* X.

Proof. Property (G1) is clear, since an automorphism of (L, ∇) is a locally constant $\mathbb{C}^*$-valued function. Let $f: Y \to X$ be a local homeomorphism, and consider two objects (L_1, ∇_1) and (L_2, ∇_2) of $\mathcal{G}(Y \xrightarrow{f} X)$. The line bundle $L_2 \otimes L_1^{\otimes -1}$ is equipped with the difference connection $\nabla_2 - \nabla_1$, which has zero curvature. So $L_2 \otimes L_1^{\otimes -1}$ is a flat line bundle. Locally on Y, it has a (flat) non-vanishing section, which gives an isomorphism $(L_1, \nabla_1) \xrightarrow{\sim} (L_2, \nabla_2)$. This proves (G2). If U is a contractible open set in X, there exists a 1-form α over U such that $d\alpha = 2\pi\sqrt{-1}\cdot\omega$; it follows that $\mathcal{G}(U)$ is not empty. This proves (G3). ∎

Now the theorem of Kostant-Weil can be restated as follows: the sheaf of groupoids $\mathcal{G}$ is trivial if and only if the cohomology class $[\omega]$ is integral. We have the following

5.5.11. Proposition. *The class in* $H^2(X, \mathbb{C}^*)$ *of the gerbe G of 5.5.10 is equal to* $\exp([\omega])$.

There is a variant of this example which is important in complex analysis and in the theory of $\mathcal{D}$-modules. Now let X be a complex manifold and let $\mathcal{D}_X$ be the sheaf of differential operators with holomorphic coefficients. Recall that $\mathcal{D}_X$ is the union of the subsheaves $\mathcal{D}_X(m)$ of differential operators of order at most m.

5.5.12. Lemma. *(Beilinson-Bernstein) The sheaf of automorphisms of $\mathcal{D}_X$ which induces the identity on $\mathcal{O}_X \subset \mathcal{D}_X$ identifies with the sheaf $\Omega_X^{1,\mathrm{cl}}$ of closed 1-forms.*

Note that $\alpha \in \Omega_X^{1,\mathrm{cl}}$ induces the unique automorphism of $\mathcal{D}_X$ which is Id on $\mathcal{O}_X$ and maps a vector field ξ to $\xi + \langle\alpha, \xi\rangle$.

5.5.13.   Definition. *(Beilinson-Bernstein) Let X be a complex manifold. A twisted sheaf of differential operators on X is a sheaf A of algebras over X, together with a monomorphism $\mathcal{O}_X \hookrightarrow A$ of sheaves of algebras, such that the pair $(A, \mathcal{O}_X)$ is locally isomorphic to $(\mathcal{D}_X, \mathcal{O}_X)$.*

For example, let L be a holomorphic line bundle over X, and let $\mathcal{D}(L)$ be the sheaf of differential operators operating on sections of L. Then $\mathcal{D}(L)$ is a twisted sheaf of differential operators. More generally, if $\frac{p}{q}$ is a rational number, we can construct the sheaf $\mathcal{D}_X(L^{\otimes \frac{p}{q}})$, even though the fractional tensor power $L^{\otimes \frac{p}{q}}$ may not exist globally; thus the literature on symplectic manifolds often mentions the algebra $\mathcal{D}_X(L^{\otimes \frac{p}{q}})$.

The set of isomorphism classes of twisted sheaves of differential operators over X is in bijection with the cohomology group $H^1(X, \Omega_X^{1\,\mathrm{cl}})$.

Given a twisted sheaf of differential operators A on X, there is a natural flat sheaf of groupoids $\mathcal{G}$, which embodies the obstruction to finding a line bundle L such that $A \overset{\sim}{\to} \mathcal{D}_X(L)$. For $f : Y \to X$ a local homeomorphism, the category $\mathcal{G}(Y \overset{f}{\longrightarrow} X)$ has for objects the sheaves L of left $f^{-1}A$-modules which are locally free of rank 1 over $\mathcal{O}_Y \subset f^{-1}A$. The morphisms are isomorphisms of sheaves of A-modules. Then we have

5.5.14. Proposition. *The class in $H^2(X, \mathbb{C}^*)$ of the above flat sheaf of groupoids is the image of $[A] \in H^1(X, \Omega_X^{1\,\mathrm{cl}})$ under the boundary map relative to the exact sequence of sheaves*

$$0 \to \mathbb{C}^* \to \mathcal{O}_X^* \overset{d\ \log}{\longrightarrow} \Omega_X^{1\,\mathrm{cl}} \to 0$$

Torsion classes in $H^2(X, \mathbb{C}^*)$ thus correspond to "fractional powers of line bundles."

The work of Asada [As] contains an interesting study of bundles whose structure group is a loop group, from a point of view which is quite different from that of this book.

Chapter 6

Line Bundles over Loop Spaces

The theme of this chapter is that a Dixmier-Douady sheaf of groupoids with connective structure over a manifold M leads naturally to a line bundle over the free loop space LM. This is in complete analogy to the well-known fact that a line bundle with connection over M leads to a function over LM, its holonomy.

6.1. Holonomy of line bundles

In this section we recall some well-known facts. Let (L, ∇) be a line bundle with connection over a smooth manifold M. For any path $\gamma : [0,1] \to M$, we defined in Proposition 2.4.6 the parallel transport along γ, which is an isomorphism $T_\gamma : L_{\gamma(0)} \xrightarrow{\sim} L_{\gamma(1)}$. In the case of a loop, this automorphism of the fiber $L_{\gamma(0)}$ is given by a number, denoted by $H(\gamma) \in \mathbb{C}^*$ and called the *holonomy* of the line bundle around the loop γ. Some of its properties are now summarized. Recall that $Diff^+(S^1)$ denotes the Lie group of orientation preserving diffeomorphisms of S^1; similarly $Diff^-(S^1)$ will denote the manifold of orientation reversing diffeomorphisms. Thus $Diff^+(S^1)$ and $Diff^-(S^1)$ are the connected components of the Lie group $Diff(S^1)$.

6.1.1. Proposition. *Let L be a line bundle over the manifold M. Let ∇ be a connection on L, with curvature K. Then the following are true:*

(1) *For $\gamma \in LM$ and $f \in Diff^\pm(S^1)$, we have $H(\gamma \circ f) = H(\gamma)^{\pm 1}$.*

(2) *The function $\gamma \mapsto H(\gamma)$ on LM is smooth; its derivative dH is the linear functional*

$$v \in T_\gamma LM \mapsto -\int_0^1 i(v) \cdot K. \qquad (6-1)$$

(3) *Let S be a smooth oriented surface, with boundary $\partial S = C_1 \coprod C_2 \cdots \coprod C_r$. Let $f_i : S^1 \xrightarrow{\sim} C_i$ be an orientation preserving diffeomorphism. Let $\phi : S \to M$ be a smooth map, and let $\gamma_i = \phi_{/C_i} \circ f_i \in LM$ be the "boundary components" of ϕ. Then we have*

$$\prod_{i=1}^r H(\gamma_i) = exp(-\int_S \phi^* K). \qquad (6-2)$$

Note that, in the case of a hermitian connection, the holonomy takes values in $\mathbb{T} \subset \mathbb{C}^*$.

A less standard fact is that the holonomy of a line bundle with connection can be expressed in terms of a transgression map in Deligne cohomology. Recall from §2.2 that the group of isomorphism classes of line bundles with connection over M is isomorphic to the hypercohomology group $H^1(X, \underline{\mathbb{C}}_M^* \xrightarrow{d \log} \underline{A}_{M,\mathbb{C}}^1)$. Note that the complex of sheaves

$$\underline{\mathbb{C}}_M^* \xrightarrow{d \log} \underline{A}_{M,\mathbb{C}}^1 \qquad\qquad (6-3)$$

is quasi-isomorphic to $(2\pi\sqrt{-1})^{-1} \cdot \mathbb{Z}(2)_{\mathcal{D}}^\infty[-1]$, where $\mathbb{Z}(2)_{\mathcal{D}}^\infty$ is the smooth Deligne complex.

6.1.2. Lemma. *There exists a natural transgression map*

$$\tau : H^1(M, \underline{\mathbb{C}}_M^* \xrightarrow{d \log} \underline{A}_{M,\mathbb{C}}^1) \to H^0(LM, \underline{\mathbb{C}}_{LM}^*), \qquad (6-4)$$

given as follows. Let $(U_i)_{i \in I}$ be an open covering of M, and let $(V_j)_{j \in J}$ be the open covering of LM, indexed by the set J of pairs (t, f), where t is a triangulation of S^1 and f is a mapping from the set of simplices of t to I. For $(t, f) \in J$, the open set $V_{(t,f)}$ of LM is the set of γ such that for any closed simplex σ of t, one has $\gamma(\sigma) \in U_{f(\sigma)}$. Let $(\underline{h}, \underline{\alpha})$ be a 1-cocycle of (U_i) with coefficients in the complex of sheaves (6-3). Then $\tau(\underline{h}, \underline{\alpha})$ is the smooth $\mathbb{C}^*$-valued function F whose restriction to the open set $V_{(t,\phi)}$ is

$$F(\gamma) = \prod_{e \text{ edge of } t} exp(-\int_e \gamma^* \alpha_{f(e)}) \cdot \prod_{v \text{ edge of } t} h_{f(e^-(v)), f(e^+(v))}(\gamma(v)).$$

$$(6-5)$$

Here for a vertex v of t, we denote by $e^+(v)$ the edge which ends at v, and by $e^-(v)$ the edge which begins at v.

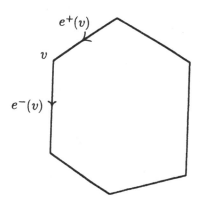

The triangulation t.

Then we have

6.1.3. Proposition. *The holonomy of (L, ∇) is equal to the function $\tau([L, \nabla])$.*

6.2. Construction of the line bundle

In this section, we construct a line bundle on the loop space LM, starting with a (Dixmier-Douady) sheaf of groupoids over M with connective structure. We show that a curving of the connective structure gives rise to a connection on the line bundle. The line bundle can be viewed as the *anomaly line bundle* associated to the sheaf of groupoids. The construction of a line bundle on loop space in this spirit was first done by Gawedzki [**Ga1**], using a transgression map on Deligne cohomology which will be discussed later. The geometric construction presented here grew out of discussions with Deligne.

It will actually be convenient to use the language of $\mathbb{C}^*$-bundles rather than line bundles. What we construct is a sort of "tautological $\mathbb{C}^*$-bundle" over LM. The construction of the $\mathbb{C}^*$-bundle is given in the following long statement.

6.2.1. Definition and Proposition. *Let M be a finite-dimensional manifold, and let $\mathcal{C}$ be a Dixmier-Douady sheaf of groupoids over M, equipped with a connective structure $P \mapsto Co(P)$. Define a smooth principal $\mathbb{C}^*$-bundle $\mathcal{P} \to LM$ as follows. The set $\mathcal{P}$ consists of equivalence classes of quadruples (γ, F, ∇, z), where $\gamma : S^1 \to M$ is a smooth loop, F is an object of the pullback sheaf of groupoids $\gamma^*(\mathcal{C})$ on S^1, ∇ is a section of $\gamma^* Co\,(F)$, and $z \in \mathbb{C}^*$.*

The equivalence relation is generated by the following two relations.

(i) The first relation identifies (γ, F, ∇, z) and (γ, F', ∇', z) if the pairs (F, ∇) and (F', ∇') of objects of $\gamma^*(\mathcal{C})$ with connection are isomorphic.

(ii) The second relation identifies $(\gamma, F, \nabla + \alpha, z)$ with $(\gamma, F, \nabla, z \cdot exp(-\int_0^1 \alpha))$ if α is a complex-valued 1-form on S^1.

$\mathbb{C}^*$ operates on $\mathcal{P}$ by $w \cdot (\gamma, F, \nabla, z) = (\gamma, F, \nabla, w \cdot z)$. Then there is a unique smooth manifold structure on $\mathcal{P}$ satisfying the following two conditions:

[a]: The projection $(\gamma, F, \nabla, z) \mapsto \gamma$ gives a smooth principal fibration $p : \mathcal{P} \to LM$;

[b]: For an open set U of M, an object F of $\mathcal{C}(U)$ and $\nabla \in Co(F)$, the

section

$$\sigma(\gamma) = (\gamma, \gamma^* F, \gamma^* \nabla, 1) \qquad\qquad (6-6)$$

of the restriction of the principal bundle $P \to LM$ to LU is smooth.

Proof. It follows from Theorem 5.2.8 that any sheaf of groupoids over S^1 is trivial, and this shows that every fiber of $p : P \to LM$ is non-empty. It is then clear that $\mathbb{C}^*$ acts transitively on every such fiber. To verify that the action is fixed-point free, we must show that the induced equivalence relation on a given fiber does not identify (γ, F, ∇, z) with $(\gamma, F, \nabla, z \cdot w)$ for $w \in \mathbb{C}^*$, $w \neq 1$. Since there is only one isomorphism class of objects of the sheaf of groupoids over S^1, we have only to consider the action of an automorphism of such an object F induced by a smooth function $g : S^1 \to \mathbb{C}^*$. Such an automorphism acts on Co (F) by $\nabla \mapsto \nabla - \frac{dg}{g}$, according to property (C2) in §5.3. So it transforms (γ, F, ∇, z) into $(\gamma, F, \nabla - \frac{dg}{g}, z)$. But this element is identified with $(\gamma, F, \nabla, z \cdot exp(\int_0^1 \frac{dg}{g}))$, which is equal to (γ, F, ∇, z), since $\frac{1}{2\pi\sqrt{-1}} \cdot \int_0^1 \frac{dg}{g}$ is the winding number of g, which is an integer.

We next define a topology and a manifold structure on P. For U an open set of M, F an object of $\mathcal{C}(U)$ and $\nabla \in Co(F)$, the section σ of the restriction of P to LU allows us to identify $p^{-1}(LU)$ with $LU \times \mathbb{C}^*$, and therefore to give a topology and a manifold structure to $p^{-1}(LU)$. Note that the image of any smooth path $\gamma : S^1 \to M$ is contained in such an open set U; hence we may define the topology on P to be that generated by the open subsets of all these $p^{-1}(LU)$. Clearly P is Hausdorff. To verify that we have a manifold structure on P, we should observe that the transition map between charts is of the type: $(\gamma, z) \mapsto (\gamma, z \cdot exp(\int_0^1 \gamma^* \alpha))$, which is a smooth map. It is then obvious that p is a locally trivial $\mathbb{C}^*$-bundle. ∎

We denote by $H_\gamma(F, \nabla)$ the element $(\gamma, F, \nabla, 1)$ of P. This should be understood as the *holonomy* of (F, ∇) around γ. The next step is to describe a connection on the $\mathbb{C}^*$-bundle $p : P \to LM$, for which we use a curving of the connective structure.

6.2.2. Proposition. *With the assumptions of 6.2.1, assume a curving $\nabla \in Co(F) \mapsto K(\nabla)$ of the connective structure on G is given. There is a natural connection D on the $\mathbb{C}^*$-bundle $p : P \to LM$, characterized by the condition that, for any open set U of M, any object F over U, and any $\nabla \in Co(F)$, we have for the local section $\sigma(\gamma) = (\gamma, \gamma^* F, \gamma^* \nabla, 1)$ of (6.2.1)[b], and for any tangent vector v to LM at γ,*

$$\frac{D_v \sigma}{\sigma} = - \int_\gamma i(v) \cdot K. \qquad\qquad (6-7)$$

Here of course $D_v\sigma$ means the covariant derivative of σ in the direction of v, and v is also viewed as a section of γ^*TM.

Let Ω be the 3-curvature of the curving; so $\Omega = 2\pi\sqrt{-1}\cdot\nu$, where ν has integral cohomology class. Then the curvature of the connection D on $\mathcal{P}$ is equal to $-2\pi\sqrt{-1}\cdot\beta$, where the 2-form β on LM is defined by transgressing ν, as in §3.1, i.e.,

$$\beta_\gamma(u,v) = \int_0^1 \nu_{\gamma(x)}(\frac{d\gamma}{dx}, u(x), v(x))dx. \qquad (6-8)$$

Proof. We want to define a connection on the $\mathbb{C}^*$-bundle by using formula (6-7) for the covariant derivative $D_v\sigma$ of the section σ along a tangent vector v. From Leibniz's rule, we then find the covariant derivative of any smooth section of p. We need to verify that the definition of D is independent of the choice of the section ∇ of $Co(F)$ that defined σ. Any other section of $Co(F)$ is of the form $\alpha + \nabla$, and gives rise to a local smooth section σ'. Formula (6-7) yields $\frac{D_v(\sigma')}{\sigma'} = -\int_\gamma i(v)\cdot K'$; here K' is the curvature of ∇', which is equal to $K + d\alpha$. On the other hand, the logarithmic form of Leibniz's formula and the equality $\frac{\sigma'}{\sigma}(\gamma) = exp(-\int_0^1 \gamma^*\alpha)$ give

$$\frac{D_v\sigma'}{\sigma'} - \frac{D_v\sigma}{\sigma} = -v\cdot(\gamma \mapsto \int_0^1 \gamma^*\alpha).$$

The derivative along v of $\int_0^1 \gamma^*\alpha$ is equal to $\int_\gamma \langle v, \alpha\rangle = \int_\gamma i(v)\cdot d\alpha$. Hence the two expressions for $\frac{D_v(\sigma')}{\sigma'}$ coincide. Since D_v depends smoothly on the tangent vector v, we obtain a connection on $p: \mathcal{P} \to LM$.

The curvature of the connection D is locally equal to $d(\frac{D\sigma}{\sigma})$, for a section σ as above, where we view $\frac{D\sigma}{\sigma}$ as a 1-form on LM. This 1-form is equal to $-\int_{S^1} ev^*K$, for K the curvature form of (F, ∇), which is a 2-form on M. We use here the notation of §1, and the fact that the assignment $\beta \mapsto \int_{S^1} ev^*\beta$ defines a morphism of complexes from the de Rham complex of M to the de Rham complex of LM, shifted by 1. It follows that $d(\frac{D\sigma}{\sigma})$ is equal to $-\int_{S^1} ev^*\Omega$, which is the 2-form we want. This concludes the proof. ∎

6.2.3. Proposition. *Let C be a sheaf of groupoids over M with connective structure. The principal $\mathbb{C}^*$-bundle $p : \mathcal{P} \to LM$, constructed in 6.2.1, is naturally $Diff^+(S^1)$-equivariant, so that, for an object F of $C(\mathcal{U})$ and $\nabla \in Co(F)$, the section $\sigma(\gamma) = H_\gamma(F, \nabla)$ of $\mathcal{P} \to LM$ defined over the $Diff^+(S^1)$-invariant open set LU is invariant. If there is a curving of the connective structure, the $Diff^+(S^1)$-action on $\mathcal{P}$ preserves the connection D.*

Proof. Locally on M, we can define the action of $Diff^+(S^1)$ on the principal bundle $p : \mathcal{P} \to LM$ by requiring that the section σ of the statement be invariant under the action. We need to verify that this action remains unchanged if we choose another section σ', corresponding to another object of $Co(F)$, say $\alpha + \nabla$. This amounts to the fact that $\dfrac{\sigma'}{\sigma}(\gamma) = exp(- \int_0^1 \gamma^* \alpha)$ is $Diff^+(S^1)$-invariant, which is obvious. ∎

As an example, consider the trivial sheaf of groupoids with trivial connective structure, but with the non-trivial curving $\nabla \mapsto K(\nabla) + \beta$, where β is some given 2-form. Then the $\mathbb{C}^*$-bundle $\mathcal{P} \to LM$ is trivial, but the connection is not the trivial connection, since the 1-form A which defines it is the transgression $A = - \int_{S^1} ev^* \beta$.

We now turn to the construction of an *action functional* on the space $Map(\Sigma, M)$ of smooth mappings $\Sigma \to M$, where Σ is an oriented surface with boundary. Let $C_1, \dots, C_r$ be the components of $\partial \Sigma$, and choose orientation preserving diffeomorphisms $f_i : S^1 \xrightarrow{\sim} C_i$ for $1 \leq i \leq r$. Then we have maps $b_i : Map(\Sigma, M) \to LM$ such that $b_i(\phi) = \phi_{/C_i} \circ f_i$, which associate to $\phi : \Sigma \to X$ its r boundary components.

6.2.4.   Theorem.   (1) *Let $\mathcal{C}$ be a sheaf of groupoids with connective structure over the manifold M. There is a canonical smooth section S of the principal $\mathbb{C}^*$-bundle $\mathcal{R} = \times_{i=1}^r b_i^* \mathcal{P}$ (a contracted product of r principal bundles), which is invariant under the natural action of $Diff^+(\Sigma)$ over $\mathcal{R}$. The following glueing axiom is satisfied. If the surface Σ is obtained by glueing together two surfaces Σ_1 and Σ_2 along some parts of their boundary, then denoting by $\mathcal{R}_j \to Map(\Sigma_j, M)$ the $\mathbb{C}^*$-bundles and by $\rho_j : Map(\Sigma, M) \to Map(\Sigma_j, M)$ the restriction maps, we have a natural isomorphism between $\mathcal{R}$ and $\rho_1^*(\mathcal{R}_1) \times^{\mathbb{C}^*} \rho_2^*(\mathcal{R}_2)$, and this isomorphism maps S to $\rho_1^*(S_1) \times \rho_2^*(S_2)$.*

(2) *Let $\mathcal{C}$ be equipped with a curving, and let D be the corresponding connection on $\mathcal{P}$. If we denote by the same symbol the induced connection on $\mathcal{R}$, we have*

$$\frac{D_v S}{S} = \int_\Sigma i(v) \cdot \Omega \qquad\qquad (6-9)$$

for a tangent vector v to $Map(\Sigma, M)$ at ϕ, viewed as a section of the pull-back $\phi^* TM$ of the tangent bundle to M.

(3) *Let Σ be the boundary of a 3-manifold V, and assume that $\phi : \Sigma \to M$ extends to a piecewise smooth continuous mapping $\Phi : V \to M$. Then we*

have

$$S(\phi) = exp(\int_V \Phi^* \Omega). \qquad (6-10)$$

Proof. The construction of S will at first be done locally on $Map(\Sigma, M)$. Denote by $ev : \Sigma \times Map(\Sigma, M) \to M$ the evaluation map. If U is a contractible open set of $Map(\Sigma, M)$, then the restriction to $\Sigma \times U$ of the pulled-back sheaf of groupoids $ev^*(\mathcal{C})$ will have a global object, since the obstruction to finding one is in $H^3(\Sigma \times U, \mathbb{Z}) = 0$. Choose such an object P, and pick an object ∇ of $Co(P)$. Then define the section S of $\mathcal{R}$ over U by

$$S(\phi) = exp(\int_\Sigma K(\nabla)) \cdot \prod_{i=1}^r H_{b_i(\phi)}(P, \nabla), \qquad (6-11)$$

where each $H_{b_i(\phi)}$ belongs to the fiber of $\mathcal{P}$ at the loop $b_i(\phi)$. This is clearly a smooth section of $\mathcal{R}$ over U, and we have to show that it is independent of the choices of P and ∇.

We will show that for P fixed, S is independent of the choice of $\nabla \in Co(P)$. Another element of $Co(P)$ is of the form $\nabla + \alpha$, for α a 1-form. If we replace ∇ with $\nabla + \alpha$, the curvature $K(\nabla)$ becomes $K(\nabla) + d\alpha$, so the term $exp(\int_\Sigma K(\nabla))$ gets multiplied by $exp(\int_\Sigma d\alpha)$. Each term $H_{b_i(\phi)}(P, \nabla)$ gets multiplied by $exp(-\int_{C_i} \alpha)$. The product of these terms, running over $1 \leq i \leq r$, is $exp(-\int_{\partial\Sigma} \alpha)$, which is equal to $exp(-\int_\Sigma d\alpha)$ by Stokes' theorem.

Next we study the effect of replacing P by another object. If this other object is isomorphic to P, S will not be changed. The set of isomorphism classes of objects of the sheaf of groupoids over $\Sigma \times U$ is $H^2(\Sigma \times U, \mathbb{Z}(1)) = H^2(\Sigma, \mathbb{Z}(1))$, which will only be non-zero if Σ has no boundary. Assuming this is the case, let $C \to \Sigma$ be a $\mathbb{C}^*$-bundle, and pull it back to $\Sigma \times U$. Choose a connection ∇' for C. Then, $P \times^{\mathbb{C}^*} C$ will be another object of our sheaf of groupoids, and to the pair (∇, ∇') there will correspond an object of $Co(P \times^{\mathbb{C}^*} C)$, which we denote by $\nabla + \nabla'$. The curvature of $\nabla + \nabla'$ is equal to $K(\nabla) + K(\nabla')$. The number $exp(\int_\Sigma K(\nabla))$ gets multiplied by $exp(\int_\Sigma K(\nabla'))$; as $\int_\Sigma K(\nabla')$ belongs to $\mathbb{Z}(1)$, its exponential is equal to 1. This proves that S is independent of all the choices, hence is a well-defined global section of $\mathcal{R} \to Map(\Sigma, M)$.

The glueing property is immediate. Locally on $Map(\Sigma, M)$, we can choose a global object over the product of Σ and an open set of this space of mappings, and use its restriction to the Σ_j to compute S_j. We need only to observe that if a boundary component of Σ_1 gets identified with a boundary component of Σ_2, the corresponding circle C appears with different orientations in the two surfaces, hence the corresponding holonomy symbols $H_{\pm C}(P, \nabla)$ cancel out. This finishes the proof of (1).

To prove formula (6-10), we may look at a one-parameter family of maps, i.e., a map $\Phi : \Sigma \times [0,1] \to M$, and consider the pull-back of C to $\Sigma \times [0,1]$. We pick an object P of this pull-back and $\nabla \in Co(P)$. With $v = \Phi_*(\frac{d}{dt})$, we have

$$\frac{D_v S}{S} = \frac{d}{dt} \int_\Sigma K(\nabla) - \sum_{i=1}^r \int_{C_i} i(v) \cdot K(\nabla).$$

The first term can be written as

$$\int \int_\Sigma \mathcal{L}(v) \cdot K(\nabla) = \int \int_\Sigma i(v) \cdot dK(\nabla) + d(i(v) \cdot K(\nabla)).$$

Using Stokes' theorem, we find:

$$\frac{Ds}{s} = \int_\Sigma i(v) \cdot dK(\nabla) = \int \int_\Sigma i(v) \cdot \Omega.$$

This proves (2). (3) also follows from Stokes' theorem. ∎

The upshot of the theorem is as follows. To any smooth mapping $\phi : \Sigma \to M$ there corresponds an *action functional* $S(\phi)$ in the fiber $\mathcal{R}_\phi$. The action functional is a smooth section of the $\mathbb{C}^*$-bundle $\mathcal{R} \to Map(\Sigma, M)$.

If $\Sigma = [0,1] \times S^1$ is a cylinder with boundary $-C_0 \coprod C_1$, then the element $S(\phi)$ of $\mathcal{P}_{b_0(\phi)}^{\otimes -1} \otimes \mathcal{P}_{b_1(\phi)}$ gives an isomorphism between $\mathcal{P}_{b_0(\phi)}$ and $\mathcal{P}_{b_1(\phi)}$. This isomorphism is equal to the parallel transport along the path $t \in [0,1] \mapsto \phi_{\{t\} \times S^1}$ in LM. If Σ has no boundary, then the action functional $S(\phi)$ is just a number, which is also called the *holonomy* of the sheaf of groupoids around the surface $\phi : \Sigma \to M$. Thus a sheaf of groupoids allows us to define a 2-dimensional version of the holonomy of a line bundle. The structure we have obtained is very analogous to the notion of *elliptic object* described by Segal [Se3]. It is a sort of 2-dimensional classical field theory (in the sense of theoretical physicists) for the manifold M.

The $\mathbb{C}^*$-bundle $\mathcal{P} \to LM$ has a multiplicativity property with respect to composition of loops. To make sense of this, we first observe that $\mathcal{P}$ can be extended to a $\mathbb{C}^*$-bundle over the manifold $L_{ps}M$ of piecewise smooth loops with respect to a given decomposition of S^1 into intervals. Then we have

6.2.5. Proposition. *Let γ_1 and γ_2 be piecewise smooth loops which may be*

composed. Then there is natural isomorphism of $\mathbb{C}^*$*-torsors*

$$\alpha : \mathcal{P}_{\gamma_1} \times^{\mathbb{C}^*} \mathcal{P}_{\gamma_2} \xrightarrow{\sim} \mathcal{P}_{\gamma_1 * \gamma_2}.$$

This isomorphism is compatible with the action functional S of Theorem 6.2.4 in the following sense. Let $\sigma_j : D^2 \to M$, for $j = 1, 2$, be a piecewise mapping which has restriction to the boundary S^1 of D^2 equal to γ_j. Then we define a map $\sigma_1 \cup \sigma_2 : D^2 \to M$ as follows: let $f : D^2 \to D^2 \vee D^2$ be the map which contracts a diameter of D^2 to a point. We put: $\sigma_1 \cup \sigma_2 = (\sigma_1 \vee \sigma_2) \circ f$. Then we have

$$\alpha(S(\sigma_1) \times S(\sigma_2)) = S(\sigma_1 \cup \sigma_2).$$

Finally we will show the geometric significance of our construction of the $\mathbb{C}^*$-bundle over the free loop space by describing it in classical differential-geometric terms for some important sheaves of groupoids.

In the first example, the manifold will be denoted by X. Let $f : Y \to X$ be a smooth fibration with $H^1(Y_x, \mathbb{C}^*) = 0$, and let β be a 2-form on Y, such that $d\beta$ is basic and the cohomology class of $\beta_x = \beta_{/Y_x}$ is integral for any $x \in X$ (this example was studied in Chapter 5). We will introduce a *relative action* $S(\gamma, \gamma')$ for two loops γ and γ' which project to the same loop δ in X. Let $F : [0,1] \times [0,1] \to Y$ be a homotopy from γ to γ' such that $f \circ F(t, u) = \delta(u)$. Then put

$$S(\gamma, \gamma') = exp(2\pi\sqrt{-1} \cdot \int_{[0,1] \times [0,1]} F^*\beta). \qquad (6-12)$$

An investigation of the spectral sequence for f shows that $S(\gamma, \gamma')$ is independent of the choice of F. In Chapter 5, we associated to this geometric situation a sheaf of groupoids over X, equipped with a connective structure. Then the total space $\mathcal{P}$ of the corresponding $\mathbb{C}^*$-bundle over LX has the following description. It is the quotient of $LY \times \mathbb{C}^*$ by the equivalence relation $\simeq$ given by

$$(\gamma, z) \simeq (\gamma', z') \iff f \circ \gamma = f \circ \gamma' \text{ and } z = z' \cdot S(\gamma, \gamma'). \qquad (6-13)$$

The other example comes from a central extension $1 \to \mathbb{C}^* \to \tilde{B} \to B \to 1$ of Lie groups, and a B-bundle $P \to M$ over a manifold M. Then we have the sheaf of groupoids $\mathcal{C}$ whose objects are local liftings of the structure group of P to $\tilde{B}$. We pick a connection on the principal bundle. Then we have a corresponding $\mathbb{C}^*$-bundle $\mathcal{P} \to LM$, which may be described as follows. For $\gamma \in LM$, and for $x \in P$ lying above the origin $\gamma(0)$ of γ, we have the holonomy $H(\gamma, x) \in B$ of the connection. This means that the horizontal lift of the loop γ which starts at x ends at $x \cdot H(\gamma, x)$. We view H as a smooth mapping $H : P \times_M LM \to B$, and we introduce the manifold $(P \times_M LM) \times_B \tilde{B}$, which consists of triples $(x, \gamma, \tilde{g})$, where $\gamma \in LM$, x is a point of P above $\gamma(0)$, and $\tilde{g} \in \tilde{B}$ is an element of $\tilde{B}$ which projects to $H(\gamma, x)$. Then $\mathcal{P}$ is the quotient of $(P \times_M LM) \times_B \tilde{B}$ by the equivalence relation which identifies $(x, \gamma, \tilde{g})$ with $(x \cdot h, \gamma, h^{-1}\tilde{g}h)$ for any $h \in B$. Here $h^{-1}\tilde{g}h$ denotes the element $\tilde{h}^{-1}\tilde{g}\tilde{h}$, for any lift $\tilde{h}$ of h to $\tilde{B}$. We see immediately that this gives a $\mathbb{C}^*$-bundle over LM, which is equivariant under the subgroup of $Diff^+(S^1)$ which fixes 0. It takes a little more work to see that it is even $Diff^+(S^1)$-equivariant, as follows from Proposition 6.2.3. This was our original example of a "tautological line bundle" over LM, before the whole theory of sheaves of groupoids with their differentiable structures was developed.

6.3. The line bundle on the space of knots

In §6.2 we constructed a $\mathbb{C}^*$-bundle on the free loop space LM of a manifold M associated with a sheaf of groupoids $\mathcal{C}$ over M. In this section we consider the case of a 3-dimensional manifold M and show how to obtain a $\mathbb{C}^*$-bundle over the space of oriented singular knots $\hat{Y}$. Recall from §3.1 that $\hat{Y}$ is equal to the quotient manifold $\hat{X}/Diff^+(S^1)$, where $\hat{X}$ is the open subset of LM consisting of smooth loops $\gamma : S^1 \to M$ which "have only finitely many singularities."

6.3.1. Theorem. (1) *Let M be a smooth 3-manifold. Let $\mathcal{C}$ be a sheaf of groupoids over M with a connective structure. The restriction to $\hat{X} \subset LM$ of the $\mathbb{C}^*$-bundle $\mathcal{P} \to LM$ descends to a $\mathbb{C}^*$-bundle $q : Q \to \hat{Y}$. Moreover, the canonical involution ι of §3.1 (reversal of knot orientation) lifts to an isomorphism $Q \xrightarrow{\sim} \iota^* Q^{-1}$ of $\mathbb{C}^*$-bundles.*

(2) *Pick a curving of the connective structure. Then the connection D descends to a connection on the bundle $Q \to \hat{Y}$; the curvature of this connection is the opposite of the 2-form on $\hat{Y}$ defined by transgression of the 3-curvature of the sheaf of groupoids.*

Proof. The action of $Diff^+(S^1)$ on the open set $\hat{X}$ of LM is fixed-point free; hence we may define Q as the quotient of $\mathcal{P} \times_{LM} \hat{X}$ by the action of $Diff^+(S^1)$. There is a principal $\mathbb{C}^*$-fibration $q : Q \to \hat{Y}$. The non-trivial element ι of $Diff(S^1)/Diff^+(S^1)$ now acts as a bundle isomorphism from Q to $\iota^* Q^{-1}$, described as follows. Given an object F of $\mathcal{C}$ over an open set of M, the corresponding local section σ of $\mathcal{P}$, which is $Diff^+(S^1)$-invariant, gives a local section of Q. Then ι transforms the section σ into the section σ^{-1} of Q^{-1}. This proves (1).

In order to see that the connection D descends to a connection on Q in the presence of a curving, we need to show that for a smooth local section s of $q : Q \to \hat{Y}$, and a vector field v on $\hat{X}$ which is vertical with respect to the projection $\pi : \hat{X} \to \hat{Y}$, the covariant derivative $D_v s$ is 0. We may assume that $s = \sigma$ as in (1), since $v \cdot g = 0$ for g a $Diff^+(S^1)$-invariant function. But the integral $\int_\gamma i(v) \cdot K$ is 0, since it is equal to $\int_0^1 i(\frac{d\gamma}{dx}) \cdot (i(v) \cdot K)$, and the iterated interior product involved is everywhere 0, since $v(x)$ is parallel to $\frac{d\gamma}{dx}$.

The curvature of the connection on Q pulls back to the curvature of D, so that it is the 2-form $-2\pi\sqrt{-1} \cdot \beta$, where β is given by (6-8). The map $Q \to \iota^* Q^{-1}$ maps the section σ to σ^{-1}. We have the connection $-D$ on Q^{-1}. Therefore

$$\left(\frac{-D_{d\iota(v)}\sigma^{-1}}{\sigma^{-1}}\right)_{\overline{\gamma}} = -\int_{\overline{\gamma}} i(d\iota(v)) \cdot (-K) = -\int_\gamma i(v) \cdot K = \left(\frac{D_v\sigma}{\sigma}\right)_\gamma;$$

hence this map is compatible with the connections. ∎

Note that we have remarkable geometric structures on the manifold $\hat{Y}$, namely:

(1) The symplectic form β.

(2) The complex structure J of §3.4, which is weakly integrable by Theorem 3.4.6 of Lempert.

(3) The $\mathbb{C}^*$-bundle $Q \to \hat{Y}$, which has a connection with curvature $-2\pi\sqrt{-1} \cdot \beta$.

(4) The Lie group G of unimodular diffeomorphisms of M acts on $\hat{Y}$ by symplectomorphisms.

Note that β is of type $(1,1)$ with respect to the complex structure J. If $\hat{Y}$ were finite-dimensional, a theorem of Griffiths (Theorem 2.2.26) asserts that the $\mathbb{C}^*$-bundle has a unique holomorphic structure such that a local section s is holomorphic if and only if $\frac{Ds}{s}$ is purely of type $(0,1)$. In the present situation, we can at least define the notion of a holomorphic section of Q, and it should not be hard to use ideas from [Pe-Sp] to show that there are non-zero holomorphic sections of $Q \to \hat{Y}$ locally.

The problem of geometric quantization would then be to construct, for each component C of the space of knots Y, an appropriate space $E(C)$ of holomorphic sections of the line bundle associated to Q (with suitable growth conditions) and to produce an action of the group G (or of some central extension of this group) on the vector space $E(C)$. Hopefully this representation would be an interesting invariant of the component C, i.e., of an isotopy type of knots. Such ideas have been advocated for some time by physicists [Ra-Re] [Pe-Sp].

This is, of course, a very ambitious goal. We turn to examining the central extension of a subgroup of G, obtained from the fundamental theorem of Kostant (Proposition 2.4.5). Let H be the subgroup of G which preserves the isomorphism class of the $\mathbb{C}^*$-bundle with connection (Q, D) over C. Then there is a natural central extension of Lie groups

$$1 \to \mathbb{C}^* \to \tilde{H} \to H \to 1. \qquad (6-14)$$

such that the action of $\tilde{H}$ on $\hat{Y}$ lifts to an action on Q. Recall that $\tilde{H}$ is the group of pairs (g, ψ), where $g \in H$ and $\psi : g_*(Q, D) \stackrel{\sim}{\to} (Q, D)$ is an isomorphism of $\mathbb{C}^*$-bundles with connections. Furthermore, if $H^1(C, \mathbb{C}^*) = 0$, then $H = G$.

There are criteria which ensure that the subgroup H is equal to G.

6.3.2. Proposition. *Let C be a connected component of $\hat{Y}$. Consider the restriction to C of the action of the connected component G^0 of G. We have $G^0 \subseteq H$, i.e., the isomorphism class of the pair (Q, D) over C is preserved by G^0, in each of the following two cases:*

(1) *LM is simply-connected;*

(2) *$H^2(M, \mathbb{C}^*) = 0$.*

Hence in either case there is a natural central extension of Lie groups

$$1 \to \mathbb{C}^* \to \tilde{G}^0 \to G^0 \to 1. \qquad (6-15)$$

Proof. In case (1), any g in G preserves the isomorphism class of the $\mathbb{C}^*$-bundle with connection $(\mathcal{P}, D)$ over LM, constructed in Proposition 6.2.1. Hence it preserves the isomorphism class of the restriction of this pair to the open set $\hat{X}$. Now we want to compare the groups of isomorphism classes of $\mathbb{C}^*$-torsors with connections on $\hat{X}$ and on $\hat{Y} = \hat{X}/Diff^+(S^1)$. These groups are $H^1(\hat{X}, \underline{\mathbb{C}}_{\hat{X}}^* \to \underline{A}_{\hat{X},\mathbb{C}}^1)$, resp. $H^1(\hat{Y}, \underline{\mathbb{C}}_{\hat{Y}}^* \to \underline{A}_{\hat{Y},\mathbb{C}}^1)$. We have a diagram with exact rows:

$$
\begin{array}{ccccccccc}
0 & \to & H^1(\hat{Y}, \mathbb{C}^*) & \to & H^1(\hat{Y}, \underline{\mathbb{C}}^* \to \underline{A}_{\hat{Y},\mathbb{C}}^1) & \to & H^2(\hat{Y}, \mathbb{Z}(1)) & \to & 0 \\
 & & \downarrow \pi^* & & \downarrow \pi^* & & \downarrow \pi^* & & \\
0 & \to & H^1(\hat{X}, \mathbb{C}^*) & \to & H^1(\hat{X}, \underline{\mathbb{C}}^* \to \underline{A}_{\hat{X},\mathbb{C}}^1) & \to & H^2(\hat{X}, \mathbb{Z}(1)) & \to & 0
\end{array}
$$

The pull-back map π^* on the H^1's is injective since $\pi_1(\hat{X})$ maps onto $\pi_1(\hat{Y})$, but the pull-back map on the groups H^2 with coefficients $\mathbb{Z}(1)$ may have a non-trivial kernel. We simply retain from this analysis the fact that the kernel B of $\pi^* : H^1(\hat{Y}, \underline{\mathbb{C}}^*_Y \to \underline{A}^1_{Y,\mathbb{C}}) \to H^1(\hat{X}, \underline{\mathbb{C}}^*_{\hat{X}} \to \underline{A}^1_{\hat{X},\mathbb{C}})$ is a discrete group. Now consider the isomorphism class of the $\mathbb{C}^*$-bundle with connection (Q, D) on $\hat{Y}$. Its pull-back to $\hat{X}$ is preserved by G, hence we have $[g^*(Q, D)] = [(Q, D)] \otimes P_g$, for an element P_g of B. Since P_g depends smoothly on $g \in G$, it must be constant on G^0, hence equal to 0. This proves the statement in case (1).

In case (2), we use the homomorphism from the group of equivalence classes of sheaves of groupoids with connective structure and curving on M, to the group of isomorphism classes of $\mathbb{C}^*$-bundles with connection over $\hat{Y}$, given by the construction of Theorem 6.3.1. This homomorphism is obviously G-equivariant. The group of equivalence classes of sheaves of groupoids with connective structure and curving over M was determined in Theorem 5.3.13: it is the second hypercohomology group of the complex of sheaves $\underline{\mathbb{C}}^*_M \xrightarrow{d \log} \underline{A}^1_{M,\mathbb{C}} \to \underline{A}^2_{M,\mathbb{C}}$, which sits in the exact sequence

$$H^2(M, \mathbb{C}^*) \to H^2(M, \underline{\mathbb{C}}^* \xrightarrow{d \log} \underline{A}^1_{M,\mathbb{C}} \to \underline{A}^2_{M,\mathbb{C}}) \to A^3(M)^{cl}_{\mathbb{C}}.$$

By assumption, $H^2(M, \mathbb{C}^*) = 0$. So the equivalence class of a sheaf of groupoids with connective structure and curving with extra structure is determined by its 3-curvature. Since G preserves the 3-form in question, it preserves the equivalence class of the sheaf of groupoids with extra structures. Hence it preserves the isomorphism class of the corresponding $\mathbb{C}^*$-bundle over $\hat{Y}$. ∎

We want to analyze to some extent the central extension of Lie algebras corresponding to the central extension of G^0 by $\mathbb{C}^*$ and to the choice of a connected component C of $\hat{Y}$. This is a central extension

$$0 \to \mathbb{C} \to \tilde{\mathfrak{g}} \to \mathfrak{g} \to 0. \tag{6 - 16}$$

We have the following result.

6.3.3. Proposition. *The central extension of Lie algebras (6-16) corresponding to a connected component C of $\hat{Y}$ is*

(1) *split if the closure of $\pi^{-1}(C)$ inside LM contains the constant loops;*

(2) *non-split if $M = S^1 \times S^1 \times S^1$, ω is $2\pi\sqrt{-1} \cdot dx \wedge dy \wedge dz$, and C contains the knot $S^1 \times \text{pt} \times \text{pt}$.*

Proof. According to Proposition 2.3.10, a Lie algebra cocycle c_γ giving this central extension may be found for any point γ of $\hat{Y}$. We have $c_\gamma(v, w) = -\beta_\gamma(v, w)$. Since the 2-form β is actually defined on LM, and since, according to [Bry2], the cohomology class of the 2-cocycle c_γ depends only on the connected component of LM containing γ, if the closure of $\pi^{-1}(C)$ in LM contains a constant loop p, we can find a path in LM from the point γ of $\pi^{-1}(C)$ to p. Hence c_γ is cohomologous to c_p, but β vanishes at p, hence c_p is zero. This proves statement (1).

In case (2), we show that the 2-cocycle c_γ associated to the knot $x \in [0, 1] \mapsto (x, 0, 0) \in S^1 \times S^1 \times S^1$ is non-cohomologous to 0 by exhibiting two commuting elements v and w of $\mathfrak{g}$ such that $c_\gamma(v, w) \neq 0$. We take $v = \frac{\partial}{\partial y}$ and $w = \frac{\partial}{\partial z}$. Then

$$c_\gamma(v, w) = \int_0^1 \nu(\frac{d\gamma}{dx}, v, w)dx = \int_0^1 \nu(\frac{\partial}{\partial x}, \frac{\partial}{\partial y}, \frac{\partial}{\partial z})dx = 1 \neq 0$$

as stated. ∎

6.4. Central extension of loop groups

In this section we give a concrete description, which follows from methods already utilized, of the central extension of a loop group. We recover geometrically the description of the central extension in terms of action functionals, as found in the physics literature [Mic]. Our treatment will be brief, as a detailed discussion, which applies also to non simply-connected groups, is to be found in [Br-ML2].

In section 6.2 we described how a sheaf of groupoids C with connective structure over a manifold M leads to a $\mathbb{C}^*$-bundle over the free loop space LM. If the sheaf of groupoids and connective structure are unitary, we actually obtain a $\mathbb{T}$-bundle over LM.

Now let G be a compact simple simply-connected Lie group. In §5.4 we described a sheaf of groupoids C with unitary connective structure and curving over G in terms of the path fibration $ev : P_1 G \to G$. The 3-curvature is $\Omega = -2\pi\sqrt{-1} \cdot \nu$ for the canonical 3-form ν on G. The free loop space LG has a Lie group structure in which the product is given by pointwise product of loops [P-S]: LG is called the *loop group* of G. The corresponding $\mathbb{T}$-bundle over LG can be identified with the central extension

$$1 \to \mathbb{T} \to \widetilde{LG} \to LG \to 1 \qquad (6-14)$$

which is constructed, for example, in [P-S]. The definition of the central extension $\widetilde{LG}$ in [P-S] consists essentially in making LG act by symplectomorphisms on some infinite-dimensional Grassmann manifold, and then taking the Kostant central extension. The Lie group $\widetilde{LG}$ is called a *Kac-Moody group*, and its Lie algebra is the affine Kac-Moody Lie algebra $\widetilde{Lg}$. One reason for the richness of the theory of affine Kac-Moody Lie algebras is that it is in some sense contained in the so-called extended Dynkin diagram, and is thus analogous to the well-known theory of finite-dimensional simple Lie algebras. Since the theory of affine Kac-Moody algebras falls outside of the scope of this book, we refer the reader to the book [**Kac**].

From our point of view, we have a sort of "tautological construction" of the $\mathbb{T}$-bundle over LG. To describe this explicitly, we need the action functional $S(\phi)$ for a continuous and piecewise smooth map $\phi : S^2 \to G$. In this case, it is simply

$$S(\phi) = exp(\int_{D^3} \Phi^*\Omega) \qquad (6-15),$$

where $\Phi : D^3 \to G$ is a continuous and piecewise smooth extension of ϕ to the 3-ball D^3 (which has S^2 as its boundary). If $\sigma_1, \sigma_2 : D^2 \to G$ are two continuous and piecewise smooth maps, agreeing on the boundary S^1, we denote by $\sigma_1 \sqcup \sigma_2 : S^2 \to G$ the continuous and piecewise smooth map which is given by σ_1 on the upper hemisphere and by σ_2 on the lower hemisphere.

The total space, say $\mathcal{E}$, of the $\mathbb{T}$-bundle over LG may be described as the set of equivalence classes of pairs (σ, z), where $\sigma : D^2 \to G$ is a smooth mapping and $z \in \mathbb{T}$. Here D^2 denotes the closed 2-dimensional disc. The boundary of D^2 is identified with S^1. The equivalence relation is called $\simeq$; we have $(\sigma_1, z_1) \simeq (\sigma_2, z_2)$ if and only if the following two conditions hold

$$
\begin{align}
&(1) \ (\sigma_1)_{/S^1} = (\sigma_2)_{/S^1} \\
&(2) \ z_2 = z_1 \cdot S(\sigma_1 \sqcup \sigma_2)^{-1} \qquad (6-16)
\end{align}
$$

(note that $S(\sigma_1 \sqcup \sigma_2)$ is well-defined, because the mapping $\sigma_1 \sqcup \sigma_2 : S^2 \to G$ is continuous and piecewise smooth). There is a natural manifold structure on $\mathcal{E}$, and the projection $(\sigma, z) \mapsto \sigma_{/S^1}$ makes $\mathcal{E}$ into a $\mathbb{T}$-bundle over LG.

It is not clear at first how to define a product on $\mathcal{E}$, because the 3-form ν on G does not satisfy the required multiplicativity property. To explain this, let $p_1, p_2, m : G \times G \to G$ denote the maps $p_1(g_1, g_2) = g_1$, $p_2(g_1, g_2) = g_2$, $m(g_1, g_2) = g_1g_2$. Then $m^*\nu$ is not equal to $p_1^*\nu + p_2^*\nu$, although their cohomology classes agree ($[\nu]$ is a so-called *primitive* cohomology class). However,(cf. [**Mic**] [**Br-ML2**]) we may introduce a 2-form ω on $G \times G$ which satisfies

(a) $$p_1^* \nu + p_2^* \nu - m^* \nu = d\omega$$

(b) Let d_0, d_1, d_2, d_3 be the "face maps" from $G \times G \times G$ to $G \times G$ given by

$$d_0(g_1, g_2, g_3) = (g_2, g_3)$$
$$d_1(g_1, g_2, g_3) = (g_1 g_2, g_3)$$
$$d_2(g_1, g_2, g_3) = (g_1, g_2 g_3)$$
$$d_3(g_1, g_2, g_3) = (g_1, g_2)$$

Then we have $d_0^* \omega - d_1^* \omega + d_2^* \omega - d_3^* \omega = 0$. The 2-form ω is left-invariant under the action of the first copy of G, and right-invariant under the action of the second copy of G. Thus ω is defined by an element of $\wedge^2(\mathfrak{g} \oplus \mathfrak{g})^*$, namely,

$$\omega(\xi_1, \xi_2) = 3 \cdot \langle \xi_1, \xi_2 \rangle \qquad (6-17)$$

where $\langle\ ,\ \rangle$ is the invariant bilinear form which describes ν (cf. §5.4).

We come to the following

6.4.1. Theorem. (*Mickelsson*) [Mic] *The product*

$$(\sigma_1, z_1) \cdot (\sigma_2, z_2) = (\sigma_1 \sigma_2, z_1 z_2 \cdot exp(-2\pi\sqrt{-1} \cdot \int_{D^2} (\sigma_1 \times \sigma_2)^* \omega)) \quad (6-18)$$

defines a Lie group structure on $\mathcal{E}$, which identifies it with $\widetilde{LG}$.

We refer the reader to [Br-ML2] for a geometric discussion of this construction from the point of view of $\mathbb{C}^*$-bundles over simplicial manifolds; the advantage is that it applies to non simply-connected compact Lie groups also. [Br-ML2] also contains a discussion of the *Segal reciprocity law* of [Se3], which says that for a surface Σ with r boundary components, the Baer sum of the r pull-backs of the central extension of LG to $Map(\Sigma, G)$ (by the r restriction maps to the boundary circles) has a canonical splitting. All of these results admit holomorphic analogs for which [Ga1] is an excellent reference.

The main open question seems to be to obtain the representation theory of the group $\widetilde{LG}$ from the canonical sheaf of groupoids on G. The method of [P-S] consists in studying the global holomorphic sections of line bundles on some infinite-dimensional flag manifolds. We ask whether some quantization method exists, based on the sheaf of groupoids (i.e., ultimately on the 3-form on G), which will produce representations of $\widetilde{LG}$. This would be of importance for a full mathematical understanding of the Chern-Simons field theory of Witten [Wi2], also discussed in [Br-ML2].

6.5. Relation with smooth Deligne cohomology

In section 2 we showed how a sheaf of groupoids with connective structure over a finite-dimensional manifold M leads by a transgression operation to a $\mathbb{C}^*$-principal bundle $\mathcal{P} \to LM$, and a curving leads to a connection on $\mathcal{P} \to LM$. Theorem 5.3.12 says that the group of equivalence classes of sheaves of groupoids with connective structure and curving over M is isomorphic to $H^2(M, \underline{\mathbb{C}}_M^* \to \underline{A}_{M,\mathbb{C}}^1 \to \underline{A}_{M,\mathbb{C}}^2)$, hence to $(2\pi\sqrt{-1})^{-2} \cdot H^3(M, \mathbb{Z}(3)_D^\infty)$; Theorem 2.2.12 says that the group of isomorphism classes of line bundles with connection over LM is isomorphic to $H^1(LM, \underline{\mathbb{C}}_{LM}^* \to \underline{A}_{LM,\mathbb{C}}^1)$, hence to $(2\pi\sqrt{-1})^{-1} \cdot H^2(LM, \mathbb{Z}(2)_D^\infty)$. So in cohomological terms the transgression in question amounts to a homomorphism

$$H^2(M, \underline{\mathbb{C}}_M^* \to \underline{A}_{M,\mathbb{C}}^1 \to \underline{A}_{M,\mathbb{C}}^2) \to H^1(M, \underline{\mathbb{C}}_{LM}^* \to \underline{A}_{LM,\mathbb{C}}^1).$$

We compare this with a homomorphism due to Gawedzki [Ga1], and defined in Čech cohomology as follows. First recall some notations from Lemma 6.1.2. Choose a triangulation t of S^1. Let $\mathcal{U} = (U_i)_{i \in I}$ be an open covering of M. Let J be the set of pairs (t, f), where t is a triangulation of S^1 and f is a mapping from the set of simplices of t to I. For $(t, f) \in J$, the open subset $V_{(t,f)}$ of LM consists of those γ such that $\gamma(\sigma) \in U_{f(\sigma)}$ for each closed simplex σ of t. Given two triangulations t_0 and t_1 of S^1, there is a common refinement $\bar{t}$ of t_0 and t_1, whose simplices are all the non-empty intersections $\sigma_o \cap \sigma_1$, for σ_j a simplex of t_j ($j = 0$ or 1). Given functions f_j from simplices of t_j to I, we define functions $\overline{f}_j$ from simplices of $\bar{t}$ to I as follows: If $\bar{e}$ is an edge of $\bar{t}$, $\bar{e}$ is of the form $\bar{e} = e_0 \cap e_1$, where e_j is an edge of t_j, and we set $\overline{f}_0(\bar{e}) = f_0(e_0)$. If $\bar{v}$ is a vertex of $\bar{t}$, there are two cases: Either $\bar{v}$ is a vertex of t_0, in which case we set $\overline{f}_0(\bar{v}) = f_0(\bar{v})$; or $\bar{v}$ is an interior point of an edge e_0 of t_0, and we set $\overline{f}_0(\bar{v}) = f_0(e_0)$. The function $\overline{f}_1$ is defined similarly.

Let $(\underline{g}, \eta, \underline{\omega})$ be a Čech cocycle of (U_i) with coefficients in the complex of sheaves $\underline{\mathbb{C}}_M^* \to \underline{A}_{M,\mathbb{C}}^1 \to \underline{A}_{M,\mathbb{C}}^2$. We define the Čech 1-cochain $(\underline{G}, \underline{A})$ of the covering $(V_{(t,f)})$ with coefficients in $\underline{\mathbb{C}}_{LM}^* \to \underline{A}_{LM,\mathbb{C}}^1$ as follows. We set

$$G_{(t_0,f_0),(t_1,f_1)}(\gamma) = exp(\sum_{\bar{e}} \int_{\bar{e}} \gamma^* \eta_{\overline{f}_0(\bar{e}),\overline{f}_1(\bar{e})}) \cdot \prod_{\bar{v} \in \partial\bar{e}} \frac{g_{\overline{f}_0(\bar{v}),\overline{f}_1(\bar{v}),\overline{f}_1(\bar{e})}}{g_{\overline{f}_0(\bar{v}),\overline{f}_0(\bar{e}),\overline{f}_1(\bar{e})}}(\gamma(\bar{v}))$$

$$(6-19)$$

where the product is over all pairs $(\bar{v}, \bar{e})$, where $\bar{v}$ is a vertex of $\bar{t}$, $\bar{e}$ is an edge of $\bar{t}$, and $\bar{v}$ is a boundary point of $\bar{e}$. We then set

$$\langle A_{t,f}, \xi \rangle_\gamma = \sum_e \int_e \gamma^*(i(d\gamma(\xi)) \cdot \omega_{f(e)}) + \sum_{v \in \partial e} \langle \xi(v), \eta_{f(v), f(e)}(\gamma(v)) \rangle \quad (6-20)$$

where ξ denotes a tangent vector to LM at γ, and $\xi(v) \in T_{\gamma(v)}M$ its value at $v \in S^1$.

6.5.1. Proposition. *The assignment* $(\underline{g}, \eta, \underline{\omega}) \mapsto (\underline{G}, \underline{A})$ *induces a group homomorphism from* $H^2(M, \underline{\mathbb{C}}_M^* \to \underline{A}_{M,\mathbb{C}}^1 \to \underline{A}_{M,\mathbb{C}}^2)$ *to* $H^1(LM, \underline{\mathbb{C}}_{LM}^* \to \underline{A}_{LM,\mathbb{C}}^1)$, *which is equal to the opposite of the transgression map from sheaves of groupoids with connective structure and curving over* M *to line bundles with connection over* LM.

Proof. We explain the relation of the map on smooth Deligne cohomology with the construction of the $\mathbb{C}^*$-bundle $\mathcal{P} \to LM$. Let $\mathcal{C}$ be a sheaf of groupoids with connective structure over M, and let P_i be an object of $\mathcal{C}(U_i)$; we choose isomorphisms $u_{ij} : (P_j)_{/U_{ij}} \xrightarrow{\sim} (P_i)_{/U_{ij}}$. Then $h_{ijk} = u_{ik}^{-1} u_{ij} u_{jk}$ is a 2-cocycle of (U_i) with values in $\underline{\mathbb{C}}_M^*$. Next, pick a section ∇_i of $Co(P_i)$, and set $\eta_{ij} = u_{ij*}(\nabla_j) - \nabla_i$ and $\omega_i = K(\nabla_i)$. We will now indicate how to find a section $s_{(t,f)}$ of $\mathcal{P}$ over the open set $V_{(t,f)}$ of LM. We describe $s_{(t,f)}$ at $\gamma \in V_{(t,f)}$. For each simplex σ of the triangulation t, we have $\gamma(\sigma) \subset U_{f(\sigma)}$. Then we obtain an object $Q_{(t,f)}$ of the sheaf of groupoids $\gamma^* \mathcal{C}$ over S^1 as follows. Over each σ we have the pull-back $(\gamma_{/\sigma})^* P_{f(\sigma)}$ of $P_{f(\sigma)}$. This object extends to a neighborhood of σ. Now we cover S^1 by open neighborhoods of the simplices of t in such a way that the intersection of any three different open sets is empty. If v is a vertex which lies in the boundary of an edge e, we glue $(\gamma_{/v})^* P_{f(v)}$ with $(\gamma_{/e})^* P_{f(e)}$ by the isomorphism $u_{f(v), f(e)}$; this gives a global object of $\gamma^* \mathcal{C}$, and in fact an object $Q_{(t,f)}$ of the pull-back of $\mathcal{C}$ to $V_{(t,f)} \times S^1$ by the evaluation map. We also have to construct an object $D_{(t,f)}$ of $Co(Q_{(t,f)})$. Actually it is much easier to work with the sheaf of 1-forms on $V_{(t,f)} \times S^1$ having smooth restriction to each $V_{(t,f)} \times \sigma$, for σ a simplex of t. Then we just take $\nabla_{f(\sigma)}$ over each simplex σ, so that we have an object $D_{(t,f)}$ of a torsor under the sheaf of piecewise smooth 1-forms of the above type.

Now if we have two different triangulations t_0 and t_1, we construct an isomorphism between the corresponding objects of the pull-back sheaf of groupoids. Precisely, for $\bar{t}$ the smallest common refinement of t_0 and t_1, we define for every simplex $\bar{\sigma}$ of $\bar{t}$ an isomorphism $\phi_{\bar{\sigma}} : P_{\bar{f}_0(\bar{\sigma})} \xrightarrow{\sim} P_{\bar{f}_1(\bar{\sigma})}$ of the form $\phi_{\bar{\sigma}} = u_{\bar{f}_1(\bar{\sigma}), \bar{f}_0(\bar{\sigma})} \cdot h_{\bar{\sigma}}$, where $h_{\bar{\sigma}}$ is a $\mathbb{C}^*$-valued function. If $\bar{v}$ is a vertex of $\bar{t}$ in the boundary of the edge $\bar{e}$, we obtain a coherency condition that gives the value of the "jump" $\frac{h_{\overline{v}}}{h_{\overline{e}}}$ at the vertex. This jump is expressed in terms of values of the 2-cocycle $\underline{h}$. Now we compute the difference between the holonomy of $(Q_{(t_0,f_0)}, D_{(t_0,f_0)})$ and that of $(Q_{(t_1,f_1)}, D_{(t_1,f_1)})$. This becomes a product of two types of terms:

(a) the product of exponentials of integrals over the edges;

(b) the product over all edges $\bar{e}$ of the exponential of the integral of $Log(\phi)$ over $\bar{e}$, which is also equal to the quotient of the product of the "jumps" for $Q_{(t_1,f_1)}$ by the similar jumps for $Q_{(t_0,f_0)}$.

In this way we get the inverse of (6-19). We verify (6-20) by taking the derivative of the holonomy of $D_{(t,f)}$, and observing that because of the discontinuity of $D_{(t,f)}$ at a vertex v of t, there is an exotic term which involves $\eta_{f(v),f(e)}$, for v in the boundary of e. ∎

Gawedzki's approach is to use this homomorphism of smooth Deligne cohomology groups in order to obtain a line bundle over the free loop space. Gawedzki also examines the holomorphic case.

It may be useful to point out another process to describe homomorphisms from $H^{k+1}(M, \mathbb{Z}(p+1)_D^\infty)$ to $(2\pi\sqrt{-1}) \cdot H^k(LM, \mathbb{Z}(p)_D^\infty)$, or equivalently from $H^k(M, \underline{\mathbb{C}}_M^* \to \underline{A}^1_{M,\mathbb{C}} \to \cdots \to \underline{A}^p_{M,\mathbb{C}})$ to $H^{k-1}(LM, \underline{\mathbb{C}}_{LM}^* \to \underline{A}^1_{LM,\mathbb{C}} \cdots \underline{A}^{p-1}_{LM,\mathbb{C}})$. First we have the pull-back map ev^* from $H^k(M, \underline{\mathbb{C}}_M^* \to \underline{A}^1_{M,\mathbb{C}} \to \cdots \to \cdots \to \underline{A}^p_{M,\mathbb{C}})$ to $H^k(LM \times S^1, \underline{\mathbb{C}}_{LM \times S^1}^* \to \underline{A}^1_{LM \times S^1,\mathbb{C}} \to \cdots \to \underline{A}^p_{LM \times S^1,\mathbb{C}})$. Then we can use the following construction of an *integration map*

$$\int_{S^1} : \; H^k(LM \times S^1, \underline{\mathbb{C}}_{LM \times S^1}^* \to \underline{A}^1_{LM \times S^1,\mathbb{C}} \to \cdots \to \underline{A}^p_{LM \times S^1,\mathbb{C}})$$
$$\to H^{k-1}(LM, \underline{\mathbb{C}}_{LM}^* \to \underline{A}^1_{LM,\mathbb{C}} \to \cdots \to \underline{A}^{p-1}_{LM,\mathbb{C}})$$

Such a map $\int_{S^1}$ exists with LM replaced by any manifold X, so we discuss $\int_{S^1}$ in that generality. Recall that we have an integration map for differential forms $\int_{S^1} : A^l(X \times S^1) \to A^{l-1}(X)$, which gives rise to a morphism of the de Rham complexes $A^\bullet(X \times S^1) \to A^{\bullet-1}(X)$. The induced homomorphism $\int_{S^1} : H^l(X \times S^1) \to H^{l-1}(X)$ is the projection to the second factor of the Künneth decomposition

$$H^l(X \times S^1) = H^l(X) \oplus [H^{l-1}(X) \otimes H^1(S^1)] = H^l(X) \oplus H^{l-1}(X).$$

The next proposition shows that this can be extended to smooth Deligne cohomology.

6.5.2. Proposition. (1) *Let X be a paracompact, smooth, possibly infinite-*

dimensional manifold. Let $p_1 : X \times S^1 \to X$ be the projection. Then there is a natural map $\int_{S^1}$ from $H^k(X \times S^1, \underline{\mathbb{C}}^*_{X \times S^1} \to \underline{A}^1_{X \times S^1, \mathbb{C}} \to \cdots \to \underline{A}^p_{X \times S^1, \mathbb{C}})$ to $H^{k-1}(X, \underline{\mathbb{C}}^*_X \to \underline{A}^1_{X, \mathbb{C}} \to \cdots \to \underline{A}^{p-1}_{X, \mathbb{C}})$, which, for $k = p$, makes the following diagram commute:

$$
\begin{array}{ccc}
H^p(X \times S^1, \underline{\mathbb{C}}^*_{X \times S^1} \to \underline{A}^1_{X \times S^1, \mathbb{C}} \to \cdots \underline{A}^p_{X \times S^1, \mathbb{C}}) & \xrightarrow{d} & A^{p+1}(X \times S^1)_{\mathbb{C}} \\
\downarrow{\scriptstyle \int_{S^1}} & & \downarrow{\scriptstyle \int_{S^1}} \\
H^{p-1}(X, \underline{\mathbb{C}}^*_X \to \underline{A}^1_{X, \mathbb{C}} \to \cdots \underline{A}^{p-1}_{X, \mathbb{C}}) & \xrightarrow{d} & A^p(X)_{\mathbb{C}}
\end{array}
$$

(2) *The induced map on Čech hypercohomology groups may be realized as follows. Take a locally finite open covering $\mathcal{U} = (U_i)_i$ of X such that all multiple intersections of U_i's are empty or contractible. Take a cyclically ordered sequence $(x_1, x_2, \ldots, x_n)$ of points of S^1, and let I_j be the closed interval $I_j = [x_j, x_{j+1}]$ (consider the indices $1, 2, \ldots n-1, n$ as elements of $\mathbb{Z}/n \cdot \mathbb{Z}$). Choose an open interval J_j containing I_j. Let $\mathcal{W}$ be the open covering of $X \times S^1$ by the $W_{ij} = U_i \times J_j$. Consider a degree k Čech cocycle $(\underline{f}, \underline{\omega}^1, \cdots, \underline{\omega}^p)$ of the covering $\mathcal{W}$ with values in $\underline{\mathbb{C}}^*_{X \times S^1} \to \underline{A}^1_{X \times S^1, \mathbb{C}} \to \cdots \underline{A}^p_{X \times S^1, \mathbb{C}}$, where $\underline{f}$ is a Čech k-cocycle with values in $\underline{\mathbb{C}}^*_{X \times S^1}$, $\underline{\omega}^1$ is a Čech $(k-1)$ cochain with values in $\underline{A}^1_{X \times S^1, \mathbb{C}}$, and so on. Then $\int_{S^1}(\underline{f}, \underline{\omega}^1, \ldots, \underline{\omega}^p)$ is represented by the degree $k-1$-Čech cochain $(\underline{g}, \underline{\alpha}^1, \cdots, \underline{\alpha}^{p-1})$ of the open covering $\mathcal{U}$ of X, with values in the complex of sheaves $\underline{\mathbb{C}}^*_{X \times S^1} \to \underline{A}^1_{X, \mathbb{C}} \to \cdots \underline{A}^{p-1}_{X, \mathbb{C}}$, where*

$$
\begin{aligned}
\underline{g}_{i_0, \ldots, i_{k-1}} = & \prod_{j=0}^{n-1} exp\left(\int_{I_j} \underline{\omega}^1_{(i_0, j), \cdots, (i_{k-1}, j)} \right) \\
& \times \prod_{j=0}^{n-1} \prod_{m=0}^{k-1} (\underline{f}_{(i_0, j), \cdots, (i_m, j), (i_m, j+1), \cdots, (i_{k-1}, j+1)})^{(-1)^{k+m}}_{/x = x_{j+1}}
\end{aligned}
$$

$$(6 - 21)$$

and

$$
\begin{aligned}
\underline{\alpha}^l_{i_0, \ldots, i_{k-l-1}} = & \sum_{j=0}^{n-1} \int_{I_j} \underline{\omega}^{l+1}_{(i_0, j), \ldots, (i_{k-l-1}, j)} \\
& + \sum_{j=0}^{n-1} \sum_{m=0}^{k-l-1} (-1)^{k-l+m} \cdot (\underline{\omega}^l_{(i_0, j), \ldots, (i_m, j), (i_m, j+1), \ldots, (i_{k-l-1}, j+1)})_{/x = x_{j+1}}
\end{aligned}
$$

$$(6 - 22)$$

We note that since the covering $\mathcal{U}$ satisfies the assumption that all intersections of open sets in the covering are empty or contractible, so does the covering $\mathcal{W}$ of the statement. This explains why the method of (2) completely describes the map $\int_{S^1}$ on smooth Deligne cohomology.

In the case $p = 2$ and $k = 2$, we obtain a map $\int_{S^1} ev^*$ from $H^2(M, \underline{\mathbb{C}}_M^* \to \underline{A}_{M,\mathbb{C}}^1 \to \underline{A}_{M,\mathbb{C}}^2)$ to $H^1(LM, \underline{\mathbb{C}}_{LM}^* \to \underline{A}_{LM,\mathbb{C}}^1)$. This coincides with the homomorphism of Gawedzki, described by (6-19) and (6-20).

It would be very interesting to have integration formulas in smooth Deligne cohomology, with S^1 replaced by a closed oriented surface. We refer to [Br-ML2] for an abstract version of such a homomorphism, in relation to degree 4 characteristic classes and line bundles over moduli spaces.

6.6. Parallel transport for sheaves of groupoids

In the case of a line bundle L with connection on a manifold M, the holonomy around a path is a special case of the parallel transport, which may be viewed as an isomorphism $T : ev_0^*L \tilde{\to} ev_1^*L$ of line bundles over the unrestricted path space PM, with $ev_t : PM \to M$ the evaluation of a path at $t \in [0,1]$.

We now consider a sheaf of groupoids $\mathcal{C}$ with connective structure over M. Construct an equivalence $T : ev_0^*\mathcal{C} \to ev_1^*\mathcal{C}$ of sheaves of groupoids over PM by first constructing the functor T at $\gamma \in PM$. Let $P \in \mathcal{C}(\gamma(0))$. In order to describe an object $T(P)$ of $\mathcal{C}(\gamma(1))$, we construct a homogeneous space $T'(P,Q)$ under $\mathbb{C}^*$ for any object Q of $\mathcal{C}(\gamma(1))$, together with isomorphisms $T'(P, Q_1) \tilde{\to} T'(P, Q_2) \times^{\mathbb{C}^*} Isom(Q_1, Q_2)$ which satisfy a natural associativity condition. Then the object $Q \times^{\mathbb{C}^*} T'(P,Q)$ of $\mathcal{C}(\gamma(1))$ will be independent of the choice of $Q \in \mathcal{C}(\gamma(1))$, and we will set

$$T(P) = Q \times^{\mathbb{C}^*} T'(P,Q) \qquad (6-23)$$

Our task therefore is to describe the $\mathbb{C}^*$-torsor $T'(P,Q)$.

6.6.1. Definition and Proposition. *Let $\gamma \in PM$, let $P \in \mathcal{C}(\gamma(0))$ and $Q \in \mathcal{C}(\gamma(1))$. Define S to be the set of quintuples $(R, \alpha, \beta, \nabla, z)$ where*

(1) *R is an object of the sheaf of groupoids $\gamma^*\mathcal{C}$ over S^1.*

(2) *$\alpha : R_{/0} \tilde{\to} P$ is an isomorphism in $\mathcal{C}(\gamma(0))$, $\beta : R_{/1} \tilde{\to} Q$ is an isomorphism in $\mathcal{C}(\gamma(1))$.*

(3) *∇ is a section of $\gamma^*Co(R)$.*

(4) *$z \in \mathbb{C}^*$.*

We put on S the equivalence relation $\simeq$ generated by the following three relations:

(a)

$$(R, \lambda \cdot \alpha, \mu \cdot \beta, \nabla, z) \simeq (R, \alpha, \beta, \nabla, \lambda^{-1} \cdot \mu \cdot z)$$

for $\lambda, \mu \in \mathbb{C}^*$.

(b) *two isomorphic quintuples are considered equivalent.*

(c)

$$(R, \alpha, \beta, \nabla + \omega, z) \simeq (R, \alpha, \beta, \nabla, z \cdot exp(- \int_0^1 \omega))$$

for a complex-valued 1-form ω on $[0,1]$. Then the quotient space $T'(P,Q) = S/ \simeq$ is a homogeneous space under $\mathbb{C}^*$, whose action is given by multiplication on the variable $z \in \mathbb{C}^*$.

Proof. The point is to show the consistency of the three relations (a), (b) and (c). According to (b), given a smooth map $g : S^1 \to \mathbb{C}^*$, $(R, \alpha, \beta, \nabla, z)$ is equivalent to $(R, g(0) \cdot \alpha, g(1) \cdot \beta, \nabla + g^{-1} dg, z)$. According to (c), this is equivalent to $(R, g(0) \cdot \alpha, g(1) \cdot \beta, \nabla, z \cdot exp(- \int_0^1 g^{-1} dg))$, which in turn equals $(R, g(0) \cdot \alpha, g(1) \cdot \beta, \nabla, z \cdot g(0)g(1)^{-1})$. According to (a), this is equivalent to $(R, \alpha, \beta, \nabla, z)$. ∎

From this we easily obtain

6.6.2. Theorem. *There exists an equivalence $T : ev_0^* C \to ev_1^* C$ of sheaves of groupoids over the unrestricted path space PM, such that for each $\gamma \in PM$ and $P \in C(\gamma(0))$, the object $T(P)$ of $C(\gamma(1))$ is canonically isomorphic to $T'(P,Q) \times^{\mathbb{C}^*} Q$, for any object Q of $C(\gamma(1))$.*

Over the free loop space $LM \subset PM$, the equivalence of the sheaf of groupoids $ev_0^* C$ with itself is given by twisting objects of $ev_0^* C$ by the $\mathbb{C}^*$-torsor $\mathcal{P} \to LM$ of §6.2. In this sense Theorem 6.6.2 is a generalization of 6.2.1.

6.6.3. Proposition. *If we have a curving of the connective structure, T extends to an equivalence of sheaves of groupoids with connective structure over PM, such that for $D_0 \in Co(P)$ and $D_1 \in Co(Q)$, the 1-form $D_1 - T_*(D_0)$ is equal to*

$$- \int_0^1 i(\frac{d\gamma}{dx}) \cdot K(\nabla)$$

for any section ∇ of $Co(R)$, where the quadruple (R, α, β, z) satisfies the assumptions of 6.6.1 as well as the further conditions: $\alpha_*(\nabla) = D_0$ and $\beta_*(\nabla) = D_1$.

Over LM, this corresponds to the connection over the $\mathbb{C}^*$-bundle $\mathcal{P} \to LM$.

The notion of parallel transport for sheaves of groupoids is used in the geometric study of the first Pontryagin class undertaken in [Br-ML1] [Br-ML2].

The last thing that we will extend from LM to PM is the construction of an action functional. We will only state it for mappings of a square to M.

6.6.4. Proposition. *Let γ, γ' be two paths in M from a point a to a point b, and let $\sigma : [0,1] \times [0,1]$ be a homotopy from γ to γ', such that $\sigma(0,y) = a$ and $\sigma(1,y) = b$. Then there is a natural transformation $S(\sigma)$ from the functor $T_\gamma : C_a \to C_b$ to the functor $T_{\gamma'} : C_a \to C_b$. This natural transformation satisfies a glueing axiom.*

The constructions of this section have an interpretation in terms of Deligne cohomology, as in §6.5. The appropriate Deligne cohomology groups are those of the diagram $PM \underset{ev_1}{\overset{ev_0}{\rightrightarrows}} M$. They are the cohomology groups of some Čech triple complex.

Chapter 7

The Dirac Monopole

7.1. Dirac's construction

We begin by reviewing Dirac's treatment of the magnetic monopole. While we will take a brief look at some of the basic ideas of quantum mechanics, very few details are actually needed for Dirac's elegant derivation. Our treatment of the basics is close to [L-L], with all the physics carefully eliminated.

We start with a particle of mass m whose motion is governed by a potential V. In classical mechanics such a particle has position $\vec{q}(t) = (x(t), y(t), z(t))$ which evolves with time, a velocity vector $\vec{v}(t) = \frac{d\vec{q}(t)}{dt}$ and a *momentum* $\vec{p}(t) = m \cdot \vec{v}(t)$. The potential $V(\vec{q})$ creates a force $\vec{F} = -\vec{\nabla} V$. The motion of the particle is given by Newton's law

$$\frac{\partial \vec{p}}{\partial t} = \vec{F} = -\vec{\nabla} V \qquad (7-1)$$

There is another formulation, in the framework of the hamiltonian formalism, of the equation of motion. In that setting, the phase space is the symplectic manifold $\mathbb{R}^6 = T^*\mathbb{R}^3$, of which a point is denoted $(\vec{q}, \vec{p})$, where $\vec{q} = (x, y, z) \in \mathbb{R}^3$ is the position vector, and $\vec{p} = (p_x, p_y, p_z)$ is the momentum vector. The symplectic form is $d\vec{q} \cdot d\vec{p} = dp_x \wedge dx + dp_y \wedge dy + dp_z \wedge dz$. Typical Poisson brackets of functions are $\{x, p_x\} = 1$, $\{y, p_y\} = 1$, and so on.

The *hamiltonian* H is the function

$$H(\vec{q}, \vec{p}) = \frac{\|\vec{p}\|^2}{2m} + V(\vec{q}) \qquad (7-2)$$

on $\mathbb{R}^6$. The corresponding hamiltonian vector field (in the sense of §2.3) is

$$X_H = \frac{\vec{p}}{m} \cdot \frac{\partial}{\partial \vec{q}} - \vec{\nabla} V \cdot \frac{\partial}{\partial \vec{p}}$$

Hence the equation of motion for the point $x = (\vec{q}, \vec{p})$ of $\mathbb{R}^6$ may be written simply as

$$\frac{\partial x}{\partial t} = X_H(x) \qquad (7-3)$$

which means that the time evolution of $x \in \mathbb{R}^6$ is the flow of the vector field X_H.

If we go from classical to quantum mechanics, the particle is no longer a purely punctual object in 3-space; instead its existence is revealed by a *wave function* $\psi(\vec{q}, t)$ which depends on the spatial variable $\vec{q} \in \mathbb{R}^3$ and on time t. The probability of presence of the particle at the point $\vec{q}$ is proportional to $\|\psi(\vec{q}, t)\|^2$, which is therefore the most relevant physical quantity. One assumes that $\|\psi\|$ has total mass 1, that is, $\int_{\mathbb{R}^3} \|\psi(\vec{q}, t)\|^2 \, dx dy dz = 1$ at all times t, but this is not essential for our purposes. The function $\vec{q} \mapsto \psi(\vec{q}, t)$ (for given t) therefore belongs to some Hilbert space E.

A fundamental tenet of quantum mechanics is that observable quantities (in our case the position and the momentum of the particle, or any function on $\mathbb{R}^6$) are incarnated into operators on the Hilbert space. This is the physical meaning of quantization of observables. If a is an observable, we let $\hat{a}$ be the corresponding operator on E. The assignment $a \mapsto \hat{a}$ has to satisfy some constraints, which relate Poisson brackets of observables to commutators of operators. We state two important constraints of the quantization procedure:

(Q1) If a is a real-valued observable, the operator $\hat{a}$ is hermitian.

(Q2) The operator $\hat{1}$ corresponding to the function 1 is the identity.

(Q3) Let a and b be observables. Then we have

$$\sqrt{-1} \cdot \hbar \cdot \widehat{\{a, b\}} = [\hat{a}, \hat{b}] \qquad (7-4)$$

where $\hbar$ is the *Planck constant*.

We are now ready to state the

7.1.1. Fundamental principle of quantum mechanics: *The equations of motion of quantum mechanics are obtained from the hamiltonian equations of classical mechanics by the following substitutions:*

(1) *One replaces the particle by the wave function $\psi(\vec{q}, t)$,*

(2) *One replaces each hamiltonian vector field ξ_a by the corresponding operator $\frac{1}{\sqrt{-1}\hbar} \cdot \hat{a}$ on E.*

If we apply this to the equation of motion (7-3) for a free particle, we obtain the *Schrödinger equation* in its abstract form

$$\sqrt{-1}\hbar \cdot \frac{\partial \psi}{\partial t} = \hat{H} \cdot \psi. \qquad (7-5)$$

To make this more concrete, we have to determine the operator $\hat{H}$. Since our Hilbert space E is a space of functions of $\vec{q} \in \mathbb{R}^3$, it is natural to associate to the function $V(\vec{q})$ the corresponding multiplication operator (which may of course be unbounded). This choice is compatible with the constraints (Q1) and (Q2). Next we need to quantize a component, say p_x of the momentum. Since we have $\{p_x, x\} = -1$, $\{p_x, y\} = 0$, $\{p_x, z\} = 0$, we obtain from the constraint (Q3) that the operator $\hat{p}_x$ commutes with multiplication by y and by z, and must satisfy the rule

$$[\hat{p}_x, x] = -\sqrt{-1}\hbar \cdot Id.$$

The only natural choice is therefore

$$\hat{p}_x = -\sqrt{-1}\hbar \cdot \frac{\partial}{\partial x}. \qquad (7-6)$$

Similarly we must have

$$\hat{p}_y = -\sqrt{-1}\hbar \cdot \frac{\partial}{\partial y},$$
$$\hat{p}_z = -\sqrt{-1}\hbar \cdot \frac{\partial}{\partial z}. \qquad (7-6')$$

In the same way we obtain the following formula for $\hat{H}$ (waving our hands a bit with respect to the quantization of the square of a function):

$$\hat{H} = -\frac{\hbar^2}{2m} \cdot \Delta + V(\vec{q}) \qquad (7-7)$$

where $\Delta = \dfrac{\partial}{\partial x^2} + \dfrac{\partial}{\partial y^2} + \dfrac{\partial}{\partial z^2}$ is the Laplace operator, and $V(\vec{q})$ is the operator of multiplication by the function V. Therefore the Schrödinger equation for the wave function $\psi(\vec{q}, t)$ of a particle moving in a potential $V(\vec{q})$ is

$$\frac{\partial \psi}{\partial t} = -\sqrt{-1} \cdot \hbar^{-1} \cdot \hat{H} \cdot \psi = \frac{\sqrt{-1}\hbar}{2m} \cdot \Delta\psi - \sqrt{-1} \cdot \hbar^{-1} \cdot V(\vec{q}) \cdot \psi. \qquad (7-8)$$

To solve this equation, the obvious method is to diagonalize the operator $\hat{H}$, which by (Q1) is hermitian. Let ψ be an eigenvector of $\hat{H}$, i.e.,

$$\hat{H} \cdot \psi = E\psi. \qquad (7-9)$$

Then the evolution of ψ in time is simply

$$\psi(\vec{q}, t) = e^{-\sqrt{-1}\hbar^{-1}Et} \cdot \psi(\vec{q}, 0). \qquad (7-10)$$

We will now complicate things somewhat by introducing a *magnetic field* $\vec{B}(\vec{q})$. To compensate, we will forget about the potential $V(\vec{q})$ altogether (its function was in fact only pedagogical). The magnetic field $\vec{B}$ comes together with an electric field $\vec{E}(\vec{q}, t)$, and we have (in vacuum) the following *Maxwell equations*, in which c denotes the speed of light:

$$\overrightarrow{curl}\ \vec{E} = -\frac{1}{c}\frac{\partial \vec{B}}{\partial t}$$

$$div(\vec{B}) = 0 \qquad (7-11)$$

This admits the following solution, parametrized by a vector field $\vec{A}$, the *magnetic potential vector*:

$$\vec{E} = -\frac{1}{c} \cdot \frac{\partial \vec{A}}{\partial t}$$

$$\vec{B} = \overrightarrow{curl}\ \vec{A} \qquad (7-12)$$

The equation of motion of a charged particle of charge e is given by

$$\frac{\partial \vec{p}}{\partial t} = e\vec{E} + \frac{e}{c}\vec{v} \times \vec{B}. \qquad (7-13)$$

This justifies the following formal operation: if we replace the momentum $\vec{p}$ by the modified momentum $\vec{P} = \vec{p} + \frac{e}{c}\vec{A}$, then taking (7-12) into account, we see that (7-13) is transformed into the equation

$$\frac{\partial \vec{P}}{\partial t} = \vec{F} \qquad (7-14)$$

where $\vec{F} = \frac{e}{c} \cdot \vec{v} \times \vec{B}$ is the *Lorentz force* (exerted by the magnetic field $\vec{B}$ on a charged particle moving through it with velocity $\vec{v}$). This is very similar to Newton's equation (7-1).

Thus when we incorporate the magnetic field $\vec{B}$ into the equation governing the wave function $\psi(\vec{q}, t)$ of a charged particle (an electron), we need to choose a magnetic potential $\vec{A}$ such that $\overrightarrow{curl}\ \vec{A} = \vec{B}$, and we replace an operator like $\hat{p}_x$ by

$$\hat{P}_x = \hat{p}_x + \frac{e}{c}A_x = -\sqrt{-1}\hbar \cdot \frac{\partial}{\partial x} + \frac{e}{c}A_x \qquad (7-15)$$

and we do the same for each component of $\vec{p}$. In vector notation we have

$$\hat{P} = -\sqrt{-1}\hbar \cdot \vec{\nabla} + \frac{e}{c}\vec{A} \qquad (7-16)$$

where both sides belong to the tensor product $\mathbb{R}^3 \otimes End(E)$, i.e., are vectors of operators. Then the hamiltonian operator $\hat{H}$ becomes

$$\hat{H} = -\frac{\hbar^2}{2m} \cdot (\vec{\nabla} + \sqrt{-1} \cdot \frac{e}{c\hbar}\vec{A})^2 \qquad (7-17)$$

in analogy with (7-7), but note that we have removed the scalar potential $V(\vec{q})$ from (7-7). This is the hamiltonian operator for an electron moving in a magnetic field.

Note that there is something artificial about the choice of the potential vector $\vec{A}$ with $\overrightarrow{curl}\ \vec{A} = \vec{B}$. So consider another choice of potential vector, necessarily of the form $\vec{A}_1 = \vec{A} - \vec{\nabla} f$, for some function f. We want the equation of motion to be invariant under this change of potential vector. However the vector-valued operator $\hat{P} = -\sqrt{-1}\hbar \cdot \vec{\nabla} + \frac{e}{c}\vec{A}$ is changed to $\hat{P}_1 = -\sqrt{-1}\hbar \cdot \vec{\nabla} + \frac{e}{c}\vec{A}_1 = \hat{P} - \frac{e}{c}\vec{\nabla}f$. It appears now that we must have a different equation of motion, putting the whole theory in jeopardy. The solution to this quandary, as described by Dirac [Di], is that we have the freedom to change the wave function ψ into a new wave function $\psi_1 = e^{\sqrt{-1}\phi} \cdot \psi$ for some real-valued function ϕ. This is reasonable from the physicists' point of view, since it is only the probability amplitude $|\psi|^2$ which has physical significance. The transformation $\psi \mapsto T(\psi) = e^{\sqrt{-1}\phi} \cdot \psi$ is called a *gauge transformation*. A simple computation then shows

7.1.2. Lemma. *Let T be the gauge transformation $T(\psi) = e^{\sqrt{-1}\frac{e}{\hbar c}f} \cdot \psi$. Then we have the following relation between the operators $\hat{P} = -\sqrt{-1}\hbar \cdot \vec{\nabla} + \frac{e}{c}\vec{A}$ and $\hat{P}_1 = \hat{P} - \frac{e}{c}\vec{\nabla}f$:*

$$T \circ \hat{P}(\psi) = \hat{P}_1 \circ T(\psi) \qquad (7-18)$$

for any function $\psi(\vec{q})$. In other words, T conjugates the operator $\hat{P}$ relative to the potential vector $\vec{A}$ into the operator $\hat{P}_1$ relative to the potential vector $\vec{A}_1$.

Now the hamiltonian operator $\hat{H}$ of (7-17) depends on the choice of the potential vector $\vec{A}$. Therefore the new potential vector $\vec{A}_1$ gives rise to a new hamiltonian operator $\hat{H}_1$.

7.1.3. Proposition. *Let ψ be a wave function, and let*

$$\psi_1 = T(\psi) = e^{\sqrt{-1}\frac{e}{\hbar c}f} \cdot \psi$$

be its transform under the gauge transformation T. Then ψ satisfies the wave equation $\hat{H} \cdot \psi = E\psi$ if and only if ψ_1 satisfies the modified wave equation $\hat{H}_1 \cdot \psi_1 = E\psi_1$.

Thus we see that the gauge transformation T transforms eigenvectors of $\hat{H}$ into eigenvectors of $\hat{H}_1$; hence the spectrum of $\hat{H}$ (which gives the "energy levels" of the particle) is independent of the choice of $\vec{A}$.

A consequence of this formalism of gauge transformations is that we must give up the idea that ψ is intrinsically a function of $\vec{q}$, since a gauge transformation would transform ψ into another function $e^{\sqrt{-1}\phi}\psi$. However, note that if ψ and ψ' are non-vanishing wave functions, then the ratio $\frac{\psi}{\psi'}$ remains unchanged under any gauge transformation. This means mathematically that we should view ψ as a section of some hermitian line bundle L. In fact, the first part of Dirac's paper is largely devoted to explaining in a beautiful way the concept of line bundle, which of course he did not have available to him. It is true that any line bundle over $\mathbb{R}^3$ is trivial, but there will be no preferred trivialization. A wave function will then be a smooth section of L. A trivialization of L (compatible with the hermitian structure) is called a *gauge* by physicists. A gauge transformation means a change of trivialization. For each given trivialization, there is a corresponding potential vector.

We must now interpret the operator $\hat{P}$ in terms of this line bundle L. First we return for a moment to a free particle and to the "vector-valued operator" $\hat{p}$ with components given by (7-6). We have $\hat{p} = -\sqrt{-1}\hbar \cdot \vec{\nabla}$. In the language of differential forms, the operator $\hat{p}$ sends a function $\psi(\vec{q})$ to the 1-form $\hat{p} \cdot \psi = -\sqrt{-1}\hbar \cdot d\psi$, where d is exterior differentiation.

If we now incorporate the potential-vector $\vec{A}$, we will get an operator $\hat{P}$ from sections of L to 1-forms with values in L. We have seen such operators in Chapter 2, namely *connections* on the line bundle L. We are consequently forced to the following conclusion, which gives the essential mathematical content of the physical situation. Here we interpret $\overrightarrow{curl}$ as the exterior differentiation, from 1-forms to 2-forms.

7.1.4. Fact: *The operator $\hat{P}$ maps sections of L to 1-forms with values in L. It is equal to*

$$\hat{P} = -\sqrt{-1}\hbar \cdot \nabla \qquad (7-19)$$

where ∇ is a connection on L. In a local trivialization (gauge) of L, ∇ becomes

$$\nabla = d + \sqrt{-1}\frac{e}{\hbar c}A, \qquad (7-20)$$

where A is the 1-form corresponding to $\vec{A}$. The curvature R of the connection

∇ *is equal to* $\sqrt{-1}\frac{e}{\hbar c}dA$ *in a local gauge; hence we have*

$$R = \sqrt{-1}\frac{e}{\hbar c}\vec{B} \qquad (7-21)$$

(if we identify 2-forms with vector-valued functions in the usual way).

So far we have assumed that the magnetic field $\vec{B}$ is defined over the whole euclidean space $\mathbb{R}^3$, in which case there exists a globally defined potential vector, and any two potential vectors differ by the gradient of some function. In terms of the connection on the line bundle L, one can trivialize the bundle, giving ∇ the expression of (7-20). However Dirac explored the idea of a magnetic field $\vec{B}$ defined on $\mathbb{R}^3 \setminus \{0\}$ which has a singularity at the origin. This singularity corresponds to the existence of a *magnetic monopole* localized at the origin. Such an object had not been considered in electromagnetism before Dirac, as everybody assumed that singularities of the electromagnetic field would always occur in pairs.

The magnetic monopole has a *strength* μ, defined as

$$\mu = \frac{1}{4\pi}\int\int_{\Sigma}\vec{B} \times d\sigma,$$

that is, the flux of the magnetic field $\vec{B}$ across a 2-sphere Σ centered at the origin, divided by the surface area of the unit 2-sphere. Since $div(\vec{B}) = 0$, this is independent of the particular surface considered. In terms of the curvature 2-form R, we have

$$\mu = -\sqrt{-1}\frac{c\hbar}{4\pi e} \cdot \int_{\Sigma} R. \qquad (7-22)$$

Now it follows from the quantization condition theorem for the curvature of a connection on a line bundle that the integral $\int_{\Sigma} R$ belongs to $\mathbb{Z}(1) = 2\pi\sqrt{-1} \cdot \mathbb{Z}$; this implies

7.1.5.   Theorem. (*Dirac* [Di]) *The strength μ of a magnetic monopole satisfies the quantization condition*

$$\mu = \frac{c\hbar}{2e} \cdot n \text{ for some } n \in \mathbb{Z} \qquad (7-23)$$

Dirac discusses the case of the radial magnetic field $\vec{B} = \mu \cdot \dfrac{\vec{q}}{\|\vec{q}\|^3}$ and finds that the lowest eigenvalue of the operator $\hat{H}$ is equal to $1/2$. I. Tamm showed that the spectrum of $\hat{H}$ consists of the rational numbers of the form $n^2 + n + \frac{1}{2}$ for $n \in \{0, 1, 2, \ldots, \}$.

7.2. The sheaf of groupoids over S^3

In Dirac's treatment of a monopole located at the origin of $\mathbb{R}^3$, the magnetic field $\vec{B}$ is proportional to the curvature R of some line bundle with connection over $\mathbb{R}^3 \setminus \{0\}$. The quantization condition for the strength of the monopole then follows from the fact that the cohomology class of $\frac{R}{2\pi\sqrt{-1}}$ is integral. In other words, the cohomology group which comes into this quantization condition is $H^2(\mathbb{R}^3 \setminus \{0\})$. It would be more natural to have a topological interpretation which involves the origin itself (where the monopole is supposed to be located) rather than the complementary subset $\mathbb{R}^3 \setminus \{0\}$. We would like to think mathematically of the monopole as some sort of singular object, in the spirit of Dirac's famous delta-function δ_0. To this end, let us reflect on the topological significance of δ_0: First, it is a distribution, that is, a continuous linear form on functions defined by

$$\delta_0(f) = f(0). \tag{7-22}$$

We may view δ_0 as a generalized 3-form on $\mathbb{R}^3$ supported at 0; the third cohomology group with support in 0, $H^3_{\{0\}}(\mathbb{R}^3, \mathbb{R}) = H^3(\mathbb{R}^3, \mathbb{R}^3 \setminus \{0\})$, may be computed as the cohomology group of the complex $D^\bullet_{\{0\}}(\mathbb{R}^3)$ of generalized differential forms with support in 0. This cohomology group is one-dimensional, and since $\delta_0(1) = 1$, the cohomology class $[\delta_0]$ is the canonical generator.

We have the exact sequence

$$H^2(\mathbb{R}^3) = 0 \to H^2(\mathbb{R}^3 \setminus \{0\}) \to H^3_{\{0\}}(\mathbb{R}^3) \to H^3(\mathbb{R}^3) = 0 \tag{7-23}$$

and the canonical generator of $H^2(\mathbb{R}^3 \setminus \{0\})$ maps to the canonical generator of $H^3_{\{0\}}(\mathbb{R}^3)$.

We present the point of view here that the topological analog of the monopole is the generator of $H^3_{\{0\}}(\mathbb{R}^3, \mathbb{Z})$, rather than that of $H^2(\mathbb{R}^3 \setminus \{0\}, \mathbb{Z})$. This may seem like a byzantine distinction, since the two groups are canonically isomorphic, however the main point is that we are talking about three-dimensional cohomology, and not two-dimensional cohomology, so that the geometric interpretation is quite different. An immediate advantage is that we can now think of moving the monopole around inside $\mathbb{R}^3$, or better, inside its one-point compactification $S^3 = \mathbb{R}^3 \cup \{\infty\}$. In fact the canonical map

$$H^3_{\{0\}}(\mathbb{R}^3) \xrightarrow{\sim} H^3_{\{0\}}(S^3) \xrightarrow{\sim} H^3(S^3) \tag{7-24}$$

is an isomorphism, so we do not have to worry any longer about cohomology with support. This will be crucial to the relation of the monopole with secondary characteristic classes, discussed in §3; this relation, however, could not even be formulated without shifting the geometric interpretation to the whole 3-sphere.

We now come to the geometric object realizing the generator of $H^3(S^3, \mathbb{Z})$. The construction that we give was suggested to us by Deligne. We will describe a Dixmier-Douady sheaf of groupoids over S^3, which will be obtained by some canonical process from the monopole itself. The monopole is given by a line bundle L with connection ∇ over the open set $\mathbb{R}^3 \setminus \{0\}$ of S^3. The Chern class of this line bundle is equal to $2\pi\sqrt{-1} \cdot u$, for the canonical generator u of $H^2(\mathbb{R}^3 \setminus \{0\}, \mathbb{Z})$. We then define a sheaf of groupoids as follows. Cover S^3 by the contractible open sets $U_0 = \mathbb{R}^3$ and $U_\infty = S^3 \setminus \{0\}$, with intersection $\mathbb{R}^3 \setminus \{0\}$. Then we set

7.2.1. Definition and Proposition. *Let $\mathcal{C}$ be the sheaf of groupoids defined as follows. For a local homeomorphism $f : X \to S^3$, let $X_0 = f^{-1}(U_0)$ and $X_\infty = f^{-1}(U_\infty)$. An object of $\mathcal{C}(X)$ is a triple $(\mathcal{L}_0, \mathcal{L}_\infty, \phi)$, where*

(1) $\mathcal{L}_0$ *is a line bundle over X_0.*

(2) $\mathcal{L}_\infty$ *is a line bundle over X_∞.*

(3) $\phi : (\mathcal{L}_0)_{/X_{0,\infty}} \xrightarrow{\sim} (\mathcal{L}_\infty)_{/X_{0,\infty}} \otimes f^*L$ *is an isomorphism of line bundles over $X_{0,\infty} = f^{-1}(U_{0,\infty})$.*

A morphism $\psi : (\mathcal{L}_0^1, \mathcal{L}_\infty^1, \phi^1) \to (\mathcal{L}_0^2, \mathcal{L}_\infty^2, \phi^2)$ is a pair (ψ_0, ψ_∞), where

(a) $\psi_0 : \mathcal{L}_0^1 \xrightarrow{\sim} \mathcal{L}_0^2$ *is an isomorphism of line bundles over X_0.*

(b) $\psi_\infty : \mathcal{L}_\infty^1 \xrightarrow{\sim} \mathcal{L}_\infty^2$ *is an isomorphism of line bundles over X_∞.*

The pair (ψ_0, ψ_∞) must satisfy the following condition

(c) *the following diagram is commutative*

$$
\begin{array}{ccc}
(\mathcal{L}_0^1)_{/X_{0,\infty}} & \xrightarrow{\phi_1} & (\mathcal{L}_\infty^1)_{/X_{0,\infty}} \otimes f^*L \\
\downarrow{\psi_0} & & \downarrow{\psi_\infty \otimes Id} \\
(\mathcal{L}_0^2)_{/X_{0,\infty}} & \xrightarrow{\phi_2} & (\mathcal{L}_\infty^2)_{/X_{0,\infty}} \otimes f^*L
\end{array}
$$

Then $\mathcal{C}$ is a Dixmier-Douady sheaf of groupoids over S^3.

Proof. It is clear that $\mathcal{C}$ is a sheaf of groupoids. Thus we only need to verify properties (G1), (G2) and (G3) of §5.2. For an object $(\mathcal{L}_0, \mathcal{L}_\infty, \phi)$ of $\mathcal{C}(X)$, as in the statement, an automorphism of this object is a pair (ψ_0, ψ_∞), where $\psi_0 \in \Gamma(X_0, \underline{\mathbb{C}}^*_{X_0})$, $\psi_\infty \in \Gamma(X_\infty, \underline{\mathbb{C}}^*_{X_\infty})$, and ψ_0, ψ_∞ satisfy $\phi\psi_0 = \psi_\infty\phi$, hence $\psi_0 = \psi_\infty$ on the intersection $X_{0,\infty}$. This means that ψ_0 and ψ_∞ glue together to give a smooth $\mathbb{C}^*$-valued function on X. This proves (G1). (G2) and (G3) will follow once we show that the restriction of C to the open sets U_0 and U_∞ is equivalent to the gerbe of all $\underline{\mathbb{C}}^*$-torsors. Over U_0, we construct an equivalence of sheaves of groupoids F from the sheaf of groupoids of line bundles to C. If $f : X \to U_0$ is a local homeomorphism, and $\mathcal{L}$ is a line bundle over X, we set $F(\mathcal{L}) = (\mathcal{L}, \mathcal{L} \otimes f^* L^{\otimes -1}, \text{can})$, where

$$\text{can} : \mathcal{L} \overset{\sim}{\to} (\mathcal{L} \otimes f^* L^{\otimes -1}) \otimes f^* L$$

is the canonical isomorphism of line bundles over $X_\infty = X_{0,\infty}$. There exists a similar equivalence between the sheaf of groupoids of line bundles and the restriction of C to U_∞. ∎

The sheaf of groupoids C is intrinsically attached to the line bundle with connection (L, ∇) over $U_{0,\infty} = \mathbb{R}^3 \setminus \{0\}$. We make the following observation

7.2.2. Proposition. *The sheaf of groupoids over S^3 has no global object.*

Proof. We will show that the existence of an object $(\mathcal{L}_0, \mathcal{L}_\infty, \phi)$ of $C(S^3)$ leads to a contradiction. $\mathcal{L}_0$ is a line bundle over $\mathbb{R}^3$, hence is trivial. $\mathcal{L}_\infty$ is a line bundle over $S^3 \setminus \{0\} \overset{\sim}{\to} \mathbb{R}^3$, so is trivial. On the other hand ϕ is an isomorphism between the trivial line bundle $(\mathcal{L}_0)_{/\mathbb{R}^3 \setminus \{0\}}$ over $\mathbb{R}^3 \setminus \{0\}$ and the non-trivial line bundle $\mathcal{L}_\infty \otimes L$; this is absurd. ∎

We will now proceed to put a connective structure and curving on the sheaf of groupoids C, and to compute the 3-curvature.

7.2.3. Definition and Proposition. *For a local homeomorphism $f : X \to S^3$ and for an object $P = (\mathcal{L}_0, \mathcal{L}_\infty, \phi)$ of $C(X)$, as in 7.2.3, let $Co(P)$ consist of pairs (D_0, D_∞), where*

(a) *D_0 is a connection on $\mathcal{L}_0$, D_∞ is a connection on $\mathcal{L}_\infty$.*

(b) *The isomorphism ϕ is compatible with the connections, i.e.,*

$$\phi_*(D_0) = D_\infty + \nabla.$$

Let $\alpha \in \underline{A}^1(X)_{\mathbb{C}}$ act on $Co(P)$ by $\alpha + (D_0, D_\infty) = (D_0 + \alpha, D_\infty + \alpha)$. Then $Co(P)$ is a torsor under $\underline{A}^1_{X,\mathbb{C}}$, and the assignment $P \mapsto Co(P)$ defines a connective structure on C.

Proof. Clearly we have a group action of the sheaf $\underline{A}^1_{X,\mathbb{C}}$ on the sheaf $Co(P)$.

Locally $Co(P)$ is not empty: for instance, if V is an open subset of X_0 over which $\mathcal{L}_0$ has a connection D_0, we can pick D_∞ to be the unique connection on $(\mathcal{L}_\infty)_{/V}$ which makes ϕ compatible with the connections. If (D_0, D_∞) and (D_0', D_∞') are two sections of $Co(P)$, then $\alpha_0 = D_0' - D_0$ is a 1-form on X_0, and $\alpha_\infty = D_\infty' - D_\infty$ is a 1-form on X_∞. The condition of compatibility of the isomorphism ϕ with both pairs of connections implies that α_0 and α_∞ have the same restriction to $X_{0,\infty}$; so they glue together to give a 1-form α over X, and we have $(D_0, D_\infty) + \alpha = (D_0', D_\infty')$. The rest of the proof is easy. ∎

In order to find a curving, we need to make the auxiliary choice of a partition of unity (f_0, f_∞) of S^3 subordinate to the open covering (U_0, U_∞) of S^3. Then we set

7.2.4. Definition and Proposition. *For $f : X \to S^3$ a local homeomorphism, for $P = (\mathcal{L}_0, \mathcal{L}_\infty, \phi)$ and for $(D_0, D_\infty) \in Co(P)$, as in 7.2.3, let $K(D_0, D_\infty)$ be the complex-valued 2-form on X defined by*

$$K(D_0, D_\infty) = f_0 \cdot K_0 + f_\infty \cdot D_\infty \qquad (7-25)$$

where K_0 is the curvature of D_0 and K_∞ is the curvature of D_∞. Then the assignment $(D_0, D_\infty) \mapsto K(D_0, D_\infty)$ gives a curving of the connective structure of Proposition 7.2.3. The 3-curvature Ω has support contained in $\mathbb{R}^3 \setminus \{0\}$ and is equal to $df_0 \wedge R$ on $\mathbb{R}^3 \setminus \{0\}$. We have

$$\int_{S^3} \Omega = -2\pi\sqrt{-1} \qquad (7-26)$$

Hence the class of the sheaf of groupoids C in $H^3(S^3, \mathbb{Z}(1))$ is equal to $-2\pi\sqrt{-1}$ times the canonical generator of $H^3(S^3, \mathbb{Z})$.

Proof. For a 1-form α on X, we have

$$K(\alpha + (D_0, D_\infty)) = K(D_0 + \alpha, D_\infty + \alpha) = K(D_0, D_\infty) + (f_0 + f_\infty) \cdot d\alpha$$
$$= K(D_0, D_\infty) + d\alpha.$$

This shows that we have a curving. We now compute the 3-curvature Ω. Over the open set $S^3 \setminus \{0\}$, we can take $\mathcal{L}_\infty$ to be the trivial line bundle and $D_\infty = d$ to be the trivial connection. We take $\mathcal{L}_0 = L$ and $D_0 = \nabla$. Then the 3-curvature reduces to $\Omega = d(f_0 R) = df_0 \wedge R$, since R is a closed 2-form. Hence the restriction of Ω to $S^3 \setminus \{0\}$ has support contained in $\mathbb{R}^3 \setminus \{0\}$. A similar reasoning shows that the restriction of Ω to $\mathbb{R}^3$ is equal to $df_\infty \wedge (-R) = df_0 \wedge R$. This proves the formula for Ω, which has support contained in $\mathbb{R}^3 \setminus \{0\}$. To compute the integral $\int_{S^3} \Omega$, we note that this integral gives the class of the sheaf of groupoids in $\tilde{H}^3(S^3, \mathbb{Z}(1))$. This class is, of course, independent of the chosen partition of unity. We may therefore assume that f_0 is a function of r^2, where r is the distance of a point of $\mathbb{R}^3$ to the origin. Now, using spherical coordinates, we have

$$\int_{S^3} \Omega = \int_0^\infty df_0(r^2) \cdot \int_{S_r} R$$

where S_r is the sphere of radius r. Since the integral $\int_{S_r} R$ is equal to $2\pi\sqrt{-1}$, we find

$$\int_{S^3} \Omega = 2\pi\sqrt{-1} \cdot \int_0^\infty df_0(r^2) = -2\pi\sqrt{-1}. \qquad \blacksquare$$

We wish however to have a curving with 3-curvature equal to $-2\pi\sqrt{-1} \cdot \nu$, for ν the unique $SO(4)$-invariant volume form on S^3 with volume 1. Then we need to choose a 2-form β with $d\beta = -2\pi\sqrt{-1}\cdot\nu - \Omega$. The curving defined by $K'(D_0, D_\infty) = K(D_0, D_\infty) + \beta$ has 3-curvature equal to $-2\pi\sqrt{-1} \cdot \nu$.

If we were working with the cohomology group $H^3_0(\mathbb{R}^3, \mathbb{Z})$ instead of $H^3(S^3, \mathbb{Z})$, we would have to develop the concept of a sheaf of groupoids on a manifold M with support in a closed subset Y; this would be a pair $(\mathcal{C}, P)$, where $\mathcal{C}$ is a sheaf of groupoids, and P is an object of $\mathcal{C}$ over $M \setminus Y$. Then the group of equivalence classes of sheaves of groupoids with support in Y identifies with $H^3_Y(M, \mathbb{Z}(1))$ and the monopole corresponds to the trivial sheaf of groupoids of line bundles over $\mathbb{R}^3$, together with the line bundle L over $\mathbb{R}^3 \setminus \{0\}$. This describes a geometric object which has support at the origin. But our purpose in the next section is to study the effect of $SU(2)$ on the sheaf of groupoids; this in itself would force us to work with a sheaf of groupoids over S^3.

7.3. Obstruction to $SU(2)$-equivariance

In §2 we used the Dirac monopole to construct a sheaf of groupoids $\mathcal{C}$ over S^3 with connective structure and curving, whose 3-curvature is equal to $-2\pi\sqrt{-1} \cdot \nu$, where ν is the standard volume form with volume 1. Recall that S^3 identifies with the group $SU(2)$. For $g \in S^3$, define $L_g : S^3 \to S^3$ to be left translation by g; the sheaf of groupoids $L_g^* \mathcal{C}$ is then equivalent to $\mathcal{C}$, since a sheaf of groupoids with connective structure and curving over S^3 is determined up to equivalence by its 3-curvature. We wish to analyze the obstruction to constructing a $SU(2) = S^3$-equivariant sheaf of groupoids over S^3. We will discuss this notion only when the group $SU(2)$ is viewed as a discrete group.

7.3.1.   Definition. *Let the discrete group G act on the manifold M by diffeomorphisms. A G-equivariant sheaf of groupoids over M is a sheaf of groupoids $\mathcal{C}$ over M, together with the following structures:*

(a) *for each $g \in G$, an equivalence $\phi_g : g_* \mathcal{C} \to \mathcal{C}$ of sheaves of groupoids.*

(b) *For each pair $(g_1, g_2) \in G \times G$, an invertible natural transformation*

$$\psi_{g_1, g_2} : \phi_{g_1 g_2} \theta_{g_1, g_2} \to \phi_{g_1} \phi_{g_2}$$

between functors from $g_{1*} g_{2*} \mathcal{C}$ to $\mathcal{C}$. Here θ_{g_1, g_2} denotes the natural transformation from $g_{1*} g_{2*}$ to $(g_1 g_2)_*$, as discussed in §5.1.

The following associativity condition should hold: For every triple (g_1, g_2, g_3) of elements of G, we should have the following equality of natural transformations from $\phi_{g_1} \phi_{g_2} \phi_{g_3}$ to $\phi_{g_1 g_2 g_3}$.

$$\psi_{g_1, g_2} \psi_{g_1 g_2, g_3} = \psi_{g_2, g_3} \psi_{g_1, g_2 g_3}. \qquad (7-27)$$

One defines similarly the notion of an equivariant sheaf of groupoids with connective structure and curving. The equivalences ϕ_g must be equivalences of sheaves of groupoids with connective structure and curving, and the natural transformations ψ_{g_1, g_2} must be understood in the same manner. We will describe the obstruction to making the sheaf of groupoids $\mathcal{C}$ over S^3 equivariant under $S^3 = SU(2)$ viewed as a discrete group. Recall that we discussed in §2.4 the obstruction to making a line bundle with connection equivariant under a group H of symplectomorphisms. The obstruction is a class in $H^2(H, \mathbb{C}^*)$ which corresponds to a central extension of H.

In the case of the sheaf of groupoids $\mathcal{C}$, the obstruction will be a class in $H^3(SU(2), \mathbb{C}^*)$. In fact, we can even replace $SU(2)$ with the larger group $Diff(S^3, \nu)$ of volume preserving diffeomorphisms of S^3. This is explained as follows. For each $g \in G$, choose ϕ_g as in 7.3.1 (a). Then for each pair (g_1, g_2) we choose ψ_{g_1, g_2} as in 7.3.1 (b). If equation (7-27) holds, we have an equivariant sheaf of groupoids. However there is no reason for this be true. Since both sides of (7-27) are horizontal sections of the same $\mathbb{C}^*$-torsor with connection, namely, the torsor of natural transformations between $\phi_{g_1} \phi_{g_2} \phi_{g_3}$ and $\phi_{g_1 g_2 g_3}$, we have

$$\psi_{g_1, g_2} \psi_{g_1 g_2, g_3} = f(g_1, g_2, g_3) \cdot \psi_{g_2, g_3} \psi_{g_1, g_2 g_3} \qquad (7-28)$$

for a unique number $f(g_1, g_2, g_3) \in \mathbb{C}^*$.

Then we have the following abstract result

7.3.2. Theorem. *The function* $(g_1, g_2, g_3) \mapsto f(g_1, g_2, g_3)$ *is a 3-cocycle of* G *with values in* $\mathbb{C}^*$, *that is, we have*

$$f(g_1, g_2, g_3) \cdot f(g_0 g_1, g_2, g_3)^{-1} \cdot f(g_0, g_1 g_2, g_3) \cdot$$
$$\cdot f(g_0, g_1, g_2 g_3)^{-1} \cdot f(g_0, g_1, g_2) = 1 \qquad (7-29)$$

for any $(g_0, g_1, g_2, g_3) \in G^4$. *The cohomology class of* f *is independent of all the above choices. It is trivial if and only if the sheaf of groupoids with connective structure and curving* $\mathcal{C}$ *can be made equivariant under* G.

We also note that $f(g_1, g_2, g_3)$ is associated to a non-commutative formal tetrahedron

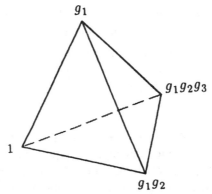

The abstract tetrahedron

In [Br-ML1], [Br-ML2] such formal tetrahedra are studied by differential-geometric means and replaced by concrete tetrahedra in the group in order to obtain a formula for a degree 4 characteristic class. Here we will follow a somewhat different path, perhaps better adapted to group cohomology.

We work now with cocycles with values in $\mathbb{T}$ rather than in $\mathbb{C}^*$. In fact, if we consider $\mathcal{C}$ as a unitary sheaf of groupoids, we obtain a degree 3 cocycle with values in $\mathbb{T}$ instead of in $\mathbb{C}^*$. From now on, we will work with this $\mathbb{T}$-valued 3-cocycle on G. So the obstruction to making $\mathcal{C}$ equivariant is a class in $H^3(G, \mathbb{T})$. To make this more concrete, we will use a variant of the approach of Sinh [Si] [Ul] to degree 3 cohomology using gr-categories. We use the terminology *groupoid with tensor product* instead of gr-category. Then following [Ul] [Bre] a groupoid with tensor product $\mathcal{C}$ is a category equipped with a composition law, which is a functor $\otimes : \mathcal{C} \times \mathcal{C} \to \mathcal{C}$, denoted by $(X, Y) \mapsto X \otimes Y$, together with an *associativity constraint*, which is a functorial isomorphism

$$c_{X,Y,Z} : X \otimes (Y \otimes Z) \xrightarrow{\sim} (X \otimes Y) \otimes Z, \qquad (7-30)$$

and a unit object I for which there are given functorial isomorphisms

$$g_X : I \otimes X \xrightarrow{\sim} X, \quad d_X : X \otimes I \xrightarrow{\sim} X. \qquad (7-31)$$

The following diagrams are required to be commutative:

$$(X \otimes I) \otimes Y \longrightarrow X \otimes (I \otimes Y)$$

$$X \otimes Y \qquad (7-32)$$

and

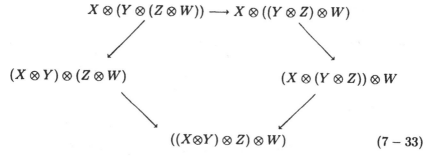

$$(7-33)$$

Diagram (7-33) is the pentagon of MacLane.

It is required that every object X admits an "inverse" X^* for which there is an isomorphism $\epsilon_X : X \otimes X^* \xrightarrow{\sim} I$. There is also, therefore, a well-defined isomorphism: $\eta_X : I \xrightarrow{\sim} X^* \otimes X$ (cf. [Bre]).

In a groupoid with tensor product, the set $\pi_0(C)$ of isomorphism classes of objects is a group (the *Picard group*) under tensor product. If $\pi_1(C)$ denotes the group $Aut_C(I)$, the tensor product map induces a group homomorphism $\pi_1(C) \times \pi_1(C) \to \pi_1(C)$, which is group multiplication. It follows that $\pi_1(C)$ is abelian. For $a \in \pi_1(C)$, and any object X of C, we denote also by a the automorphism of $X \simeq X \otimes I$ induced by $1 \otimes a$. In this way $\pi_1(C)$ acts on any object of C. The group $\pi_0(C)$ operates on $\pi_1(C)$ as follows: for X an object of C and γ an automorphism of I, let $[X] \cdot \gamma$ denote the automorphism of $I \simeq X \otimes (I \otimes X^*)$ given by $Id_X \otimes (\gamma \otimes Id_{X^*})$. A theorem of Sinh says that there is a canonical class in the cohomology group $H^3(\pi_0(C), \pi_1(C))$ which represents the obstruction to finding an assignment $g \in \pi_0(C) \mapsto P_g \in Ob(C)$, together with isomorphisms $\psi_{g_1, g_2} : P_{g_1 g_2} \simeq P_{g_1} \otimes P_{g_2}$ for all g_1, g_2 in $\pi_0(C)$, in such a way that, for three elements of this group, a natural associativity diagram of the type (c) in 7.3.1 commutes. This obstruction is found as follows. One chooses objects P_g and isomorphisms ψ_{g_1, g_2}. Then we have, with $c = c_{P_{g_1}, P_{g_2}, P_{g_3}}$, the equality

$$c \circ (Id \otimes \psi_{g_2, g_3}) \circ \psi_{g_1, g_2 g_3} = f(g_1, g_2, g_3) \circ (\psi_{g_1, g_2} \otimes Id) \circ \psi_{g_1 g_2, g_3} \quad (7-34)$$

for a unique $f(g_1, g_2, g_3) \in \pi_1(C)$. The cohomology class of f in $H^3(\pi_0(C), \pi_1(C))$ is independent of all choices, and is called the *Sinh invariant* of C.

We need the notion of equivalence between groupoids with tensor product C_1 and C_2, with given π_0 and π_1. That is, we need a functor $F : C_1 \to C_2$ together with functorial isomorphisms $\lambda : F(X \otimes Y) \xrightarrow{\sim} F(X) \otimes F(Y)$ and $\mu : F(I) \xrightarrow{\sim} I$, which are compatible with the associativity isomorphisms and the identity isomorphisms in C_1 and C_2; it is also required that F induces the identity maps on π_0 and π_1.

We have the following theorem:

7.3.3. Theorem. (*Sinh* [Si], *see* [Bre] [Ul]) *By attaching to a groupoid with tensor product its Sinh invariant, we obtain an isomorphism between the group of equivalence classes of groupoids with tensor product C, for which $\pi_0(C) = G$ and $\pi_1(C) = M$, with given action of G on M, and the cohomology group $H^3(G, M)$.*

The similarity with 7.3.1 is no accident. Indeed there is a groupoid with tensor product whose objects are pairs (g, ϕ), where $\phi \in G$ and $\phi : g_* C \to C$ is an equivalence of sheaves of groupoids. Morphisms are natural transformations between such equivalences, and the tensor product is given by composing equivalences. This, however, is just a way of rephrasing the constructions which led us to the cohomology class in $H^3(G, \mathbb{C}^*)$. It is much more interesting to find an equivalent groupoid with tensor product involving only differential geometry without the sheaf of groupoids C. First we have a natural groupoid associated to the action of G on S^3. The objects are pairs (g, γ), where $g \in G$, and γ is a path from 1 to $g \cdot 1$. A morphism from (g, γ_1) to (g, γ_2) is a homotopy class of maps $\sigma : [0,1] \times [0,1] \to S^3$ such that $\sigma(0, y) = 1$, $\sigma(1, y) = g \cdot 1$, $\sigma(x, 0) = \gamma_1(x)$ and $\sigma(x, 1) = \gamma_2(x)$. The composition of morphisms is given by vertical juxtaposition of squares, and the tensor product by horizontal juxtaposition of squares.

The groupoid C that we want is a sort of "quantization" of the groupoid of paths in S^3: the objects are quantizations of paths and the arrows are quantizations of squares. This is somewhat reminiscent of the quantization of symplectic groupoids discussed in [W-X]. To give the precise construction, we let the objects of C be pairs (g, γ), where $g \in G$ and $\gamma : [0, 1] \to S^3$ is a piecewise-smooth path from 1 to $g \cdot 1$. There will only be morphisms in C from a pair (g, γ_1) to a pair (g, γ_2). The set $Hom_C((g, \gamma_1), (g, \gamma_2))$ will then be the fiber at the loop $\gamma_1 * (\gamma_2)^{-1}$ of the $\mathbb{T}$-bundle $\mathcal{P}$ over LS^3. Composition of arrows is defined as follows: For $v \in \mathcal{P}_{\gamma_1 * \gamma_2^{-1}}$ and $w \in \mathcal{P}_{\gamma_2 * \gamma_3^{-1}}$, note that, by Proposition 6.2.5, we have an isomorphism $\alpha : \mathcal{P}_{\gamma_1 * \gamma_2^{-1}} \times^{\mathbb{T}} \mathcal{P}_{\gamma_2 * \gamma_3^{-1}} \simeq \mathcal{P}_{\gamma_1 * \gamma_3^{-1}}$; the composition $w \cdot v$ is then defined to be $\alpha(v \times w)$. It follows from the compatibility of the isomorphism m with composition of arrows that this composition law is associative; it is also clear that it is given by a smooth map. The category C is obviously a groupoid.

We introduce another equivalent description of the set $Mor(C)$ of morphisms of C, as consisting of equivalence classes of triples (g, σ, λ), where $g \in G$, $\sigma : [0, 1] \times [0, 1] \to S^3$ is a smooth map, with $\sigma(0, y) = 1$ and $\sigma(1, y) = g \cdot 1$, and $\lambda \in \mathbb{T}$. The equivalence relation is as follows. If σ and σ' have the same restriction to the boundary of the square $[0, 1] \times [0, 1]$, we may glue two copies of the square to get a compact oriented surface Σ, and obtain from σ and σ' a smooth mapping $\Phi = \sigma \sqcup \sigma' : \Sigma \to S^3$. There is, according to §6.2, a number $S(\Phi)$ associated to Φ. This is defined purely in terms of the 3-form ν in view of equation (6-10). Then we identify (g, σ, λ) with $(g, \sigma', \lambda \cdot S(\Phi))$. The source (or beginning) of the arrow (g, σ, λ) is $(g, \sigma(*, 0))$, and its target is $(g, \sigma(*, 1))$. The composition of arrows is obtained by *vertical* composition of mappings from squares. More precisely, the arrows $(g, \sigma, \lambda) \circ (g, \sigma', \lambda')$ are composable if and only if $\sigma'(x, 1) = \sigma(x, 0)$. In that case, their composition is the class of the triple $(g, \sigma' *_v \sigma, \lambda \cdot \lambda')$, where the map $\sigma' *_v \sigma : [0, 1] \times [0, 1] \to S^3$ is defined as usual by

$$\sigma' *_v \sigma(x, y) = \left\{ \begin{array}{ll} \sigma'(x, 2y) & \text{if } 0 \leq y \leq 1/2 \\ \sigma(x, 2y - 1) & \text{if } 1/2 \leq y \leq 1 \end{array} \right\}.$$

The link with the previous description of $Mor(C)$ is obtained by associating with the triple (g, σ, λ) the arrow $\lambda \cdot T_\sigma$ from $(g, \sigma(*, 0))$ to $(g, \sigma(*, 1))$, where $T_\sigma : \mathcal{P}_{\sigma(*, 0)} \simeq \mathcal{P}_{\sigma(*, 1)}$ is the isomorphism of parallel transport associated with σ as in §6.6 (taking into account the fact that the vertical sides of the square are being mapped to a point).

Using this description of arrows in C, we define a tensor product operation in the groupoid C. The tensor product $(g, \gamma) \otimes (h, \delta)$ of two objects is the pair $(g \cdot h, \gamma * (g \cdot \delta))$. Here $g \cdot \delta$ is the transform by the diffeomorphism g of the path δ; so it is a path from $g \cdot 1$ to $gh(1)$.

We next describe the tensor product of two arrows described as equivalence classes of (g, σ, λ) and (g', σ', λ'). Their tensor product is the arrow $(g \cdot g', \sigma *_h (g \cdot \sigma'), \lambda \cdot \lambda')$, where $*_h$ denotes *horizontal* composition of mappings from the square. The fact that the tensor product is a functor reflects the relation $(\sigma_1 *_v \sigma_2) *_h (\sigma_3 *_v \sigma_4) = (\sigma_1 *_h \sigma_3) *_v (\sigma_2 *_h \sigma_4)$ between the horizontal and the vertical compositions. One first has to show that the tensor product operation is compatible with the equivalence relation on triples. Using the glueing axiom for the action functional of a mapping of a closed surface into S^3, as explained in §2, one sees that if σ and σ_1 (resp. σ' and σ_1') have the same restriction to the boundary, then

$$S((\sigma *_h (g \cdot \sigma')) \sqcup (\sigma_1 *_h (g \cdot \sigma_1')) = S(\sigma \sqcup \sigma_1) \times S(g \cdot (\sigma' \sqcup \sigma_1')).$$

We now observe the following:

7.3.4. Lemma. *Let g be a volume-preserving diffeomorphism of S^3. For $\Phi : \Sigma \to S^3$ a smooth mapping, with Σ a closed oriented surface of genus zero, we have $S(\Phi) = S(g \circ \Phi)$, for any $g \in G$.*

The lemma implies that the tensor product is well-defined on equivalence classes of triples (g, σ, λ).

We next need an associativity isomorphism

$$c : (g_1, \gamma_1) \otimes ((g_2, \gamma_2) \otimes (g_3, \gamma_3)) \simeq ((g_1, \gamma_1) \otimes (g_2, \gamma_2)) \otimes (g_3, \gamma_3).$$

To define the isomorphism, we describe a mapping $\sigma : [0, 1] \times [0, 1] \to S^3$, such that the vertical sides are mapped to 1 (resp. $g_1 g_2 g_3(1)$), and the horizontal sides are the compositions $\gamma_1 *_h (g_1 \cdot (\gamma_2 *_h g_2 \cdot \gamma_3))$, resp. $(\gamma_1 *_h g_1 \cdot \gamma_2) *_h (g_1 g_2 \cdot \gamma_3)$. There exists a canonical such mapping σ, obtained by the process which shows the associativity of the product on the fundamental group. The isomorphism is then the class of the triple $(g_1 g_2 g_3, \sigma, 1)$. This is a functorial isomorphism in each of the three variables. The commutativity of the pentagonal diagram (7-33) is a consequence of the fact that the diffeomorphisms of S^1 which realize the essential associativity of composition of loops make a similar pentagonal diagram commute up to homotopy.

The identity object I for the tensor product is the pair $(1, 1)$, where 1 is the constant path. The isomorphism $(1, 1) \otimes (g, \gamma) \simeq (g, \gamma)$ corresponds to the square which describes the homotopy between a path and its composition with a constant path. There is an inverse to (g, γ), namely, $(g^{-1}, g^{-1} \cdot \gamma^{-1})$.

Thus we have found a groupoid with tensor product C. It is clear that there is an arrow between (g, γ) and (g', γ') if and only if $g = g'$, so we have $\pi_0(C) = G$. The group $\pi_1(C)$ is the group of automorphisms of the object $(1,1)$; any such automorphism is equivalent to a triple $(1,1,\lambda)$ for some unique $\lambda \in \mathbb{T}$. Hence $\pi_1(C) = \mathbb{T}$. Since the action of G on $\mathbb{T}$ is trivial, we have constructed a class in the group cohomology $H^3(G, \mathbb{T})$.

The following proposition shows that C is a concrete model for the groupoid associated to the problem of making C equivariant under G.

7.3.6. Proposition. *The obstruction to making the sheaf of groupoids C on S^3 equivariant under G is the Sinh invariant of the groupoid C with tensor product.*

Proof. We will construct an equivalence t of groupoids with tensor product from C to the abstract groupoid associated with C and the G-action on S^3. For this purpose, to any object $P_g = (g, \gamma)$ of C, we associate an equivalence $t(g, \gamma) : g_*C \to C$ of sheaves of groupoids with connective structure and curving. We note that such an equivalence is unique up to tensor product with a flat line bundle, and a flat line bundle is specified by its fiber at one point. Therefore there is a unique such equivalence $t(g, \gamma) : g_*C \to C$, whose value at the base point 1 is given by parallel transport along the path γ of S^3. A morphism $\sigma : (g, \gamma_1) \to (g, \gamma_2)$ is a class of maps $\sigma : [0,1] \times [0,1] \to S^3$. $t(\sigma)$ is, therefore, the natural transformation between $t(g, \gamma_1)$ and $t(g, \gamma_2)$, whose value at 1 is given by the generalized action functional of §6.6. The functor t is seen to be compatible with tensor products. ∎

The groupoid with tensor product C has the advantage of being very concrete. We can in fact compute easily its Sinh invariant. We will arrange things such that the 3-cocycle is differentiable in a neighborhood of 1 in G^3. Choosing a function $p : U \to PU$ such that $p(g)(1) = g$ (so $p(g)$ is a path from $1 \in G$ to g), and p is smooth in a neighborhood U of 1, we may then define the object $P_g = (g, p(g) \cdot 1)$ of C.

Next, we define an isomorphism $\psi(g_1, g_2)$ from $P_{g_1 \cdot g_2}$ to $P_{g_1} \otimes P_{g_2}$, for g_1, g_2 in U, which, in a neighborhood of 1, depends smoothly on g_1 and g_2. We do this by choosing a map $\sigma_{g_1, g_2} : [0,1] \times [0,1] \to G$ such that $\sigma_{g_1, g_2}(*, 0) = p(g_1 g_2)$ and $\sigma_{g_1, g_2}(*, 1) = p(g_1) * (g_1 \cdot p(g_2))$, and σ_{g_1, g_2} is constant on the vertical sides of the square. Putting $\psi(g_1, g_2) = (g_1 g_2, \sigma_{g_1, g_2} \cdot 1)$, we then have the $\mathbb{T}$-valued cocycle f of Sinh, smooth on $U \times U \times U$, which measures the fact that $\psi(g_1, g_2)$ need not commute with associativity isomorphisms.

In the following lemma, we will use the standard homeomorphism between a square with the two vertical sides collapsed, and the standard 2-simplex. Hence we view σ_{g_1, g_2} as a map from Δ_2 to G.

7.3.7. Lemma. *For (g_1, g_2, g_3) in U, we have*

$$f(g_1, g_2, g_3) = exp(\int_{T_{g_1, g_2, g_3} \cdot 1} 2\pi\sqrt{-1} \cdot \nu),$$

where T_{g_1, g_2, g_3} is a smooth map from the standard tetrahedron Δ_3 to G, whose four faces are σ_{g_1, g_2}, $\sigma_{g_1 g_2, g_3}$, $\sigma_{g_1, g_2 g_3}$ and $g_1 \cdot \sigma_{g_2, g_3}$, respectively.

Proof. The number $f(g_1, g_2, g_3)$ is equal to the action functional for the restriction to $S^2 \xrightarrow{\sim} \partial\Delta_3$ of the map $T_{g_1, g_2, g_3} \cdot 1 : \Delta_3 \to G$. One then uses equation (6-10). ∎

If we restrict this 3-cocycle to $SU(2) \subset G$, we get exactly the explicit 3-cocycle written down by Cheeger and Simons [**Chee-S**] for the secondary characteristic class $\hat{c}_2 \in H^3(SU(2), \mathbb{C}^*)$. We refer to [**Cher-S**] [**Chee-S**] for the construction of the secondary Chern class $\hat{c}_2$ attached to a $SU(2)$-bundle with connection $P \to M$, belonging to the group $\hat{H}^2(M, \mathbb{Z}(2))$ of differential characters discussed in §1.5. In the case of flat connections, this gives an invariant in $H^3(M, \mathbb{T})$. For the universal flat bundle over $B\ SU(2)$ (with $SU(2)$ viewed as a discrete group), one gets a class $\hat{c}_2 \in H^3(B\ SU(2), \mathbb{T}) = H^3(SU(2), \mathbb{T})$, yielding

7.3.8. Theorem. *The obstruction to making the monopole sheaf of groupoids C over S^3 equivariant under $SU(2)$ coincides with the Chern-Cheeger-Simons class $\hat{c}_2 \in H^3(SU(2), \mathbb{T})$.*

A technical point to note is that [**Chee-S**] actually gives a 3-cocycle in measurable group cohomology, but any two extensions of this densely defined 3-cocycle to an everywhere defined 3-cocycle are cohomologous by a theorem of C. Moore.

We wish to relate this to a class in the Lie algebra cohomology $H^3(\mathfrak{g}, \mathbb{R})$, where $\mathfrak{g}$ is the Lie algebra of G. In [**Bry2**], we gave a map from $H_p(\mathfrak{g})$ to $H_p(S^3)$ (in fact, the construction worked for any finite-dimensional manifold, using homology with closed supports). Dually, this gives a map from $H^p(S^3)$ to $H^p(\mathfrak{g})$. For $p = 3$, the cohomology class of a point of S^3 maps to the cocycle

$$(\xi_1, \xi_2, \xi_3) \mapsto \nu_1(\xi_1, \xi_2, \xi_3) \tag{7 - 35}$$

which consists simply in evaluating a volume at the identity. On the other hand, we get a Lie algebra 3-cocycle from the group 3-cocycle f by differentiation at the origin, as follows:

7.3.9. Lemma. *If f is a 3-cocycle for G with values in $\mathbb{C}^*$, which is differentiable near $(1,1,1)$, the following formula*

$$F(\xi_1,\xi_2,\xi_3) = \frac{\partial^3}{\partial t_1 \partial t_2 \partial t_3} \Big(\sum_{\sigma \in S_3} \epsilon(\sigma) \cdot f(exp(t_{\sigma(1)}\xi_{\sigma(1)}),$$

$$exp(t_{\sigma(2)}\xi_{\sigma(2)}), exp(t_{\sigma(3)}\xi_{\sigma(3)}))_{t_i=0} \qquad (7-36)$$

defines a Lie algebra 3-cocycle on $\mathfrak{g}$.

Then we have

7.3.10. Proposition. *The Lie algebra 3-cocycle associated by Lemma 7.3.9 to the Sinh cocycle f is equal to $-2\pi\sqrt{-1} \cdot \nu_1$, where ν_1 is defined in (7-35). Its restriction to $\mathfrak{su}(2) \subset \mathfrak{g}$ is the element of $\wedge^3 \mathfrak{su}(2)^*$ corresponding to the invariant volume form ν, multiplied by $-2\pi\sqrt{-1}$.*

Thus the obstruction to making the monopole sheaf of groupoids equivariant is a sort of "quantization" of the degree 3 cohomology class (7-35) of $\mathfrak{g}$.

It would be hard not to believe that a similar theory could be developed for other secondary characteristic classes.

Bibliography

[Ar] V. I. Arnold, *Mathematical Methods of Classical Mechanics*, Graduate Texts in Math, 2nd ed. Springer Verlag (1989)

[As] A. Asada, *Characteristic classes of loop groups and bundles and generalized string classes*, Coll. Math. Soc. J. Bolyai **56**. Diff. Geometry and its Appl., North-Holland (1989), 33–66

[At] M. F. Atiyah, *Circular symmetry and stationary phase approximation*, in "Colloque en l'honneur de Laurent Schwartz vol. I," Astérisque **131** (1985), Soc. Math. Fr., 45–59

[A-M-R] R. Abraham, J. E. Marsden and T. Ratiu, *Manifolds, Tensors and Applications*, Applied Mathematical Sciences vol. 75, 2nd ed., Springer Verlag (1988)

[Be1] A. A. Beilinson, *Higher regulators and values of L-functions*, J. Soviet Math. **30** (1985), 2036–2070

[Be2] A. A. Beilinson, *Notes on absolute Hodge cohomology*, in Contemp. Math. **55** Part I, Amer. Math. Soc. (1986)

[B-K] A. A. Beilinson and D. Kazhdan, *Projective flat connections*, in preparation

[Bl] S. Bloch, *The dilogarithm and extensions of Lie algebras*, in Algebraic K-theory, Proceedings Evanston, Lecture Notes in Math. **854** (1981), 141–167

[Bo] A. Borel, *Seminar on Transformation Groups*, Annals of Math Studies **46** (1960), Princeton Univ. Press

[Bott1] R. Bott, *On some formulas for the characteristic classes of group-actions*, in Differential Topology, Foliations, and Gelfand-Fuks cohomology, Proc., Rio de Janeiro 1976, Lecture Notes in Math. **652** (1978), Springer Verlag

[Bott2] R. Bott, *An application of Morse theory to the topology of Lie groups*, Bull. Soc. Math. Fr. **84** (1956), 251–281

[B-T] R. Bott and L. Tu, *Differential Forms in Algebraic Topology*, Graduate Texts in Math (1986), Springer Verlag

[B-V] D. Burghelea and M. Vigué, *Cyclic homology of commutative algebras I*, Lecture Notes in Math. **1318** (1986), Springer Verlag, 51–72

[Bre] L. Breen, *Théorie de Schreier supérieure*, Ann. Sci. Ec. Norm. Sup., to appear

[Bry1] J-L. Brylinski, *Representations of loop groups, Dirac operators on loop space, and modular forms*, Topology **29** (1990), 461–480

[Bry2] J-L. Brylinski, *Non-commutative Ruelle-Sullivan type currents*, in Grothendieck Festschrift vol. I, Birkhäuser Boston (1990), 477–498

[Bry3] J-L. Brylinski, *The Kaehler geometry of the space of knots in a smooth threefold*, preprint Penn State PM 93 (1990)

[Br-ML 1] J-L. Brylinski and D. McLaughlin, *A geometric construction of the first Pontryagin class*, preprint (1991)

[Br-ML 2] J-L. Brylinski and D. McLaughlin, *Geometry of the first Pontryagin class*, in preparation

[Bu1] D. Burghelea, *Cyclic homology and the algebraic K-theory of spaces I*, in Proc. Summer Institute on algebraic K-theory, Boulder (1983), Contemp. Math. vol. **55** , Amer. Math. Soc. (1986), 89–115

[Bu2] D. Burghelea, *The free loop space and automorphisms of manifolds*, in Algebraic Topology, Evanston (1988), Contemp. Math. vol. **96**, Amer. Math. Soc. (1989), 59–85

[Bu-Vi] D. Burghelea and M. Vigué-Poirrier, *Cyclic homology of commutative algebras I*, J. Diff. Geom. **22** (1985), 243–253

[Ca] H. Cartan, *Idéaux et modules de fonctions analytiques de variables complexes*, Bull. Soc. Math. Fr. **78** (1950), 29–64

[Car] E. Cartan, *La Topologie des Espaces Représentatifs des Groupes de Lie*, Actual. Sci. et Industr., no. 358, Hermann (1936), in Oeuvres Complètes, Part I vol. 2, Gauthier-Villars (1952), 1307–1330

[C-E] H. Cartan and S. Eilenberg, *Homological Algebra*, Annals of Math. Studies vol.19, Princeton Univ. Press (1956)

[Ch] K-T. Chen, *Reduced bar constructions on de Rham complexes*, in "Algebra, Topology and category Theory," A. Heller and M. Tierney, eds., Academic Press (1976), 19–32

[Chee-S] J. Cheeger and J. Simons, *Differential characters and geometric invariants*, Lecture Notes in Math. **1167** (1985), Springer Verlag, 50–80

[Cher-S] S. S. Chern and J. Simons, *Characteristic forms and geometric invariants*, Ann. Math. **99** (1974), 48–69

[Co] T. J. Courant, *Dirac manifolds*, Trans. Amer. Soc. **319** (1990), 631–661

[Ded] P. Dedecker, *Algèbre homologique non-abélienne*, Colloque de Topologie Algébrique, Centre Belge de Recherche Mathématique, Bruxelles (1964)

[De1] P. Deligne, *Théorie de Hodge II*, (1971), Publ. Math. IHES **40**, 5–58

[De 2] P. Deligne, *Le symbole modéré*, Publ. Math. IHES, **73** (1991), 147–181

[Den] C. Deninger, *On the Γ-factors attached to motives*, Invent. Math. **104** (1991), 245–261

[D-D] J. Dixmier and A. Douady, *Champs continus d'espaces hilbertiens et de C*-algèbres*, Bull. Soc. Math. Fr. **91**(1963), 227–284

[D-H-Z] J. Dupont, R. Hain and S. Zucker, *Regulators and characteristic classes of flat bundles*, preprint Aarhus Univ. (1992)

[D-I] P. Donato, P. Iglesias, *Examples de groupes difféologiques: flots irrationnels sur le tore*, C.R. Acad. Sci. Paris **301**, ser. 1 (1985), 127–130

[D-M-O-S] P. Deligne, J. S. Milne, A. Ogus and K-y Shih, *Hodge Cycles, Motives and Shimura Varieties*, Lecture Notes in Math. **900** (1982), Springer-Verlag

[E-M] D. G. Ebin and J. E. Marsden, *Groups of diffeomorphisms and the motion of incompressible fluids*, Ann. Math. **92**(1970), 102–163

[E] H. Esnault, *Characteristic classes of flat bundles*, Topology **27** (1988), 323–352

[E-V] H. Esnault and E. Viehweg, *Deligne-Beilinson cohomology*, in [R-S-S], 43–92

[F-H] M. H. Freedman and Z-X. He, *Divergence-free fields: energy and asymptotical crossing number*, Ann. Math. **134** (1991), 189–229

[F-T] B. L. Feigin and B. Tsygan, *Additive K-theory*, in "K-Theory, Arithmetic and Geometry," Lecture Notes in math. **1289** (1987), Springer Verlag, 97–209

[Fr] J. Frenkel, *Cohomologie non-abélienne et espaces fibrés*, Bull. Soc. Math. Fr. **85** (1957), 135–218

[Fu] S. Fukuhara, *Energy of a knot*, in A fête of Topology (1988), Academic Press, 443–451

[Ga1] K. Gawedzki, *Topological actions in two-dimensional quantum field theory*, in Nonperturbative Quantum Field Theories, ed. G't Hooft, A. Jaffe, G. Mack, P. K. Mitter, R. Stora, NATO Series vol. 185, Plenum Press (1988), 101–142

[Ga2] K. Gawedzki, *Classical origins of quantum group symmetries in Wess-Zumino-Witten conformal field theory*, Comm. Math. Phys. **139** (1991), 201–213

[G-F] I. M. Gelfand and D. B. Fuks, *Cohomology of Lie algebras of vector fields on the circle*, Funct. Anal. Appl. **2**:3 (1968), 32–52

[G-H] M. Greenberg and J. Harper, *Algebraic Topology. A first Course*, Math. Lecture Note Series, Benjamin (1981)

[G-J-P] E. Getzler, J. Jones and S. Petrack, *Differential forms on loop spaces and the cyclic bar complex*, Topology **30** (1991), 339–371

[Gi] J. Giraud, *Cohomologie non-abélienne*, Grundl. **179**, Springer Verlag (1971)

[Go] T. Goodwillie, *Cyclic homology, derivations and the free loop space*, Topology **24** (1985), 187–215

[G-S] V. Guillemin and S. Sternberg, *Geometric Asymptotics*, Math. Surveys no. 14, Amer. Math. Soc. (1977)

[G-L-S-W] M. Gotay, R. Lashof, J. Sniatycki and A. Weinstein, *Closed forms on symplectic bundles*, Comment. Math. Helv. **58**(1963), 617–621

[Gri1] P. A. Griffiths, *Periods of integrals on algebraic manifolds I. Construction and properties of the modular varieties*, Amer. J. Math. **90** (1968), 568–626

[Gri2] P. A. Griffiths, *The extension problem in complex analysis II. Embeddings with positive normal bundle*, Amer. J. Math. **88** (1966), 366–446

[Gr1] A. Grothendieck, *A general theory of fibre spaces with structure sheaf*, Univ. of Kansas (1955)

[Gr2] A. Grothendieck, *Sur quelques points d'algèbre homologique*, Tohoku Math. J. **9** (1957), 119–221

[Gr3] A. Grothendieck, *Techniques de descente et théorèmes d'existence en géométrie algébrique*. Séminaire Bourbaki Exposé no. 221

[Gr4] A. Grothendieck, *Le groupe de Brauer*, in Dix Exposés sur la Cohomologie des Schémas, North-Holland, Masson (1968), 46–66

[Gr5] A. Grothendieck et al., *Revêtements Étales et Groupe Fondamental* (SGA 1), Lecture Notes in Math. **224**(1971), Springer Verlag

[Ham] R. Hamilton, *The inverse function theorem of Nash and Moser*, Bull. Amer. Soc. **7**(1982), 65–222

[Han] P. Hanlon, *Cyclic homology and the MacDonald conjecture*, Inv. Math. **86** (1986), 131–159

[Has] R. Hasimoto, *A soliton on a vortex filament*, J. Fluid Mechanics, **51**(1972), 477–485

[Hat] A. Hatcher, *A proof of the Smale conjecture*, Ann. Math. **117** (1983), 553–607

[He] S. Helgason, *Differential Geometry and Symmetric Spaces*, Academic Press, 1962

[Hir] M. Hirsch, *Immersions of manifolds*, Trans. Amer. Soc. **93** (1959), 242–276

[Hi1] N. J. Hitchin, *Complex manifolds and Einstein equations*, Lecture Notes in Math., **970** (1982), Springer Verlag, 73–98

[Hi2] N. J. Hitchin, *Monopoles and geodesics*, Comm. Math. Phys. **83** (1982), 579–602

[I] B. Iversen, *Cohomology of Sheaves*, Universitext (1986), Springer Verlag

[Jo] J. Jones, *Cyclic homology and equivariant homology*, Inv. Math. **87** (1987), 403–423

[K] M. Kashiwara, *Quantization of contact manifold*, preprint R.I.M.S. Kyoto Univ. (1992)

[Kac] V. Kac, *Infinite Dimensional Lie Algebras*, Progress in Math, Birkhaüser (1983)

[Ka] C. Kassel, *A Künneth formula for the cyclic homology of* $\mathbb{Z}/2$-*graded algebras*, Math. Ann. **275** (1986), 683–699

[K-K-V] H. Knop, H. Kraft and T. Vust, *The Picard group of a G-variety*, in Algebraische Transformationsgruppen und Invariententheorie, DMV Seminar Bd. 13, Birkhaüser Verlag (1989), 77–87

[Ki1] A. A. Kirillov, *Geometric quantization*, in Encycl. of Math. Sciences, Dynamical Systems vol. 4, V. I. Arnold and S. P. Novikov, eds., Springer Verlag(1990), 138–172

[Ki2] A. A. Kirillov, *Kähler structure on K-orbits of the groups of diffeomorphisms of a circle*, Funct. Anal. Appl. **21**: 2 (1987), 122–125

[K-Y] A. A. Kirillov and D. V. Yurev, *Kähler geometry of the infinite-dimensional space* $M = Diff^+(S^1)/Rot(S^1)$, Funct. Anal. **21**:4 (1987), 284–294

[Kl] W. Klingenberg, *Lectures on Closed Geodesics*, Grundl. **230**, Springer Verlag (1978)

[K-N] S. Kobayashi and K. Nomizu, *Foundations of Differential Geometry vol. 1*, Interscience Tracts in Pure and Appl. Math., Wiley (1963)

[Ko1] B. Kostant, *Quantization and unitary representations*, in Lectures in modern analysis and applications III. Lecture Notes in Math. **170** (1970), Springer Verlag, 87–208

[Ko2] B. Kostant, *On the definition of quantization*, in Géométrie Symplectique et Physique Mathématique, Colloques Intern. CNRS, vol. 237, Paris (1975) 187–210

[Ku] N. Kurokawa, *On some Euler products I*, Proc. Japan Acad. **60 A** (1984), 335–338

[Lak] M. Lakshmanan, *Continuum spin systems as an exactly solvable dynamical system*, Phys. Letters **61 A**:1 (1977), 53–54

[Lan] S. Lang, *Introduction to Differentiable Manifolds*, Academic Press (1962)

[Law] B. Lawson, *Lectures on Minimal Submanifolds*, Publish or Perish, vol. 9 (1980)

[LB] C. LeBrun, *Spaces of complex null-geodesics in complex riemannian geometry* , Trans. Amer. Math. Soc. **278** (1983), 209–231

[L-L] L. D. Landau and E. M. Lifschitz, *Quantum Mechanics-Non-Relativistic Theory*, Pergamon Press (1965)

[L-P] J. Langer and R. Perline, *Poisson geometry of the filament equation*, J. Nonlinear Sci. **1** (1991), 71–93

[Le] J. Leray, *L'anneau spectral et l'anneau filtré d'homologie d'un espace localement compact et d'une application continue*, J. Math. Pures Appl. **29** (1950), 1–139

[Lem] L. Lempert, *Loop spaces as complex manifolds*, preprint (1992)

[Les] J. Leslie, *On a differential structure for the group of diffeomorphisms*, Topology **6** (1967), 263–271

[Li] A. Lichnérowicz, *Algèbre de Lie des automorphismes infinitésimaux d'une structure unimodulaire*, Ann. Inst. Fourier Grenoble **24** (1974), 2190–2226

[Lo] J-L. Loday, *Cyclic Homology*, Grundl. Springer Verlag (1992)

[M-M] B. Mazur and W. Messing, *Universal Extensions and One-Dimensional Crystalline Cohomology*, Lecture Notes in Math. **370** (1974), Springer Verlag

[M-W] J. Marsden and A. Weinstein *Coadjoint orbits, vortices, Clebsch variables for incompressible fluids*, Physica **7 D** (1983), 305–323

[McLau] D. McLaughlin, *Orientation and string structures on loop spaces*, Pac. J. Math. **155**:1 (1992), 143–156.

[McCl] J. McCleary, *User's Guide to Spectral Sequences*, Math. Lecture Series **12**, Publish or Perish (1985)

[McL1] S. MacLane, *Categories for the Working Mathematician*, Graduate Texts in Math., Springer Verlag (1971)

[McL2] S. MacLane, *Homology*, Grundl. **114**, 3rd ed., Springer Verlag(1975),

[Mic] J. Mickelsson, *Kac-Moody groups, topology of the Dirac determinant bundle, and fermionization*,Comm. Math. Phys. **110** (1987), 173–183

[Mil] J. S. Milne, *Étale Cohomology*, Princeton Math. Series **33**, Princeton Univ. Press (1980)

[Mo] J. Moser, *On the volume elements of a manifold*, Trans. Amer. Math. Soc. **120** (1965), 286–294

[Mu1] D. Mumford, *Abelian Varieties*, Tata Studies in Math. vol. 5, Oxford Univ. Press, 2nd ed. (1974)

[Mu2] D. Mumford, *On the equations defining abelian varieties*, Inv. Math. **1** (1966), 287–354

[Mun] J. Munkres, *Topology: A First Course*, Prentice-Hall (1975)

[N-N] A. Newlander and L. Nirenberg, *Complex analytic coordinates in almost complex manifolds*, Ann. of Math. **65**(1959), 391–404

[Ok] K. Oka, *Sur quelques notions arithmétiques*, Bull. Soc. Math. Fr. **78** (1950), 1–27

[Om] H. Omori, *Infinite-dimensional Lie transformation Groups*, Lecture Notes in Math. **427** (1975), Springer Verlag

[O'H] J. O'Hara, *Energy of a knot*, Topology **30** (1991), 241–247

[O'N] B. O'Neill, *The fundamental equations of a submersion* , Mich. Jour. Math. **13** (1966), 459–469

[P-S] A. Pressley and G. Segal, *Loop groups*, Oxford Univ. Press (1986)

[Pe] J-P. Penot, *Sur le théorème de Frobenius*, Bull. Soc. Math. Fr. **98** (1970), 47–80

[Pe-Sp] V. Penna and M. Spera, *A geometric approach to quantum vortices*, J. Math. Phys. **30 (12)**, Dec. 1989, 2778–2784

[Po] V. Poenaru, *Extension des immersions combinatoires en codimension 1*, Exposé Séminaire Bourbaki no. 342, February 1968

[R-S-S] M. Rapoport, N. Shappacher and P. Schneider, ed., *Beilinson's Conjectures on Special values of L-Functions*, Perspectives in Math., Academic Press (1988)

[R-R] I. Raeburn and J. Rosenberg, *Crossed products of continuous-trace C^*-algebras by smooth actions*, Trans. Amer. Soc. **305** (1988), 1–45

[Ra-Re] M. Rasetti and T. Regge, *Vortices in He II, current algebras and quantum knots*, Physica **80A** (1975), 217–233

[R-S] T. Ratiu and R. Schmid, *The differentiable structure of three remarkable diffeomorphism groups*, Math. Z. **177** (1981), 81–100

[S-W] G. Segal and G. Wilson, *Loop groups and equations of KdV type*, Publ. Math. IHES **61** (1981), 5–65

[Sa] P. Samuel, *Méthodes d'algèbre abstraite en géométrie algébrique*, Ergeb. der Math. Bd. 4, Springer-Verlag (1955)

[Sc1] U. Schäper, *Geodesics on loop spaces*, J. Geom. Phys. **11** (1993), 553–557

[Sc2] U. Schäper, *Geometrie von Loopräume*, Inaug. Diss. Fak. Phys. A. Lud. Univ. Freiburg March 1993

[Se1] G. Segal, *Equivariant K-theory*, Publ. Math. IHES **34** (1968), 129–151

[Se2] G. Segal, *The definition of conformal field theory*, Differential Geometrical Methods in Theoretical Physics (Como, 1987), NATO Adv. Sci. Inst. Ser. C: Math. Phys. Sci., 250, Kluwer Acad. Publ. (1988), 165–171

[Se3] G. Segal, *Elliptic cohomology*. Exposé no. 695, Séminaire Bourbaki 1987/88. Astérisque **161–162**, Soc. Math. Fr., 187–201

[Ser] J-P. Serre, *Homologie singulière et espaces fibrés*, Ann. Math. **54** (1951), 425–505

[Si] H. X. Sinh, Thèse de Doctorat d' Etat, Univ. Paris 7 (1975)

[Sm] S. Smale, *The classification of immersions of spheres in euclidean space*, Ann. of Math. **69** (1959), 327–344

[Sou] C. Soulé, *Régulateurs*. Séminaire Bourbaki, Exposé no. 644, Astérisque **133–134**, Soc. Math. Fr., 237–253

[So1] J-M. Souriau, *Structure des Systèmes Dynamiques*, Dunod (1970)

[So2] J-M. Souriau, *Un algorithme générateur de structures quantiques*, in Elie Cartan et Les Mathématiques d'Aujourd'hui, Astérisque (1985), 341–399

[S-W] G. Segal and G. Wilson, *Loop groups and equations of K dV type*, Publ. Math. IHES **80**(1985), 301-342

[Spa] E. Spanier, *Algebraic Topology* (1989), Springer Verlag

[St] R. Steinberg (with J. Faulkner and R. Wilson), *Lectures on Chevalley groups*, mimeographed notes, Yale Univ. (1967)

[T] F. Trèves, *Topological Vector Spaces, Distributions and Kernels*, Academic Press (1967)

[Ul] K. Ulbrich, *Group cohomology for Picard categories*, J. Alg. **91** (1984), 464-498

[Va] V. A. Vassiliev, *Cohomology of knot spaces*, preprint (1990)

[Weil1] A. Weil, *Sur les théorèmes de de Rham*, Comm. Math. Helv. **26** (1952), 17-43

[Weil2] A. Weil, *Variétés Kaehlériennes*, Hermann (1957)

[Weil3] A. Weil, *Sur certains groupes d'opérateurs unitaires*, Acta Math. **111** (1976), 143-211

[Wein] A. Weinstein, *Cohomology of symplectomorphism groups and critical values of hamiltonians*, Math. Z. **201** (1989), 75-82

[Wi1] E. Witten, *The index of the Dirac operator on loop space*, in Elliptic Curves and Modular Forms in Algebraic Topology, Lecture Notes in Math. **1326** (1988) Springer Verlag, 161-181

[Wi2] E. Witten, *Quantum field theory and the Jones polynomial*, Comm. Math. Phys. **121** (1989), 351-399

[W-X] A. Weinstein and P. Xu, *Extensions of symplectic groupoids and quantization*, J. Rein. Angew. Math. **417** (1991), 159-189

[Y] C. N. Yang, *Fibre bundles and the physics of the magnetic monopole*, Intern. Symp. on Diff. Geom. in honor of S.S. Chern, Berkeley, 1979, W-Y. Hsiang, ed. Springer Verlag (1980), 247-253

[Z-S] V. E. Zakharov and A. B. Shabat, *Exact two-dimensional self-focusing and one-dimensional self-modulation of waves in non-linear media*, Soviet Physics JETP **34**, no. **1** (1972), 62-79

List of Notations

d', d''	components of the exterior differential, 14
D	total differential in a double complex, 14
$Tot(K^{\bullet\bullet})$	total complex of $K^{\bullet\bullet}$, 14
F_H	first filtration, 15
F_V	second filtration, 15
$Gr_F^p(K^{\bullet})$	graded subquotients of F, 15
Z_r^p	groups used in the spectral sequence, 15
$E_r^{p,q}$	spectral sequence, 15
d_r	differential in a spectral sequence, 15
E_∞^p	abutment of a spectral sequence, 16
H_H	horizontal cohomology, 18
H_V	vertical cohomology, 18
$\underline{A}_M^{\bullet}$	de Rham complex of M, 19
$\underline{A}_M^p$	sheaf of p-forms on M, 19
d	exterior differential, 19
$H^p(X, K^{\bullet})$	hypercohomology group, 21
$U_{i_0,\cdots,i_p}$	intersection of $U_{i_0},\cdots,U_{i_p}$, 25
$C^p(\mathcal{U}, A)$	group of Čech cochains, 25
δ	Čech differential, 25
$\underline{\alpha} = (\alpha_{i_0,\cdots,i_p})$	Čech cochain, 25
$\mathcal{O}_S$	sheaf of holomorphic functions, 26
$H^p(\mathcal{U}, A)$	Čech cohomology for the covering $\mathcal{U}$, 25
$\mathcal{C}^p(\mathcal{U}, A)$	sheaf of Čech p-cochains, 26
$\mathcal{C}^{\bullet}(\mathcal{U}, A)$	complex of Čech cochains, 26
$F \otimes G$	tensor product of sheaves, 29
$\cup$	cup-product, 29
$\check{H}^p(X, A)$	Čech cohomology groups, 31
$\check{H}^p(X, K^{\bullet})$	Čech hypercohomology groups, 32
$\mathcal{V} = (V_j)_{j \in J}$	open covering, 32
$\mathcal{U} \prec \mathcal{V}$	$\mathcal{V}$ is a refinement of $\mathcal{U}$, 32
$A^p(M)$	space of p-forms on M, 35
$\phi_i, \phi_j,$	local charts, 35-36
$H_{DR}^p(M)$	de Rham cohomology, 36
$\mathbb{R}_M$	constant sheaf, 36
$f^{-1}A$	inverse image sheaf, 36-37
$Supp(s)$	support of a section s of a sheaf, 39
(f_i)	partition of unity, 39
E	ILH space, 40
$\varprojlim$	inverse limit, 40
$\mathbb{Z}(p)$	subgroup $(2\pi\sqrt{-1})^p \cdot \mathbb{Z}$ of $\mathbb{C}$, 46
B	subgroup of $\mathbb{R}$, 46

$B(p)$	subgroup $(2\pi\sqrt{-1})^p \cdot B$ of $\mathbb{C}$, 46
$B(p)_D^\infty$	smooth Deligne complex, 47
κ	homomorphism from Deligne to ordinary cohomology, 48
$\sigma_{\leq p-1}$	Deligne truncation, 48
$A^{p-1}(M)_{\mathbb{C},0}$	group of closed $\mathbb{C}$-valued $p-1$-forms with integral periods, 48
$S_p(M)$	group of smooth singular chains, 50
$S_p(M)^*$	group of smooth singular cochains, 50
∂	boundary map on smooth chains, 50
∂^*	coboundary map on smooth cochains, 50
$\hat{H}^p(M, \mathbb{Z}(p))$	group of differential characters, 50
deg	degree, 52
$\tilde{d}$	differential in the Deligne complex, 52
$B(p)_D, \mathbb{Z}(p)_D$	Deligne complex, 53
$\mathcal{O}_X$	sheaf of holomorphic functions, 53
$\mathcal{O}_X^*$	sheaf of invertible holomorphic functions, 53
Ω_X^p	sheaf of holomorphic p-forms, 53
$F^p\Omega_X^\bullet$	truncation of the holomorphic de Rham complex, 53
$Hdg^p(X)$	group of Hodge cycles, 54
J^p	Griffiths intermediate jacobian, 54
$f: Y \to X$	continuous map, 54
Y_x	fiber of f, 54
$f_*(A)$	direct image of the sheaf A, 54-55
$R^q f_*(K^\bullet)$	higher direct image sheaves, 55
ρ	monodromy representation, 58
$\underline{H}^q(F, A)$	locally constant sheaf, 58
$\pi_1(X, x)$	fundamental group, 58
F^p	Cartan filtration, 58
$A^n(Y/X)$	space of relative n-forms on Y, 58
$\underline{A}^n(Y/X)$	sheaf of relative n-forms on Y, 58
$i(\xi)$	interior product, 59
E_C	Cartan-Leray spectral sequence, 59
$C^\infty(X)$	algebra of smooth functions on X, 59
$\underline{C}_X^\infty$	sheaf of smooth functions on X, 59
$\int_f$	fiber integration, 61
Eu	Euler class, 61
$p: L \to M$	line bundle, 62
L_x	fiber of the line bundle L, 62
$\phi: L_1 \tilde{\to} L_2$	isomorphism of line bundles, 62-63
s	section of a line bundle, 63
$p_+: L^+ \to M$	$\mathbb{C}^*$-bundle associated to L, 63
$F: C_1 \to C_2$	equivalence of categories, 64

f^*L	pull-back of a line bundle, 64
$\mathbb{P}$	projective space, 64
$\mathbb{CP}_n$	complex projective space, 64
$(z_1,\ldots,z_n)$	complex coordinates, 64
$Q_1 \times_M Q_2$	contracted product of two $\mathbb{C}^*$-bundles, 65
$Pic^\infty(M)$	group of isomorphism classes of line bundles over M, 65
g_{ij}	transition functions, 65
$c_1(L)$	first Chern class, 66
h	hermitian metric, 67
L^1	circle bundle associated to a hermitian line bundle, 68
$Pic(M)$	Picard group, 69
$(A^1 \otimes L)(U)$	space of 1-forms on U with values in L, 70
∇	connection, 70
∇_v	contraction of ∇ with the vector field v, 72
A	connection 1-form, 72
$K = K(\nabla)$	curvature of ∇, 74
$L_{\nabla=0}$	sheaf of horizontal sections of L, 74
$\underline{g}$	Čech cocycle (g_{ij}), 75
$\underline{A}^2_M{}^{cl}$	sheaf of closed 2-forms, 78
$A^2(M)^{cl}_\mathbb{C}$	space of closed 2-forms, 79
$\Re$	real part, 79
(f,g)	Deligne line bundle, 82
$Li_2(z)$	dilogarithm function, 83
$\mathcal{L}(v)$	Lie derivative, 85
ω	closed 2-form, 85
$Symp(M)$	Lie algebra of symplectic vector fields, 85
I	mapping $TM \xrightarrow{\sim} T^*M$ of bundles, 86
$\mathcal{H}_M$	Lie algebra of hamiltonian vector fields, 86
$Vect(M)$	Lie algebra of vector fields on M, 87
$C^\infty(M/K)$	space of hamiltonian functions, 88
$\{\,,\,\}$	Poisson bracket, 88
$\tilde{\mathcal{H}}_M$	central extension of $\mathcal{H}_M$, 89
X_f	hamiltonian vector field of f, 89
c	Lie algebra 2-cocycle, 89
$\mathcal{H}_{M,x}$	Lie subalgebra of $\mathcal{H}_M$, 90
v_h	horizontal lift of v, 90
$\mathcal{V}_L$	Lie algebra of connection-preserving vector fields, 91
$Diff(M,\omega)$	group of diffeomorphisms of (M,ω), 94
$\tilde{H}$	central extension of H, 95
T_γ	parallel transport along γ, 95
$\gamma * \gamma'$	composition of paths, 95

PM	path space of M, 96
s, t	source and target maps $PM \to M$, 96
T^*M	cotangent bundle, 97
$\Gamma(L)$	space of smooth sections of L, 98
$E = E_+ \oplus E_-$	Hilbert spaces, 99
$H.S.(E_1, E_2)$	space of Hilbert-Schmidt operators, 99
$Gr(E)$	infinite-dimensional Grassmann manifold, 99
$GL_{res}(E)$	restricted linear group, 99
$U_{res}(E)$	restricted unitary group, 99
$LU(n)$	loop group of $U(n)$, 100
$Diff^+(S^1)$	group of orientation-preserving diffeomorphisms of S^1, 100
$\widetilde{Diff}^+(S^1)$	central extension of $Diff^+(S^1)$, 100
$[\omega]$	cohomology class of ω, 103
Λ	subgroup of $\mathbb{C}$, 104
$\mathcal{T}$	sheaf of admissible trivializations of a $\mathbb{C}/\Lambda$-bundle, 104-105
D	connection on a $\mathbb{C}/\Lambda$-bundle, 106
M	manifold of dimension n ($n = 3$ often), 110
$H^n(S^1)$	Sobolev space, 110
$\|f\|_n$	norm on $H^n(S^1)$, 110
$*$	convolution product, 110
LM	free loop space of M, 111
exp	exponential map in the riemannian manifold M, 111
T_ϵ	tubular neighborhood, 111
$\hat{X}$	open subset of LM, 112
X	open subset of $\hat{X}$, 112
$\hat{Y}$	space of singular knots, 112-113
$\pi : \hat{X} \to \hat{Y}$	projection map, 112
Y	space of knots, 113
$\tilde{C}$	normalization of C, 113
$l(C)$	length of the curve C, 114
κ	curvature of a curve, 114
H	holonomy, 115
$Imm(S^1, M)$	space of immersions of S^1 into M, 116
SM	sphere bundle, 116
$A^\bullet$	differential graded algebra, 117
$B(A^\bullet)$	bar complex, 117
b	Hochschild differential, 117
$\overline{B}(A^\bullet)$	reduced bar complex, 117
$\mathcal{M}$	Sullivan model, 117
$\hat{Y}_b$	space of based singular knots, 118
$ES^1 \to BS^1$	universal S^1-bundle, 118-119

$\mathbb{P}(E)$	projective space of lines in E, 165
μ_n	group of nth roots of unity, 166
$PGL(n, \mathbb{C})$	projective linear group, 166
H	Heisenberg group in the sense of A. Weil, 166
E^∞	Fréchet space $C^\infty(S^1)$, 168
L^∞	algebra of continuous endomorphisms of E_∞, 168
$\tilde{B}$	central extension of the Lie group B by $\mathbb{C}^*$, 168
$p : Q \to M$	principal B-bundle, 169
$\mathfrak{b}$	Lie algebra of B, 169
$\tilde{\mathfrak{b}}$	Lie algebra of $\tilde{B}$, 169
$\tilde{p} : \tilde{Q} \to M$	principal $\tilde{B}$-bundle, 169
$\tilde{\theta}$	connection on $\tilde{Q} \to M$ compatible with θ, 169
$\tilde{\Theta}$	curvature of $\tilde{\theta}$, 170
$\mathcal{V}(W)$	vector bundle associated to a group representation W, 170
l	splitting $\mathcal{V}(\tilde{\mathfrak{b}}) \to \mathbb{C}$, 170
$K(\tilde{\theta})$	scalar curvature of $\tilde{\theta}$, 170
Ω	3-curvature, 171
$\underline{h} = (h_{ijk})$	Čech 2-cocycle with values in $\mathbb{C}^*_M$, 172-173
$p : Y \to X$	family of symplectic manifolds, 179
ω	relative symplectic form on Y, 179
$\coprod$	disjoint union, 183
$Y \times_X Y$	fiber product, 183
$A(Y \xrightarrow{f} X)$	value of the sheaf A on $f : Y \to X$, 184
ϕ	descent isomorphism, 185-186
p_1, p_2	projection maps from $Y \times_X Y$ to Y, 185
p_{12}, p_{13}, p_{23}	projection maps from $Y \times_X Y \times_X Y$ to $Y \times_X Y$, 185-186
p_1, p_2, p_3	projection maps from $Y \times_X Y \times_X Y$ to Y, 185-186
f^{-1}	inverse image functor, 187
$Sh(Y)$	category of sheaves of sets over Y, 187
$\theta_{g,h}$	natural transformation from $h^{-1}g^{-1}$ to $(gh)^{-1}$, 187
$\underline{Hom}(A, B)$	sheaf of homomorphisms from A to B, 187
$\underline{Isom}(A, B)$	sheaf of isomorphisms from A to B, 188
$Sets$	category of sets, 188
$Desc(Z \xrightarrow{g} Y)$	descent category, 189
$\underline{Aut}(F)$	sheaf of automorphisms of F, 190
$f^{-1}T$	pull-back of a torsor, 190
f^*T	pull-back of an $\underline{A}^1_{X,\mathbb{C}}$-torsor, 191
$\mathcal{C}$	presheaf of categories, 191
$Bund_G(X)$	category of G-bundles over X, 192
$Desc(\mathcal{C}, g)$	descent category, 193
$f^{-1}\mathcal{C}$	inverse image of the sheaf of categories $\mathcal{C}$, 195

G_1, G_2, G_3	axioms for gerbes, 196
$P \times^H I$	twist of an object P of $\mathcal{C}$ by a torsor I, 198
Φ	equivalence of gerbes, 199
$\mathcal{G}$	sheaf of groupoids, 202
f^*C	pull-back of a Dixmier-Douady sheaf of groupoids, 203
Co	connective structure on a sheaf of groupoids, 206-207
$\tilde{\nabla}$	connection compatible with ∇, 208
$Conn(Q)$	torsor of connections on the $\mathbb{C}^*$-torsor Q, 210
$K(\nabla)$	curving of a connective structure, 211-212
f^*K	pull-back of a curving, 213
Ω	3-curvature of a curving, 214
$\pi_2(M)$	second homotopy group, 219
$P_1(M)$	based path space, 220
ev	evaluation map $P_1(M) \to M$, 220
ΩM	based loop space, 220
G	Lie group, 222
$\mathfrak{g}$	Lie algebra of G, 222
$H^\bullet(\mathfrak{g})$	Lie algebra cohomology, 222
B	Killing form, 222
ν	3-form on G, 223
$\mathfrak{h}$	Cartan subalgebra, 224
α	root, 224
$G_{\mathbb{C}}$	complexification of G, 226
M/G	quotient space, 228
$Pic_G^\infty(M)$	equivariant Picard group, 228
$\mathcal{X}$	sheaf on M/G, 229
$\kappa(L)$	associated character, 229
$\mathcal{D}_X$	sheaf of differential operators on X, 232
$\mathcal{D}_X(L)$	sheaf of twisted differential operators, 232
$Diff^-(S^1)$	orientation reversing diffeomorphisms of S^1, 234
t	triangulation of S^1, 235
$\mathcal{P}$	$\mathbb{C}^*$-bundle over LM, 236-237
$H_\gamma(F, \nabla)$	holonomy of (F, ∇) as section of $\mathcal{P}$, 237
D	connection on $\mathcal{P} \to LM$, 237
$Map(\Sigma, M)$	space of smooth mappings $\Sigma \to M$, 239
b_i	restriction to a boundary copmponent, 239
S	action functional, 239
$\mathcal{R}$	$\mathbb{C}^*$-bundle over $Map(\Sigma, M)$, 239
$\mathcal{Q}$	$\mathbb{C}^*$-bundle over $\hat{Y}$, 243
C	connected component of $\hat{Y}$, 244
LG	loop group of G, 247

$\widetilde{LG}$	central extension of LG, 247
$\int_{S^1}$	fiber integration map in Deligne cohomology, 252
$\vec{q} = (x, y, z)$	position vector, 257
$\vec{p} = (p_x, p_y, p_z)$	momentum vector, 257
$\vec{F}$	force, 257
H	hamiltonian, 257
$V(\vec{q})$	potential function, 257
$\vec{\nabla}$	gradient, 257
ψ	wave function, 258
$\hat{a}$	quantized operator for an observable a, 258
$\hbar$	Planck constant, 258
$\vec{E}$	electric field, 260
$\vec{B}$	magnetic field, 259-260
$\vec{A}$	magnetic potential vector, 260
c	speed of light, 260
$\vec{P}$	modified momentum vector, 260
$\hat{P} = (\hat{P}_x, \hat{P}_y, \hat{P}_z)$	quantization of $\vec{P}$, 260
R	curvature of the connection on L, 263
μ	strength of a monopole, 263
δ_0	delta function of Dirac, 264
$H^3_{\{0\}}(\mathbb{R}^3)$	cohomology with support in $\{0\}$, 264
(U_0, U_∞)	open covering of S^3, 265
$SU(2)$	special unitary group, 268
$H^3(G, -)$	degree 3 cohomology of the group G, 269-270
f	degree 3 group cocycle, 270
C	groupoid with tensor product, 270
I	identity object of C, 271
X^*	inverse of the object X, 271
$\pi_0(C)$	group of isomorphism of objects of C, 271
$\pi_1(C)$	automorphism group of I, 271
ψ_{g_1, g_2}	isomorphisms in C, 272
$*_v$	vertical composition of mappings from the square, 273
$*_h$	horizontal composition of mappings from the square, 273
$\hat{c}_2$	secondary Chern class, 276
$\mathfrak{su}(2)$	Lie algebra of $SU(2)$, 277

Index